Technology & American History

A Historical Anthology from **Technology & Culture**

Technology & American History

A Historical Anthology from **Technology & Culture**

Edited by Stephen H. Cutcliffe and Terry S. Reynolds

The University of Chicago Press

CHICAGO AND LONDON

The essays in this volume originally appeared in various issues of *Technology and Culture*.

The University of Chicago Press, Chicago 60637
The University of Chicago Press, Ltd., London

Printed in the United States of America
01 00 99 98 97   5 4 3 2 1

Library of Congress Cataloging-in-Publication Data

Technology and American History / edited by Stephen H. Cutcliffe and Terry S. Reynolds.
p. cm.
Anthology of essays from Technology and culture.
Includes bibliographical references and index.
ISBN 0-226-71027-0 (cloth). — ISBN 0-226-71028-9 (pbk.)
1. Technology—United States—History. I. Cutcliffe, Stephen H. II. Reynolds, Terry S. III. Technology and culture.
T21.T426 1997
609.73—dc21 97-5925
CIP

The paper used in this publication meets the minimum requirements of American National Standard for Information Science—Permanence of Paper for Printed Library Materials, ANSI Z39.48-1984.

Technology and American History

For Katie and Linda

Introduction

> The locomotive and the steamboat, like enormous shuttles, shoot every day across the thousand various threads of national descent and employment and bind them fast in one web. (Ralph Waldo Emerson, "The Young American" [1844])

Technology—material culture—has always been an integral strand in the fabric of human existence, but perhaps especially so in our contemporary "technological society." Without making claims for undue "uniqueness," there is an American technological story to be told. Almost exactly 150 years ago, Ralph Waldo Emerson had come to that recognition with his observation (above) regarding the integral role that the then new transportation technologies of steamboats and railroads were playing in American society. As do historians today, he recognized that technology was intricately woven into the cultural cloth that is American society. If one tried to pick out the technological strand to examine it in isolation, the "seamless web" that is society would quickly unravel. Yet, by examining technology within its societal context, it is possible, without unraveling the fabric, to understand the complex and changing patterns of our technological heritage.

Technology and American History contains fifteen essays that place the history of technology within the fabric of its cultural context. It is one-half of a two-volume set of anthologies consisting of articles selected from *Technology and Culture,* the official journal of the Society for the History of Technology (SHOT). Its companion volume (*Technology and the West*) looks at technology in the context of Western civilization. In this volume our intent is to provide a series of high-quality essays that cover, in roughly chronological order, the scope of the history of technology in American society. To this end, we have selected articles representative of key time periods, themes, and technologies without trying or claiming to be all-inclusive. An intro-

Stephen H. Cutliffe is a member of the Department of History and the director of Lehigh University's Science, Technology and Society Program, where he edits their *Science, Technology and Society Curriculum Development Newsletter.* He is also the co-editor with Robert C. Post of *In Context: History and the History of Technology—Essays in Honor of Melvin Kranzberg* (Lehigh University Press, 1989). Terry S. Reynolds is professor of history and chair of the Department of Social Sciences at Michigan Technological University. He has written several books on aspects of the history of water power and edited *The Engineer in America* (Chicago, 1991).

ductory overview essay sets the selection of essays in broad historical context, identifies common themes and issues, and offers brief synopses of each article. It is meant to be read as an essay in its own right; however, it is intended that readers will also refer back to appropriate sections prior to reading specific articles themselves.

We found the task of selecting fifteen articles to cover the scope of the history of technology in America to be a difficult one. It was, however, an even more daunting task for the companion volume on the history of technology in Western civilization. Often the best article from the viewpoint of academic scholarship was not the best article from the viewpoint of the general reader. Often good pieces had to be eliminated simply because we had space for only one article in a particular time period, if we were to cover all of American history in a single volume.

Although the primary criterion used in selecting essays was appropriateness for a general readership and for the undergraduate classroom, as well as "fit" with the focus of this anthology, a significant number of the selections have been singled out for recognition for the quality of their scholarship by SHOT. The Society awards its Usher prize annually for the best essay appearing in *Technology and Culture* selected from the previous three years' issues. Among the Usher Prize–winning pieces found in this anthology are Norman B. Wilkinson's "Brandywine Borrowings from European Technology" (1963), John G. Burke's "Bursting Boilers and the Federal Power" (1967), George Wise's "A New Role for Professional Scientists in Industry" (1983), and Robert Gordon's "Who Turned the Mechanical Ideal into Mechanical Reality?" (1991). We have also used Ruth Schwartz Cowan's "The 'Industrial Revolution' in the Home," the research for which later found expanded publication in her 1984 Dexter Prize–winning book, *More Work for Mother: The Ironies of Household Technology from the Open Hearth to the Microwave,* and Claude S. Fischer's "'Touch Someone': The Telephone Industry Discovers Sociability," which expanded into his 1995 Dexter Prize–winning book, *America Calling: A Social History of the Telephone to 1940.* Finally, we have included David Jardini's essay on the Jones and Laughlin workforce between 1885 and 1896, which won the 1992 Samuel Eleazar and Rose Tartakow Levinson Prize.[1] Taken together then,

[1] Ruth Schwartz Cowan, *More Work for Mother: The Ironies of Household Technology from the Open Hearth to the Microwave* (New York: Basic, 1983); Claude S. Fischer, *America Calling: A Social History of the Telephone to 1940* (Berkeley and Los Angeles: University of California Press, 1992). The Dexter Prize, founded and supported by the Dexter Chemical Corporation, annually recognizes an outstanding book in the

the essays assembled here represent some of the best scholarship in the field and should serve to introduce the general reader and students to the profession and craft of research and writing in the history of American technology.

In compiling this anthology, we have accumulated a number of debts. Our original stimulus came from the Executive Council of the Society for the History of Technology, which suggested this project in 1993. We would particularly like to thank Bob Post, former editor of *Technology and Culture,* for his encouragement; without it, we might very well have given up on this project. We also thank Richard Hirsh of Virginia Tech, Charles Hyde of Wayne State University, Bruce Seely of Michigan Tech, and Roger Simon and John Smith of Lehigh University for reviewing and commenting on our proposal and article selections for this reader. Finally, we are indebted to the fine staff at the University of Chicago Press Journals Division for help in bringing our early vision to final reality.

STEPHEN H. CUTCLIFFE AND TERRY S. REYNOLDS

history of technology published within the previous three years. The Levinson Prize is awarded annually for an original essay in the history of technology that explicitly examines a technological artifact or process within the framework of social or intellectual history.

Technology in American Context

STEPHEN H. CUTCLIFFE AND
TERRY S. REYNOLDS

Technology is an inherently human activity, and as such it is deeply embedded within society. Thus, whereas "location, location, location" might be the tongue-in-cheek answer given by real estate agents when advising prospective clients on the most important considerations in buying or selling a house, the not-so-tongue-in-cheek characterization of what is most important for understanding of technology in American history is "context, context, context." That is, while the technical or "nuts-and bolts" elements of an artifact or process are extremely important, ultimately it is the contextual or societal setting in which a technology rests that explains its history. A "contextual" appreciation of technology recognizes both the manner in which cultural, political, and economic forces shape technological innovation and the ways that technology has influenced society and shaped its values. Thus, technology is neither an "autonomous" juggernaut nor merely a "neutral" tool ready for just any use; it is, rather, a value-laden, interactive set of social processes. It is with an emphasis on societal context that this set of essays from *Technology and Culture* has been selected.

As a nation that has passed through an industrial revolution and since moved on to an age of "information" and a service economy, it is sometimes difficult to remember that the preindustrial era was in its own way just as technologically based as our own time. Indeed, Native Americans of the precontact period were highly dependent on technological skills for their survival and well-being. Long before Europeans set foot ashore the Americas, Native Americans had evolved sophisticated agricultural techniques including extensive irrigation systems, had developed building construction skills that in some sections of the New World resulted in cities far larger than anything the Europeans would be able to duplicate for many decades, and had achieved high levels of artistic beauty and functional form in domestic technologies like pottery making and basket weaving. Thus, by no means did European settlers step into a technological "wilderness," although they often failed to recognize the situation for what it really entailed; nor were the technologies they brought with them necessarily better than those they encountered on arrival.

European settlers brought with them Old World technologies of agriculture, building construction, transportation, and craft manufacturing that were essentially medieval. The American colonial era has been characterized as an age of wood and water, for most things were made from wood, although iron was not unknown, and beyond the animate sources of power found in humans and animals, waterpower was the major energy source for operating those few early manufacturing processes that did exist, such as saw and grain milling. Most colonial-era technology was of a craft sort, entailing the use of hand tools with little division of labor, and with most innovative change emerging out of empirical practice, rather than arising from theoretical scientific knowledge.

By the end of the 18th century, the American colonies had grown both demographically and economically and had matured politically with the Declaration of Independence and the subsequent Revolution and Constitution of 1787. Nonetheless, the new United States remained a net importer of technology. Looking back from the late 20th century, when Americans often complain and worry about other nations unfairly "borrowing" and then capitalizing on our scientific knowledge and technological expertise, it is easy to forget that the United States was in a very similar situation around 1800 when it drew heavily on Europe for the new manufacturing skills and technologies it so desperately needed if it was to survive in the rapidly industrializing Anglo-European world.

One of the early clusters of American industrial development was centered along the Bradywine Creek, near present-day Wilmington, Delaware. Norman B. Wilkinson, in the first essay in this volume, "Brandywine Borrowings from European Technology," describes the turn-of-the-century transfer of a wide range of European technologies to the New World and by implication suggests the large extent of American technological "indebtedness" to the "other side of the Atlantic." He thus relates the not atypical story of the Manchester mechanics George and Isaac Hodgson, who in 1811, before embarking for the New World themselves, disguised their tools as fruit trees and sent them on a separate ship to avoid being apprehended under British law, which made it illegal for certain classes of skilled workers to emigrate. The Hodgson experience testified to an American need for skilled technicians to develop its own industrial base.

E. I. du Pont, himself a transplanted Frenchman, considered diversifying beyond gunpowder manufacture, which was based on French-derived methods, by expanding into textiles. To that end, in 1809, he wrote to a cousin in Rouen, France, soliciting information on the latest textile machines then in use there. Du Pont also wrote to his father to inquire about advanced tanning processes. In both cases du Pont established American enterprises based on French procedures and utilizing French workers, although both ultimately failed because of depressed economic conditions of the 1810s and early 1820s.

Other industrial ventures that drew on advanced British and Continental scientific and technological understanding were the paper mills first established on the Brandywine in 1787 by the Philadelphia merchants Joshua and Thomas Gilpin. In subsequent journeys to England and the Continent, Joshua visited numerous manufacturing and production sites, making copious notes and sketches of what he heard and saw for subsequent transfer to the family's Brandywine works.

Wilkinson thus deftly shows us that we are as much indebted to our British and European forebears for their technological ideas as for their political, legal, and philosophical ideals. Americans often improved on European inventions, adapting them to their particular needs and environment, much as today we malign our international competition for doing the same with our technological expertise. To assume this international industrial transfer and adoption is but a recent, late-20th-century, one-way phenomenon is to overlook its 18th-century origins and the direction of its flow.

Further to the north in 1790, another, perhaps more famous, case of "industrial espionage" took place, one that initiated the mechanization of America's first major industry, textiles. It was in Pawtucket, Rhode Island, that Samuel Slater, a young British spinner, helped establish the nation's first waterpowered spinning mill based on knowledge of Richard Arkwright's spinning machinery. James Conrad, in an essay entitled "'Drive That Branch': Samuel Slater, the Power Loom, and the Writing of America's Textile History," is less interested in the story of the transfer of Slater's technological knowledge, per se, than he is in the subsequent role that Slater played in the development of this important industry. Some historians have argued that Slater might have been more productive if he had only adopted power-loom weaving after it became available in 1815 instead of surrendering industry leadership to the large-scale mills located first in Waltham and later Lowell, Massachusetts.

Conrad seeks to revise Slater's post-1815 career, by first examining what has been written about Slater by earlier generations of historians. These works developed the notion of two distinctive manufacturing systems: the Slater or Rhode Island system, focusing on small spinning mills utilizing child and family labor, with diversified production, and the Waltham (later Lowell) system of large integrated mills, utilizing young women workers supervised by professional managers and specializing in standard inexpensive goods woven on power looms. Later historians furthered this dichotomy by emphasizing Slater's supposed "failure" to adopt power looms when they became available. It was not until the 1960s and the emergence of the "new social history," which began to focus on the contributions of hitherto un- or underrecognized Americans, that this simplistic dichotomy began to undergo transformation.

By itself, Conrad's review of the historical construction of the Slater story is very revealing of the historian's craft, but he goes further than this. Drawing on previously unutilized source materials, Conrad finds that a number of Rhode Island mechanics had, in fact, developed power looms at least two years earlier than Lowell's 1815 Waltham loom. Slater knew of these loom experiments and encouraged similar efforts in his own mills. By 1818, the Almy, Brown, Slater mill at Smithfield, Rhode Island, was able to ship to Philadelphia "water loom shirting of Superior Quality." Conrad suggests that, encouraged by this success, within two years the Slater brothers may have been operating as many as 110 power looms, which, while not as extensive as the 175 power looms at Waltham, nonetheless still represented some 15% of the total looms in operation in the United States in 1820. Three years later the Slaters purchased the Jewett City Cotton Manufacturing Company in Connecticut, where they almost immediately installed power looms with the clear intent of fully mechanizing the cloth production process. Conrad thus argues that Slater "aggressively supported and effectively utilized the power loom and other post-1815 technologies long before most of his contemporaries," which leads us to a much needed reconfiguration of the Slater image itself and textile history more generally.

Textiles was not the only industry undergoing significant transformation in the early 1800s, for iron production was also increasing in importance, especially as the country entered into its nascent period of industrialization and required materials with the strength that only iron and, later, steel could provide. Although ironworks had been established as early as 1619 near Jamestown, Virginia, and in Saugus, Massachusetts, in the 1640s, by the late 18th century the center of the industry had shifted to the middle states, especially Pennsylvania. Most iron was produced on rural, small-scale, and largely self-sufficient "plantations" located close to forest lands that provided the charcoal fuel necessary to the process. In general, such operations were wasteful, especially of fuel, and thus iron commanded high prices, although the quality was often quite good. Historians have traditionally argued that the American iron industry only slowly shifted to coal- or coke-based production processes because of the abundance of relatively inexpensive charcoal fuels and that the nature of that charcoal meant American ironworks remained small and inefficient.

Richard Schallenberg and David Ault challenge some of the central assumptions of this traditional view in the essay "Raw Materials Supply and Technological Change in the American Iron Industry." They maintain that charcoal fuel cost was not the primary factor limiting large-scale improvements and enlargement of iron production. Instead, they suggest that accessibility to large-scale deposits of inexpensive iron ore was far more crucial and point out that where such deposits were available, the charcoal iron industry adopted techniques that allowed it to operate on the same scale and level of efficiency as coke-iron producers.

In contrast with older mid-Atlantic charcoal furnaces, Schallenberg and Ault discovered that the post–Civil War furnaces widely adopted in iron ore–rich areas of Alabama, Michigan, and Missouri were more heavily capitalized and utilized a variety of new technical innovations, including steam engines in lieu of waterpower. These technological improvements all contributed to the reduction of charcoal consumption. Equally important was more efficient conversion of wood to charcoal. With improved technologies, production increased from 35 bushels of charcoal per cord of wood early in the 19th century to 60–65 bushels per cord by the 1880s.

More efficient blast furnaces and charcoal production enabled iron producers to keep their fuel costs down, such that they were producing iron with approximately one-fifth the forest acreage per ton of iron as had been required in the antebellum period. This made it cost-effective to transport charcoal over distances of up to 100 miles, first along rivers and canals and later on the railroads, reducing dependence on nearby fuel supplies and making ready access to rich, low-cost ore in large quantities far more important. In the absence of such deposits, traditional low-output antebellum furnace operations could not justify the capital investment necessary to compete with coal- and coke-produced iron. Schallenberg and Ault have thus provided a much richer, more contextual, analysis of the iron industry's technical evolution over the course of the 19th century than some earlier, single-factor interpretations.

Dramatic developments in American transportation systems during the half-century prior to the Civil War were essential to industrialization. They helped to knit the new nation together by enhancing the ease, and dramatically reducing the cost, of shipping agricultural goods and raw materials, such as in the case of charcoal already noted, to increasingly urban and industrial areas, while at the same time facilitating the distribution of finished manufactured goods. Government, especially at the state level, had from the beginning been involved in the issuance of charters for turnpikes, canals, and railroads and, later, especially in the case of railroads, offered financial support by providing land grants along projected right-of-ways, which could then be sold or developed to offset the cost of building new lines. It was the steamboat, however, that first got the federal government into the business of regulating technology, which, while common in the late 20th century, was largely unknown in the early history of the United States. This is the story told by John Burke in his essay, "Bursting Boilers and the Federal Power."

Early Newcomen and Watt steam engines, utilized primarily for stationary mine pumping and factory applications, had been complemented by the development of smaller, high-pressure engines more suited to transportation applications very early in the 19th century. Oliver Evans's development of high-pressure engines in the United States in 1804 led quickly to the building of large numbers of steamboats, which plied both eastern and especially western rivers. By 1820 some seventy steamboats worked the Mississippi–Ohio–Missouri River systems, a number which rose to well over 700 by the outbreak of the Civil War. Low-pressure engines traditionally operated at 7–15 psi (pounds—of steam pressure—per square inch), while the newer high-pressure engines operated at levels upward of 100 psi. Increased steam pressures, in particular, led to explosions, many of which resulted in the tragic loss of life. The inappropriate use of traditional low-pressure boiler designs, the inferior materials of the day, inadequate safety equipment and procedures, and increased competition, which in turn led to excessive speed and overloaded operating levels, were major contributing factors.

In 1824, the year a steamboat explosion led to the first aborted call in Congress for regulation, Philadelphia manufacturers founded the Franklin Institute to promote technological research and industrial advancement. One of its first concerns was boiler explosions. The tragic explosion in 1830 of the *Helen McGregor,* which resulted in more than fifty deaths, led the Franklin Institute to authorize an extensive six-year experimental investigation into steam boiler explosions. Congress, prompted by the same explosion, also requested an investigation, and through the secretary of the treasury, the federal government authorized the expenditure of funds to support the Franklin Institute study, the first federally funded "scientific research grant" in the nation's history. The Franklin Institute research disproved several widely held, but erroneous, beliefs regarding the behavior of steam and boilers and uncovered new data about the strength of materials. Congress received two reports in 1836. After two years of wrangling and debate, it finally passed the first piece of regulatory legislation. Unfortunately, the law contained no inspection criteria and required only weak steam engineer qualifications, with the result that boiler explosions continued nearly unabated. Between 1841 and 1848 seventy marine explosions resulted in the deaths of over 600 people. By 1852 Congress was ready to take further action, which it did under the technical leadership of an Illinois engineer, Alfred Guthrie, whose report echoed the earlier 1836 Franklin Institute recommendations. This time the new legislation required far more stringent design and operating requirements and established a board of inspectors, with the consequent result that steamboat explosions declined sharply. By passing such legislation, Congress had for the first time asserted its power to regulate technology and thereby established a precedent for regulatory legislation, which today is far more extensive, reaching into almost every corner of American life.

As the transportation sector and the iron and steel industries grew and matured during the course of the 19th century, so too did American manufacturing and industrialization; indeed, in many ways, they evolved in tandem. A central feature in the emergence of large-scale American manufacturing was the ability to mass-produce large numbers of items or parts of a "uniform" or interchangeable nature. Often referred to as the "American System of Manufacturing" (so named by the British in deference not so much to its origins as to the extent to which it was ultimately perfected in the United States), such production relied heavily on sophisticated and increasingly accurate machine tools. Many scholars have assumed that with the increased use of such machines, which by 1880 were able to achieve accuracies of 1/1000th of an inch, that artisanal craft skills declined. Historical reports of trials in which similar-model firearms were disassembled and reassembled at random with all units perfectly functional have led scholars to assume that the mechanical ideal of interchangeability had been largely achieved, at least in the military arms industry, by roughly midcentury. From here, the ideal, and the techniques for achieving it, spread to other industries such as sewing machines, bicycles, and in the 20th century to automobiles.

As might be expected, turning the mechanical ideal of interchangeability into reality was neither a straightforward linear path, nor was it as easy as some accounts might have us believe. Neither did it have the "deskilling" impact that might be anticipated, which is precisely the argument put forward by Robert Gordon in his essay, "Who Turned the Mechanical Ideal into Mechanical Reality?" Indeed, Gordon, based on the close physical examination of a series of tumblers in musket lock mechanisms produced between 1820 and 1880, argues exactly the opposite. That is, while machines tools usefully removed large amounts of metal, thereby reducing much physical labor, precision filing for final fit by "artificers" was still required to attain the necessary tolerances for interchangeability. Skilled workers had four types of skills—dexterity, judgment, planning, and resourcefulness. Gordon posits that, while planning skills may have been circumscribed by division of task and the increased use of filing jigs to hold parts in place, the other skill requirements, in fact, increased in order to attain needed levels of precision. Thus, throughout most of the 19th century, many of the skills necessary to interchangeability were not "built into the machines" but remained in the hands of increasingly skilled artisans.

The effects of technological change on the industrial workforce were equally complex during the late 19th century, an era sometimes referred to as the Second Industrial Revolution because of its rapid and extensive mechanization, the diffusion of mass production, and the emergence of new organizational procedures. As with the earlier industrial period, historians have debated whether labor's skills, work experiences, and socioeconomic status were enhanced by the new capital-intensive technology or undercut by deskilling, mass homogenization, and a corresponding loss of bargaining power. In an insightful case study entitled "From Iron to Steel: The Recasting of the Jones and Laughlins Workforce between 1885 and 1896," David Jardini analyzes the changing skills, wages, and productivity characteristics of workers in a single plant, that of Jones and Laughlins in Pittsburgh, as it underwent rapid and extensive technical change. In particular, he looks at the shift from the puddling process for iron refining to the Bessemer process of steel production, the "quintessential technological development of the period."

Following a brief discussion of the technologies involved, Jardini looks comparatively at the wage records, skill training requirements, and productivity characteristics for iron refining and steel production and the primary-stage rolling operations associated with each process. Interestingly, he finds both expected and unexpected changes. A major expected effect was "a vast improvement in the productive efficiency of the plant." For example, it required 29.7 man-hours/ton to produce wrought iron in 1887, but only 3 man-hours/ton for steel in 1892. Even allowing for the higher capital investment in the Bessemer process, the steel plant was some three times more efficient than the puddling operation it replaced. Again, not unexpectedly, the skill levels required in the earlier puddling process were generally higher, with perhaps a third of the workforce requiring four or more years training. In contrast, three-quarters of all jobs involved in the Bessemer process required less than a year of training. Such findings would seem to support the deskilling arguments of many labor historians; however, a wage analysis reveals that after an initial period of upward fluctuation, Bessemer plant wages settled back to a level in constant dollars nearly double that of the iron department a decade earlier.

To explain the disparity between apparent deskilling and presumed loss of bargaining power and unexpected increases in real wages, Jardini offers several suggestions. He uses the idea of "strategic workers" with "diagnostic skills." He points out that in a continuous process like that involved in Bessemer steel making, an error or breakdown in even a small part of the process could result in extensive delays, massive damage, and costly losses. Thus, even less skilled workers may well have had an important level of control and bargaining power, recognized by management, which compensated them for their higher levels of responsibility and as a hedge against disgruntlement. At the same time, Jardini argues, a workforce that lost specific craft skills related to puddling may well have learned new integrative diagnostic skills associated with the faster-paced continuous process. Such a "reskilling" would thereby strengthen workers in the face of managerial efforts to control the work process. Jardini concludes by asking "to what extent can a particularly powerful or well-organized interest group manipulate the course of technological change so as to best serve its own narrowly defined interests?" His evidence runs counter to much traditional interpretation that would favor management's power over labor and suggests once again that we must probe well below the surface of technological change to examine the "unanticipated yet profound social changes" it occasions.

Not all technological developments in the 19th century occurred in basic industries like textiles and steel. As the century drew to a close, important developments also took place in consumer-oriented technology and products. Among them were the mass production and sale of such items as sewing machines, bicycles, and, as described by Reese Jenkins in "Technology and the Market: George Eastman and the Origins of Mass Amateur Photography," cameras. Prior to about 1880, photography was so technically complex that only a handful of amateurs tried to master the necessary skills, by and large leaving the field to professionals. During the last two decades of the century, however, George Eastman, a Rochester, New York, entrepreneur, revolutionized photography by enabling almost any novice to "snap" the shutter on a simple box camera and have the new gelatin-coated roll film developed and printed at a remote factory. Jenkins examines the traditional heroic inventor myths surrounding Eastman for a deeper understanding of the emergence of this new technology.

Originally a bank clerk and bookkeeper who became interested in amateur photography, George Eastman involved himself in the manufacture of photographic plates coated with dry gelatin as the carrier of photosensitive halogen silver salts as a replacement for the then-standard wet nitrocellulose or collodion process. Increased competition and lower profit margins shortly led Eastman and a local camera and plate manufacturer, William Walker, to substitute flexible roll film for their dry plates. Focusing their attention on the roll holder, the film itself, and the production machinery for manufacturing the film, Eastman and Walker developed a system based on mass-produced interchangeable units. Despite initial successes with the roll holder, the film itself remained extremely difficult to use even for professional photographers, which led to the system's rejection even in that quarter. Drawing on the experience gained from this false start, Eastman reoriented his market focus from the professional to the amateur.

The new approach necessitated developing an easily manufactured, simple-to-use roll film camera, which by late 1887 had been finalized as the "Kodak," and centralized film processing facilities. The camera, which came loaded with a 100 exposure roll of film, initially cost twenty-five dollars, and by returning the whole camera to the factory with ten dollars, the customer received back his processed pictures, the camera, and a fresh roll of film. The ease of "You press the button—we do the rest" caught the public's imagination, and demand for the new cameras grew rapidly. Subsequent improvements refined both camera and film processing, such that by the late 1890s Eastman and Walker's camera had transformed the photographic field from one dominated by small numbers of professionals to a mass market of amateurs. Through an aggressive patent protection policy Eastman was able to establish a major business enterprise, not only in the United States, but in the international market as well. This enterprise reflected a shift from a decentralized, craft-oriented mode of production to one characterized by centralization, mechanization, and mass production, traditional hallmarks of the modern period of American industrial history.

By the dawn of the 20th century, it had become increasingly clear in many areas that the old empirical, cut-and-try methods of technological innovation were no longer completely adequate. Thus, many corporations, including eventually Eastman Kodak, turned to the establishment of more scientifically based research and development (R&D) laboratories as a source of new ideas and product lines. Although there were clear predecessors, such as Thomas Edison's research lab at Menlo Park, the first real "industrial" R&D lab was that established at General Electric in 1900, which is the focus of George

Wise's essay, "A New Role for Professional Scientists in Industry: Industrial Research at General Electric, 1900–1916."

Prior to 1900 most scientists served industry as either part-time consultants or, if full-time, in applied positions such as involved with routine testing, but not original research. As businesses at the turn of the century underwent extensive consolidations and those that remained became increasingly competitive, it became clear to industry leaders that something more than part-time consulting scientists would be required to maintain technical leadership. To this general end, and as a specific response to technical threats to its dominance in the electric lighting business, General Electric decided to establish a "scientific" research lab, the key element of which was isolation from day-to-day production responsibilities. In doing so, GE established a new role, that of the industrial research scientist, entailing a blend of "research freedom" with "practical usefulness." Such employment appealed to people who wanted to be professional scientists, but not in the traditional academic setting, or to those with inventive inclinations but who were unwilling to take the risks of entrepreneurship.

Wise briefly focuses on the careers of three industrial scientists to illustrate the developmental period of GE's industrial research efforts. Willis Whitney, a young assistant professor of chemistry at MIT, was lured away by the opportunity to do more extensive independent research than his university teaching position and responsibilities allowed. The not inconsequential starting salary of $2,400, which corresponded to the pay of a full professor, was also a factor. Whitney's particular talent and role as the laboratory's director from its founding until 1932 was to inspire effective research in the team of scientists he would subsequently assemble. Among them was William Coolidge, a German-trained physical chemist and a former student and colleague of Whitney's at MIT, who came to GE in 1905. Coolidge was a talented scientist with a bent for inventiveness, but he had neither the resources nor the interest to enter the business on his own. His initial work on tungsten filaments contributed importantly to the corporation's ability to compete effectively in the lamp business. While neither Whitney nor Coolidge were devoted to knowledge for its own sake, preferring instead to focus on the usefulness of science, Irving Langmuir more closely represented the professional scientist. Also trained in Germany, Langmuir found his extensive teaching duties, limited research opportunities, and meager salary at Stevens Institute of Technology to be overly constraining. Moving to Schenectady in July of 1909, ostensibly for a summer position, Langmuir quickly agreed to stay on in a full-time capacity. While his work certainly focused on practical applications, most notably the gas-filled incandescent lamp, he pursued his research in such a way that he continued to make significant scientific contributions, on which he frequently published in the leading journals of the period.

By 1916 GE's lab was fully institutionalized within the corporation, in large part because it had achieved real results in fending off the competition. The lab was by no means a carbon copy of an academic scientific research laboratory. It entailed ambiguous distinctions among research, development, and engineering, and certainly its scientific production was lower. However, as Wise puts it, "the important thing is not that GE scientists published so little, but that they published at all." Other research labs were influenced by GE's experiment, including Eastman Kodak's and General Motors's, but while the lab was influential, it was not without faults and limitations, often born of its inward-looking focus. Neither was GE's success in industrial research typical, for other efforts such as that at Westinghouse in 1917 failed. Nonetheless, the lab's establishment and successes did signal a major shift in industrial research and development.

While mass production and science-based technical expertise had come to characterize much of America by the early 20th century, they were by no means ubiquitous. Gail Cooper's "Custom Design, Engineering Guarantees, and Unpatentable Data: The Air Conditioning Industry, 1902–1935" offers an important counterpoint to the assumption that mass production has totally dominated 20th-century technology. Factory air conditioning first appeared in 1902 with machinery not mass produced but custom designed for each specific installation. Cooper argues that far from dying out, custom design remained a vital system of production for at least three decades. Although it did draw on manufactured subassemblies, custom design involved "making" or "building" individually tailored systems. Cooper examines the experiences of Willis Carrier, often referred to as the "father of air conditioning," and the Carrier Engineering Corporation (CEC) founded in 1914, to illustrate the central elements of custom design.

Factory air conditioning emerged more out of a need to control humidity levels than to lower temperatures. Known as "process air conditioning," it helped keep cigarette machines from jamming, candy from changing color, and cotton fibers from breaking due to inelasticity. Air conditioning companies such as Carrier thus "sold air" in that they guaranteed specified seasonal humidity and temperature ranges for each installation, in contrast to machine guarantees. What gave CEC its edge in the business was Carrier's and later his engineers' ability to analyze the complex "dynamic" relationships between factory buildings, the machines they contained, and the processes being conducted therein. From this knowledge base Carrier and his engineers designed and built appropriate systems that "manufactured weather." Because the air conditioning field was expanding rapidly, such individual "experience" quickly became "expertise," which was guarded very carefully as proprietary knowledge. Although subsequently turned into corporate knowledge through the creation of a "Confidential Data" file, air conditioning expertise was as much an art as it was an engineering science. The success of CEC's business thus rested on three mutually supporting elements—custom design itself, engineering guarantees of specific climates, and a body of unpatentable data.

To be successful, custom design engineers needed to control not only the buildings themselves but increasingly the activities that went on within them, causing conflicts with both management and workers over the operation of both production processes and the air conditioning system itself. Early experimental failures at product drying in the macaroni and tobacco industries were a result of CEC's failure to recognize the complex relationships between natural materials and environmental conditions, something that the more "sensual" craft approach to production handled better. One response to these failures was to fall back on science, with the result that the American Society of Heating and Ventilating Engineers established and supported a research laboratory in Pittsburgh in 1918. The appeal to science and the lab enhanced the influence of engineers in their struggle for control over factory production within the context of the artificial climate ideal.

By the late 1920s, however, the air conditioning industry shifted increasingly to mass production, especially through the manufacture of equipment for comfort cooling, exemplified by the development of standardized, mass-produced window units, the antithesis of "manufactured weather." Under this new approach, custom engineering companies like CEC were no longer able to define and control the technology. Cooper concludes that the history of custom design suggests that we must beware of simplistic linear histories of industrialization that would have us see a straight line path from craft to mass to today's flexible production. The history of any system of production must reflect the tangled web of relationships among management, labor, and consumers and, in this case, also the engineers who helped to shape the system.

Promoters of a given technology do not always know precisely how the end users will actually put it to use, despite often times having very strong ideas regarding proper use. This was an issue for air conditioning engineers who constantly had to deal with factory operatives who opened windows to regulate humidity and heat in direct conflict with system design and operation criteria. Such "misuse" was also an issue in the telephone industry. In light of today's telephone advertising, it is perhaps hard to conceive that the industry did not always say "reach out and touch someone," but that is precisely the argument set forth in Claude Fischer's "'Touch Someone': The Telephone Industry Discovers Sociability."

Within roughly a decade and a half of Alexander Graham Bell's invention of the telephone in 1876, there were over a quarter million units in service, a figure that grew especially rapidly immediately after the original Bell patents expired in 1893–94. By 1929 some 42 percent of American homes had phone service. Nonetheless, as is often the case with new technologies of a somewhat "revolutionary" nature, how to use the telephone was not self-evident, and industry executives felt they had to "educate" the public on its proper use. Coming out of the telegraph industry for the most part, early phone executives believed the proper use of the new technology would be similar to the predecessor technology's "serious" uses: news services, business needs, ordering of commercial goods and services, train schedules, and, of course, emergencies. What they did not encourage, in fact even discouraged, was "sociability"—keeping in touch with friends and family. Such "conversation" was deemed "unnecessary," "frivolous," and "trivial." Industry executives often associated such use with women who had nothing better to do than tie up valuable phone lines from more serious business. Only in the mid-1920s, first for long-distance, when calls were metered for time, and later in the decade for local calls, did the phone industry begin to stress "comfort and convenience" in its advertising. Suddenly sociable conversation was promoted with such injunctions as "Call the folks now!" "No girl wants to be a wallflower" held out alluring promises of enhanced social life if only one had a telephone. Why was the phone industry suddenly reaching out?

Fischer analyzes several possible reasons for the industry's lag in adjusting to actual customer use patterns. He rejects economic and technical explanations, such as rate structures, the continuance of party lines, and limited numbers of rural toll lines. Instead, he attributes the industry's recalcitrance to deeply entrenched ideas carried over by executives weaned on the telegraph system. Only after actual customer use, which in fact spurred phone subscriptions, especially in rural areas, increased to the point where it could no longer be ignored did the industry become "sociable." In addition to the obvious counterintuitive enjoyment of the story, Fischer suggests that the history of telephone use contains several valuable lessons for understanding technology today. Among them are the recognition that developers and sponsors of new technology do not always understand the final use patterns, which will as often as not be determined by the consumer rather than solely constrained by technical and economic attributes.

Although historians of technology have tended to focus on the process of industrialization and its consequences for workers and society, for much of American history the country was more agricultural than it was industrial. This is not to say that farmers did not utilize technology, nor that inventors and entrepreneurs ignored their needs or market potential. Indeed, from Thomas Jefferson's early experiments with plowshares and moldboards, to Cyrus McCormick's development of mechanical reapers, to gasoline-powered tractors, farmers have depended on and generally welcomed technological innovations. Most such innovations were piecemeal, extending older skills and combining modernity with a tradition of self-sufficiency; however, the development and subsequent adoption of scientifically engineered hybrid corn in the late 1930s was fundamentally different. As described by Deborah Fitzgerald in "Farmers Deskilled: Hybrid Corn and Farmers' Work," hybrid corn was "an all-or-nothing proposition" that did not allow combining old and new skills. In fact, with hybrid corn farmers no longer fully understood their own operations without the aid of experts.

Prior to the introduction of hybrids, corn farmers were intimately involved in corn improvement efforts. Beyond the physical effort, farming required mental skills—what to do, how to do it, and when—skills that were far more important and "infinitely complex." Farmers' livelihoods depended on their ability to "read" their corn for which varieties would grow and yield best under what soil and climatic circumstances and, perhaps most important, what ears to save for next year's seed stock. Nonetheless, by the 1910s, it was becoming evident to farmers and seed breeders alike that the farmers needed more help in selecting corn for higher yields than traditional methods of judging corn allowed, especially in the complex areas of disease and drought resistance.

Corn breeding, although by no means new or particularly scientifically complex, quickly developed to a point that the scale and complexity of record keeping necessary in formal breeding programs effectively "locked farmers out of the process." The rapid adoption of hybrids in the decade after their introduction in the mid-1930s meant that farmers became even more "deskilled" in this area of their work. Varieties were no longer visually distinguishable from each other, nor were their growing properties self-evident. Since hybrids cannot be successfully reproduced from year to year, farmers had to purchase new seed on an annual basis, thereby furthering their dependence on geneticists and seed dealers. Fitzgerald concludes that the issue is not whether hybrid corn is "good" or "bad" but rather "understanding how such technologies usurp the skill and knowledge of existing workers."

As should be evident from the preceding discussion, the themes of labor and work are intimately entwined with that of technological change. Whether it be in the factory, at the foundry, or on the farm, technological change was very complex and contextually bound. We cannot simplistically assume that it always deskilled workers and thus deprived labor of its power and control over the work process or that the industrialists who introduced technological changes were always able to dictate how customers used new technologies. Thus, it should not come as a surprise that the changes women encountered during the interwar years with the adoption of household appliances were just as complex and socially mediated as those of the Bessemer process, the telephone, and hybrid corn.

Much technology is developed, promoted, and even adopted by end users as a way to eliminate drudgery and reduce labor. The traditional view of household technology holds that beginning about 1920 fairly widespread middle-class adoption of such appliances as electric irons, washing machines, and vacuum cleaners removed much of the drudgery of housework, while simultaneously reducing the time required to perform such tasks. In turn, middle-class housewives found themselves free for more rewarding activities or, somewhat later, to enter the paid work force outside the home. Certainly such machines did remove a great deal of physical drudgery. For example, a housewife no longer needed to take the rugs outside for their seasonal "cleaning" with a carpet beater, nor did she need to hand wash clothes in a tub or heat up the wood or coal stove on a hot August day in order to warm her iron. What is perhaps counterintuitive, and therefore surprising, however, is that, on balance, for women who did not also work outside the home, such laborsaving devices did not reduce the overall time spent doing housework. How this could be the case is the substance of Ruth Schwartz Cowan's essay "The 'Industrial Revolution' in the Home: Household Technology and Social Change in the 20th Century."

The answer to this seeming paradox lies partly in the technologies themselves, but more broadly in elements of the societal context within which they were embedded. First, the availability of domestic servants declined, especially due to limitations on immigration imposed by World War I and later in the early 1920s by more restrictive immigration laws. This meant that housewives took on additional, previously delegated tasks such as child rearing and cooking. At the same time, the frequency with which one did tasks increased. Thus, the availability of washing machines meant families changed clothes more frequently, often daily, with a corresponding increase in the workload of washing and ironing. While the vacuum cleaner eliminated the need to beat carpets, a semiannual chore, it led to more frequent, perhaps weekly, vacuuming. Also, the rapid middle-class diffusion of the automobile, which coincided with the decline of delivery services, often meant that housewives spent upward of a full day per week shopping and running errands, tasks, which while not necessarily physically onerous, were no less demanding of time.

In conjunction with the changing technologies and associated tasks of housework was an accompanying ideological change, one which suggested household chores should no longer be viewed as "trials," but rather as "emotional trips" that expressed a wife's and mother's love for her family. Thus, scrubbing the bathroom was more than mere cleaning, it was "an exercise in protective maternal instincts." Through an analysis of advertisements in women's magazines from the interwar years, Cowan deftly shows how promoters of new household appliances and products played on women's "guilt" regarding dirty clothes, dirty sinks, or bad breath. As with hybrid corn, the issue is not whether changes in household technology were necessarily "good" or "bad," nor the enthusiasm with which most were readily adopted, but rather clarification of the societal complexities engendered by their adoption.

Implicit throughout the essays in this volume, even if not always explicitly discussed, has been the central role engineering has played in the industrialization and modernization of America. Whether it be in the construction of transportation networks, the electrification of the nation, or the mass production of consumer goods, the building of modern technological systems was highly dependent on engineering knowledge and skills. By 1900 older forms of acquiring and utilizing technological knowledge such as the on-the-job training that built the Erie Canal or that was learned within an industrial "shop culture" and carried the nation through most of its first industrial revolution were no longer adequate. Engineering knowledge was institutionalized largely after the Civil War through the formation of professional engineering societies and college-level courses of study, such that by the turn of the century, older engineering apprenticeship systems had largely disappeared.

Even in 1900, however, most engineering educators still sought primarily to train "good practical engineers," suggesting that, despite engineering's professionalization, there was still a strong element of the practical involved. By 1960, however, as Bruce Seely chronicles in "Research, Engineering, and Science in American Engineering Colleges," engineering education looked quite different, with a much stronger emphasis on "scientifically derived theory," the language of which was mathematics. Seely argues that the shift in engineering education from practical to theoretical was largely a result of changes occurring in the funding of engineering research during the same period.

What little academic engineering research there was around 1900 tended to come out of faculty summer consulting practices, for during the school year the same extensive teaching loads that had encouraged scientists to gravitate to GE's new research laboratory mitigated against much serious engineering research. Nonetheless, at about this time, many of the land grant colleges, such as the University of Illinois in 1903, established engineering experiment stations. Their intent was somewhat parallel with the agricultural experiment stations of the previous generation. As such they had a public service mission, and their research emphasis focused on very practical problems—drainage systems, culvert design, materials performance—and often involved routine testing functions. By 1931 there were forty such engineering experiment stations in place. Despite the needs of the newer science-based fields of electrical and chemical engineering and the prompting of a few European-trained faculty, only a handful of American engineering schools conducted research that went much beyond routine experiment station testing. Putting it kindly, Seely suggests the pace of change was "deliberate."

Following World War II, the federal government began to provide research funds of unprecedented magnitude to universities, especially for military related work that required a more scientific focus. While by no means purely scientifically based, much more theory was involved than in the prewar era. State universities such as Illinois and Georgia Tech and private institutions such as MIT and Caltech, seeking to capture the new funds available, quickly transformed their research focus from practical applied work to more fundamental theoretical research. A synergistic relationship developed between the schools that embraced scientific engineering and received federal grants and the federal government, allowing the development of large research programs.

The changes in engineering research were paralleled by related changes in curricula. Academic engineering research generally requires graduate students, so schools desiring to ride the federal gravy train began to alter their curricula to prepare students for graduate school and "scientific" research, instead of for the practical needs of industry. Seely offers an interesting tabular comparison by decade from 1920 to 1970 of the evolving civil engineering curriculum at Texas A&M, which reflects this shift to an increased scientific emphasis. Despite the shift among academic engineers toward the values of science, they did not become scientists, for they still had very real connections to the world of doing, with an emphasis on the requirements of design.

Nonetheless, engineers in industry frequently came to different conclusions regarding the value of science-based education and academic research, complaining that its theoretical focus had little practical application. Clearly the boundaries between industrial engineering and engineering science are blurred. In explaining the timing of the shift in emphasis, Seely suggests we must look beyond the argument that the increasingly scientific needs of mid-20th-century technology required the reorientation of engineering education, to the social factors, such as the large postwar infusion of research funds, that actually occasioned it.

Somewhat analogously, the technical performance of a given technology is not always the determining factor in its "success" or "failure." Just such an example can be found in Gregory Kunkle's essay, "Technology in the Seamless Web: 'Success' and 'Failure' in the History of the Electron Microscope." Microscopes are often perceived as embodying the essence of the search for objective knowledge and scientific truth, and the electron microscope is by far the most powerful. Yet, in its commercial development, the electron microscope reveals a fascinatingly complex history in which societal factors played a central role in the final shaping of the artifact. At issue was the competition between two different competing lens systems—*electromagnetic* and *electrostatic*—produced by two different firms—RCA and GE—using two different approaches to developing and marketing the instrument.

The major problem GE encountered in designing an electrostatic lens was to control the aberrations or blurring caused by variations in either lens field strength or in electron velocity. GE scientists recognized that voltage variations in the electron "gun" would be offset proportionately by variations in the lens system, thereby eliminating any need for elaborate voltage regulation. This would make the device easy-to-use and keep the price low. Despite a magnification capability ten times that of the light microscope, GE's early attempts to market the new instrument in 1944 met with little success. Kunkle argues, based on internal corporate evidence and on a comparison with German and Japanese success with similar instruments, that had

GE committed the necessary funds and effectively marketed the device, it would have found applications in industrial quality-control testing, which in the 1940s and early 1950s constituted perhaps two-thirds of the total microscope market.

RCA, in contrast to GE, very early in the development of its instrument, established close institutional and individual ties with research scientists who needed the higher resolution capabilities promised by the electromagnetic version. Through these ties, which included the establishment of a special electron microscopy research laboratory at RCA, the company was able to establish the credibility of its instrument within the scientific community. As scientific research done at RCA's lab became increasingly well-known, the company, in Kunkle's words, "became identified with state-of-the-art microscopy," which further ensured commercial success. GE, in contrast, did not seek out similar ties with the scientific community and hence failed to "sell" practitioners on the workings and viability of the electrostatic instrument. For GE, the electron microscope was just another product like electric irons and refrigerators; if it did not yield immediate profits, it could be scrapped. This, indeed, was precisely its fate after a second unsuccessful attempt at production in the late 1940s and early 1950s. Drawing on sociology of technology terminology, Kunkle concludes by suggesting it was RCA's "inclusion" of a widespread "community of practitioners" or relevant and supportive "actors" in the social network in which the electron microscope was placed that ultimately accounts for the company's "success" rather than any technical superiority per se.

Although debates over the final form a technological artifact or process will take can sometimes continue over long periods of time and even involve extensive controversy, once resolution or "closure" has been reached, it can sometimes appear that a technology almost takes on a life of its own. This seeming autonomy is often what gives the public a feeling of helplessness, especially in the face of large-scale technological systems like those associated with the automobile or electric power generation and distribution. While historians and sociologists of technology have disavowed any notion of technology as being "autonomous," they have recognized that large technical systems, once they become deeply embedded within society, do seem to take on a certain "momentum." Because the social, economic, political, and technical elements of such systems are so closely intertwined, almost as if they were in a "seamless web," only a major, perhaps catastrophic, confluence of forces can seemingly alter the course of events.

It is precisely just such a societally embedded technological system faced with the possibility of catastrophic change that Richard Hirsh and Adam Serchuk, utilizing a framework of analysis suggested by historian Thomas P. Hughes, examine in "Momentum Shifts in the American Electric Utility System: Catastrophic Change—or No Change at All?" Hirsh and Serchuk focus on deregulatory changes following the oil crises of 1973 and 1979. In particular, they look at California's application of the 1978 federal Public Utility Regulatory Policies Act (PURPA), which encouraged enhanced energy efficiency by promoting "cogeneration"—traditionally the utilization of waste heat from industrial processes to generate electricity to be used locally. The act, however, also required utilities to link all producers to their grids and purchase their excess power. Although unintended, PURPA began a process of utility deregulation, one that brought various new "actors," such as California's wind farms, into the utility industry, which threatened to destabilize the entire centralized industry.

A second stress was caused by the environmental movement, which instigated legal challenges to new power plant construction. Environmentalists encouraged the use of renewable sources coupled with energy conservation and efficiency measures, sometimes referred to as "demand-side management" (DSM), to counter both pollution and projected increases in energy consumption. After an initial period of litigious tension, many utilities came to recognize the value of cogeneration, windfarms, and DSM, especially with regard to their potential to delay costly new power plant construction.

While Hirsh and Serchuk accept the incorporation of cogeneration, decentralized wind power, and DSM techniques as "significant departures from past practice," they see them as more akin to "tactical responses" to system stress than a reflection of "radical" or deep structural change within the utility industry. They view utility company incorporation of wind turbines and employment of DSM as ways "to conserve the utility system with as little change as possible." They also suggest there was nothing inherent in the alternative technologies themselves dictating their use to either destroy or conserve the existing system, which lends further support to the argument for the contextual nature of technology.

* * *

Throughout this overview a number of important themes and topics have recurred. Most important, and what holds the collection of essays together in many ways, is the argument that understanding the nature of technology and technological change over time requires that technologies be viewed in their broader cultural context.

Whether one is analyzing steam boilers, air conditioners, or electron microscopes, it is the cultural embeddedness they reflect that fully explains a technology's historical development and its implications.

Several other more specific topics have provided recurring threads to the fabric of this narrative as well. Chief among them has been the process of industrialization, which has come to characterize much of modern American society since the 19th century. Closely linked to industrialization have been the implications it has had for workers, not just in the factory setting, but also down on the farm and in the idealized middle-class household. Also implicit within the history of technology and industrialization, even if not always discussed explicitly in these essays, has been the nature and role of engineering as distinct from, but related to, science. While these topics by no means exhaust important issues and questions relevant to the history of technology, they do constitute some of the major themes historians of technology have emphasized.

Finally, most of the authors of these essays place their scholarship within the broader historiographic context of the history of technology. Thus, the student of technology has here an opportunity to view historians of technology working within the broader context of a preexisting body of scholarship to which they are both responding and contributing, much in the same way that technologists work within and add to a preexisting sociotechnical context.

Brandywine Borrowings From European Technology

NORMAN B. WILKINSON

One day in 1797 the French traveler Rochefoucault-Liancourt was shown around a large, modern merchant flour mill on Brandywine Creek in Wilmington, Delaware, by Thomas Lea, one of the owners. The visitor's attention was called to a number of recently installed automatic devices that moved the grain and processed the flour through successive operations with scarcely a touch of human hand. Lea commented with satisfaction on these improvements, the work of the Delawarean Oliver Evans, an "ingenious mechanician," and he apparently indulged in some quiet Quaker boasting about the advances being made in American milling technology.

When Liancourt put down his impressions of this visit he wrote:

> Like a true American patriot, he [Thomas Lea] persuades himself, that nowhere is any undertaking executed so well, or with so much ingenuity, as in America; that the spirit, invention, and genius of Europe, are in a state of decrepitude . . . whilst the genius of America, full of vigour, is arriving at perfection.

Such patriotic enthusiasm, characteristic of many Americans he had met, continued Liancourt, did not prevent them, however, from adopting all the good inventions developed by Europeans by which they could improve their mills.[1]

It is our purpose here to illustrate how certain industries of the lower Brandywine Valley, in the quarter century between Hamilton's *Report on Manufactures* of 1791 and the passage of the first protective tariff in 1816, were under obligation to Europe for the methods, the machin-

At the time of publication of this article, DR. WILKINSON was Research Associate at the Hagley Museum of the Eleutherian Mills-Hagley Foundation in Wilmington, Delaware. The author wishes to acknowledge the research assistance of his colleagues at the Hagley Museum in the preparation of this article, which was first presented as a paper at the December 1961 meeting of the Society for the History of Technology in Washington, D.C.

[1] Duc de la Rochefoucault Liancourt, *Travels Through the United States of North America, 1795, 1796, 1797* (London, 1800; 2nd edn., 4 vols), Vol. 3, pp. 496-97. Hereafter cited as Liancourt, *Travels.*

ery, and the "know-how" in the operations of their mills. This can only be a sampling, for by Liancourt's own estimate, there were some sixty to eighty mills on the stream in the 1790's, "almost all of different descriptions, such as paper, powder, tobacco, sawing, fulling, and flour mills. . . ." [2]

We are inclined to regard this figure of sixty to eighty mills as an exaggeration, but that the four or five miles from Wilmington upstream was one of the nation's most concentrated industrial areas is borne out by the remarks of many other visitors and by the attention given it in such promotional publications as Morse's *Geography* and *Gazeteers* and in Niles' *Register*. In his *American Gazeteer* of 1804 and 1810 Morse noted that "Wilmington and its neighborhood are probably already the greatest seat of manufactures in the United States." [3]

1. Textiles and Textile Machinery

Although we cannot identify him, it seems likely that Samuel Slater had a counterpart on the Brandywine: another English textile workman with a keen memory and mechanical talent, employed by Jacob Broom, a signer of the Constitution from Delaware. Broom was a wealthy entrepreneur, a banker and land owner, with a hand in many commercial ventures; in 1795 he established a carding and spinning mill about four miles north of Wilmington. His factory was described as "very Compleat, all the carding, woofing and spinning is done by water and machines which are excellent in their performance . . . very few excellent Cotton Stuffs are imported as those made here. . . ." [4] All the machines and implements were well made and constructed on Arkwright's plan. Broom employed fifteen workmen, all Englishmen, and fifty more were expected.[5]

This effort to duplicate what Samuel Slater and Moses Brown were achieving at Pawtucket, begun only a few years earlier, was short lived. Broom's mill was totally destroyed by fire early in February, 1797, with a loss of $10,000. The Delaware legislature granted him the right to conduct a lottery to raise money for rebuilding, but after some unsuccessful efforts Broom abandoned the venture. In 1802 his mill property was sold to E. I. du Pont, and on the site du Pont erected his

[2] Liancourt, *Travels*, Vol. 3, pp. 492-93.

[3] Jedidiah Morse, *The American Gazeteer* (Boston, 1804, 1810). Under *Delaware*.

[4] Thomas Rodney, "Propositions for new & Usefull Inventions & Improvements by Thomas Rodney Esq. Philosophical Tracts and Journals, July 17 to August 20, 1795." Historical Society of Delaware.

[5] Liancourt, *Travels*, Vol. 3, p. 502.

powder factory between 1802 and 1804. Not long after his own mills began producing powder, du Pont, writing to his father, Pierre Samuel du Pont de Nemours, in France in October, 1804, explained that he was compelled to write in code because

> the greatest danger to my business is that of attracting the attention of the English. . . . They employ all possible means to prevent the establishment of manufactures here. They burned my predecessor's cotton mill, and might easily try to do the same to my mills.[6]

If du Pont's charge was true—and we have come across no corroborating evidence in other sources—the English were adding arson and peacetime sabotage to their restrictions on the migration of mechanics and the export of machines and machine drawings in their efforts to retain their industrial leadership and monopoly. But du Pont was an Anglophobe with good cause to dislike the English.

The ease with which English artificers evaded the ban on migration is illustrated in the careers of George and Isaac Hodgson, mechanics from Manchester, who established a cotton spinning machine works a short distance below the du Pont mills in 1811. Declaring themselves to be farmers, an occupation not on the banned list, they wrapped their tools, marked them as fruit trees, and then took passage to Ireland, presumably their new home. From Belfast they sailed to Philadelphia, sending their "fruit trees" by another vessel, however, to lessen the danger of being apprehended by port officers as illegally migrating mechanics. The Hodgsons made it safely, claimed their tools in Philadelphia, and were soon operating the Brandywine Foundry making spinning machinery.[7]

The year 1811 was an opportune time to get into this business, and the Hodgsons quickly earned a reputation as manufacturers of textile machines. John Vaughan of Philadelphia and E. I. du Pont recommended them to General Thomas Pinckney when he was considering entering the textile business in South Carolina in 1812. And Charles Willson Peale, artist, museologist, and manufacturer, at du Pont's suggestion, apprenticed two of his sons, Titian and Franklin, to the Hodgsons to learn the machine trade in preparation for careers as

[6] B. G. du Pont, trans. and ed., *Life of Eleuthère Irénée du Pont, from Contemporary Correspondence* (Newark, Delaware: 1923-1927; 11 vols. and index), Vol. 7, p. 15. Hereafter cited as *Life of E. I. du Pont.*

[7] Memorandum Book of Deborah Hodgson Jones, privately owned. For details on English restrictions and the industrial immigrant see Herbert Heaton, "The Industrial Immigrant in the United States, 1783-1812," *Proceedings of the American Philosophical Society*, Vol. 95, No. 5 (October 17, 1951), pp. 519-27.

cotton manufacturers in the Peale factory at Belfield, near Philadelphia.[8]

Nearby, a little distance upstream, and using some of the Hodgson machines, was the spinning mill of Duplanty and McCall. This was one of several Brandywine textile mills established at the instigation of E. I. du Pont during the War of 1812. It was erected just outside his lower powder mill yard and was managed by a Frenchman, Raphael Duplanty, a former du Pont accountant, and Robert McCall of the Philadelphia merchant and banking family of that name.

For several years prior to the founding of this business du Pont had contemplated both a cotton mill and a woolen mill as parts of a diversified industrial community he hoped to create in the vicinity of his powder mills. In 1809 he had written to a relative in France, one Pierre Abraham Pouchet-Belmare, owner of a textile mill in Rouen, seeking information about some new French spinning and weaving machines which were reputedly very efficient and economical to operate. He asked Belmare to send him drawings or models of these machines and to give his opinion of them. Du Pont could not afford to experiment; he wanted machines that had already proved satisfactory in the French mills. He assured his cousin that the prospects for going into the textile business here in America were tremendous, particularly for one so well versed in the business as was Belmare. " What would you think of using your own knowledge of this industry for starting a factory here for yourself or one of your children? I am sure that such an establishment would be most successful." [9]

To his father, his confidante and adviser, du Pont remarked that most of the machines used for carding and spinning wool in America were English types, but he wanted to learn more about the newly-invented French machines, which might prove preferable. Their adoption in America would not harm the French textile industry, for French textiles did not sell in this country; the preference was for English merchandise, from habit, and because English prices were lower. He believed,

> It is only the English commerce that American manufactures can hurt. This country is now ready for manufacturing; with the industry and activity natural to Americans they will soon make for themselves whatever they most need; if we do not begin others will; in any case the manufactures will be established.[10]

---

[8] John Vaughan to E. I. du Pont, November 12, 1812, Eleutherian Mills Historical Library; hereafter cited as E. M. H. L. Charles W. Peale to John de Peyster, August 8, 1813, Calendar to C. W. Peale Letter Books, Letter Book XII, 102, American Philosophical Society.

[9] *Life of E. I. du Pont*, Vol. 8, p. 130. [10] *Ibid.*, pp. 144-45.

When his cousin replied, Pouchet-Belmare graciously declined the invitation to come to the Brandywine to set up a textile mill on French principles. His father had built up the family business in Rouen over twenty years; Belmare was now sole owner, and "When a man is forty-five years old he should not leave his own country or start new enterprises." But he did give a careful opinion of some of the machines about which du Pont had inquired, negative and discouraging. He recommended other types, widely used in both France and England, and offered to send several to du Pont. Also, he would try to find a good experienced textile man willing to go to America as the foreman in du Pont's contemplated mill. "And," he offered, "if you . . . want to send any of your children to France to learn our methods of manufacturing and you will trust them to me, be quite sure that I will do all in my power to help them. . . ." [11]

One outcome of this correspondence, delayed some years, was the founding of the Duplanty and McCall cotton factory. A number of French textile workers were employed at spinning fine yarn, from number 80 thread upwards to 110, and a French couple was placed under contract to construct, operate, and teach others to operate French stocking frame machines. The firm planned to do its own bleaching and for this purpose tried to engage persons in France to come to America "to take charge and establish a Berthaleinne laundry," which we believe meant the installation of Berthollet's bleaching process, first announced in 1785.[12] Post-war conditions and the competition from imported textiles forced Duplanty and McCall into bankruptcy in 1819 owing over $82,000.[13]

2. *The Tanning Industry*

A long established Delaware industry that E. I. du Pont believed could be improved by the application of European processes was the leather industry. After learning something of the methods used in the local tanneries he declared them very crude and the labor very dear. In some French technical journals he had read accounts of Armand

[11] *Ibid.*, pp. 160-66.

[12] Du Pont Family Manuscripts, V, E. M. H. L. Anthony and Marie Ravigneaux were engaged in May, 1816, to work "at the construction of stocking machinery, to direct the operation of the said machines and to train for the operation of the said machines apprentices," and, "to reel, double, sew, and to hot-press the stockings as well to instruct the apprentices in this work." The combined wage paid to husband and wife was $2.00 a day; they did not remain long in the employ of Duplanty and McCall.

[13] Duplanty, McCall Co., Final Accounts with E. I. du Pont de Nemours & Co., Old Stone Office Records, E. M. H. L.

Seguin's new process of tanning that purported to reduce the time required to tan sole leather from two years to three or four months. The descriptions were rather general, so du Pont wrote his father requesting him to investigate the Seguin process, to find out how widely it was being used in France, whether it was economical, whether any recent improvements had been made in it, and whether Seguin had made a fortune out of it.[14] His father complied, sent him several bulky memoranda, and offered this opinion about American leather:

> It is certain that your Americans make a great deal of leather and make it badly: it absorbs water like a sponge and does not wear at all. That is why it pays to send shoes, even poor shoes, from Europe as cargo.[15]

His son, however, thought it would pay him to set up a tannery using the French process, somewhere adjacent to his powder mills.

It was not until 1815, when the elder du Pont fled France upon Napoleon's return and came to live with his son at Eleutherian Mills, that the tannery was begun. It was located between the powder mills and the Duplanty, McCall Cotton Factory, under the direction of Alexandre Cardon de Sandrans, a young Frenchman who had been companion and secretary to the senior du Pont in Paris. M. Chenou, a tanner experienced in the Seguin process, supervised the operations for the first year until Cardon became thoroughly acquainted with the procedure.

The secret of the Seguin process was the addition of carefully measured amounts of sulphuric acid to the tanning solution. This accelerated the penetration of the tannin into the hide, thus sharply reducing the number of months the hides had to lie soaking in the tanning vats. Another aspect of the process was a different arrangement of the hides and the bark, and the periodic shifting of the hides in the vats.

In the first flush of successful production Cardon and the du Ponts felt assured their fortunes would be made by the tremendous advantage this saving of time and labor costs gave them. But tanning by the Seguin method sacrificed quality; though the outer part of the hide was tanned, not enough time was allowed for the tannic acid thoroughly to penetrate and tan the inner part. The result was an inferior grade of leather. There were markets for cheaper leather, but in the 1820's the Cardon tannery fell upon hard times, induced by general depressed

[14] *Life of E. I. du Pont*, Vol. 6, pp. 211-12.
[15] *Ibid.*, Vol. 8, p. 258.

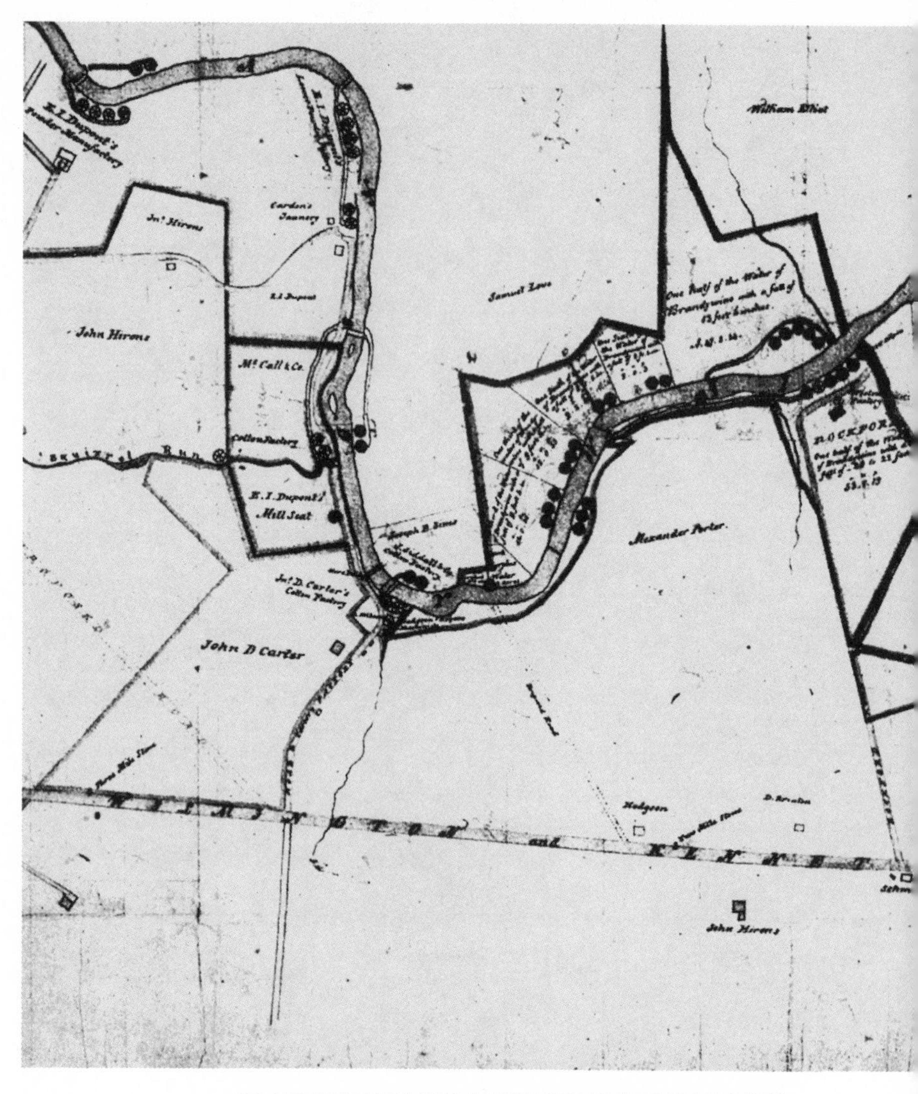

BRANDYWINE MILL SEATS COMPANY MAP (1822)

The purpose of this map is explained in a note from E. I. du Pont, a member of the company, to Bradford and Cooch, Baltimore powder agents, March 1, 1822: ". . . the map of the Brandywine you have seen has been drawn in order to promote the sale of a number of mill seats at present unoccupied. The Brandywine Mill Seats Company intends to have this map engraved shortly, when we will furnish you with a copy." Jonas P. Fairamb, a local surveyor, drew the map and made five parchment copies, one for each member of the Mill Seats Company.

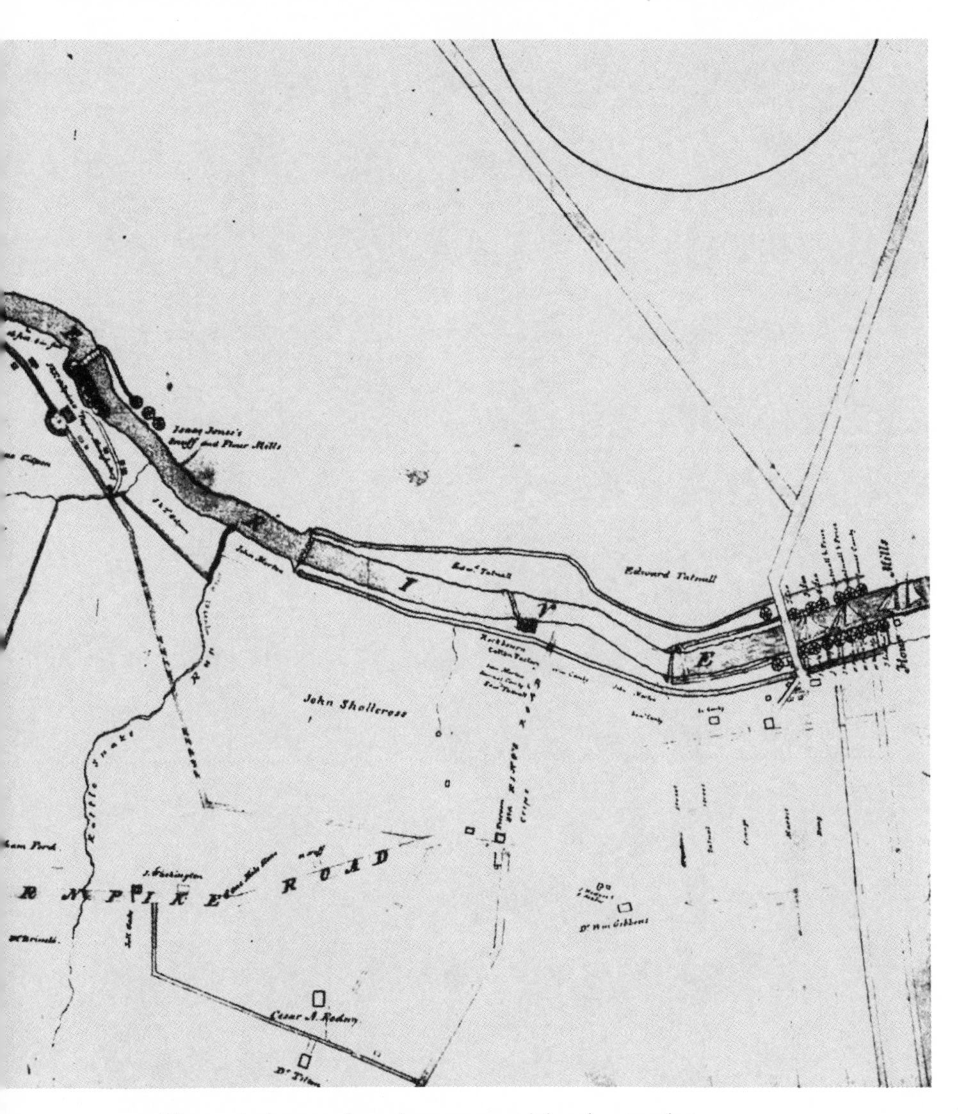

The particular area along the stream receiving the attention of the company was that heavily outlined on the eastern side of the stream, opposite the lands of Alexander Porter. All of the operating mills, from Du Pont's powder mills at upper left to the cluster of flour mills in Wilmington, are identified, and in some instances the fall of water available to each establishment is noted. (In the four to five mile stretch shown on the map, the fall of water was about 125 feet, an attractive amount of water power.) The map also pays particular attention to the roads and proposed roads as transportation facilities. The Delaware River was about one-half mile distant from the flour mills shown at the extreme right of the map.

(By permission of the Eleuthrian Mills Historical Library)

business conditions and by mismanagement within the firm. The enterprise had a spotty career, bolstered from time to time by fresh capital supplied by E. I. du Pont, but it ceased operations, bankrupt, in 1826.[16]

3. The Powder Mills

The black powder mills that E. I. du Pont erected between 1802 and 1804 were in every respect modeled after those in France. Explosives manufacture in France was a government monopoly under the direction of the Régie des Poudres, with headquarters and laboratory at the Arsenal in Paris and mills in various locations throughout France. While still in his teens, from 1787 to 1790, du Pont had been employed in the Paris Arsenal, then directed by his father's friend, Antoine Lavoisier, after which he had worked for about a year in the mills at Essonne.

From 1791 to 1799 du Pont had no connections with powdermaking; he was assisting his father in the management of a printing and publishing business in Paris. Briefly, in 1794, he was called upon to direct the gathering and refining of saltpeter in one of the districts of Paris, but this was a wartime emergency assignment of brief duration.

When the du Pont family, father and two sons and their families, thirteen in all, first came to the United States in 1799 their plans included neither powdermaking nor publishing as means of livelihood. The father had ambitions to found a colony somewhere on the near-frontier, in western Virginia or Kentucky, based on the philosophy of the Physiocrats, a school of economists of which he was a most eloquent exponent. To this end he raised capital among French sources and organized the Compagnie d'Amerique, with himself at its head, and his sons Victor and Eleuthère as his assistants.

Upon Jefferson's advice the settlement project was shortly given up, and among the dozen or so alternate ventures contemplated for the fruitful investment of their limited funds the " eighth plan " called for the establishment of a powder factory to be directed by the younger son, Eleuthère.[17]

English and Dutch gunpowder nearly monopolized the American market in 1800. The few American mills produced small amounts, and the quality was reputedly not very good. After visiting the mills of Lane and Decatur in Frankford, north of Philadelphia, du Pont noted that the head workman was a Dutchman who was using a process a half

[16] See Peter C. Welsh, " A. Cardon and Company, Brandywine Tanners, 1815-1826," in *Delaware History*, Vol. VIII, No. 2 (September, 1958), pp. 121-47.

[17] *Life of E. I. du Pont*, Vol. 5, pp. 191-92.

century or more out of date, a wasteful method that produced costly but poor quality powder. Du Pont was certain that using the most recent techniques developed in France, and with French equipment, he could make powder superior to the American and equal in quality to any of the imported.[18]

To prepare himself, du Pont returned to France in 1801 and spent three months "brushing up" on powdermaking technology. He consulted with the heads of the Régie des Poudres, MM. Bottée and Riffault, and the superintendent of the Essonne mills, M. Robin. Methods were discussed; drawings of mill structures and machinery were made available to him for copying; some equipment was obtained from the Arsenal, and orders were left for other items to be made and sent on later. Approximately $4,000 were spent for drawings, castings, eprouvettes, utensils, sieves, presses, laboratory items, and small quantities of French powder. The French authorities also promised they would try to persuade one of their more experienced powdermen to come to America to assist du Pont in getting his mills established.[19]

In this manner du Pont refreshed himself on the improvements and newer processes that the French had developed since he had left Essonne back in 1791. The willingness of the government officials to aid him so generously was due in part to the prestige and associations of his father in high places, but possibly more—as du Pont himself expressed on several occasions—to the realization by the French that a flourishing American powder producer could hurt the English in their gunpowder trade with the United States.[20] The U. S. Navy was also of assistance, for some of the heavier machinery ordered in Paris was brought over later in the year in the sloop-of-war *Maryland*, commanded by Captain John Rodgers.[21]

During the months he was constructing his mills du Pont corresponded frequently with Bottée, Riffault, and Robin, getting a great deal of additional technical information and encouragement. His appreciation was shown by sending them boxes of seeds, shrubs, plants,

[18] *Ibid.*, pp. 199-200.

[19] *Ibid.*, pp. 212-25, 230, 233-34.

[20] *Ibid.*, Vol. 8, pp. 102, 145. In a letter to his father, October 1, 1808, du Pont said, "The greatest harm that can be done to the English is to destroy their trade; the only way to accomplish that in this country is to establish manufactures that will rival theirs. The French manufacturers will never be able to overcome American prejudice and habit and it is only by American industries that England can be fought. This truth was felt in France before my journey in 1801 and secured for me all the help that I found there."

[21] Shipping Manifest, dated July 13, 1801, in "Receipted Bills, 1802-1803," E. M. H. L.

and small trees for propagation in France.[22] Du Pont had a keen interest in horticulture, and if the choice of vocation had been wholly his own there is good reason to think he would have been a botanist—the occupation he entered on his passport when he had left France in 1799.

Du Pont de Nemours, the father, returned to France in 1802, and by him Eleuthère was kept supplied with technical journals containing articles on explosives—*Bulletins des Sciences, Annales de Chimie*, the *Lycée des Arts*, and the *Journal de L'Ecole Polytechnique.* Treatises, memoranda, and government reports dealing with refining techniques, the making of charcoal, and trials of new machinery were sent to him. In 1811 Bottée and Riffault published a two-volume study entitled *L'Art de Fabriquer la Poudre à Canon;* at his son's urgent request,—"its many details of the improvements made in France may be most useful to me"—the elder du Pont sent him two copies of this highly prized work. Without doubt it served as his manual of instructions in his own Brandywine mills. Among du Pont's papers is a lengthy bibliography of articles on gunpowder and gunnery that appeared in the *Transactions* of the Royal Society of London from 1665 to 1800, an indication of his interest in the earlier and the contemporary developments in British powdermaking.[23]

In 1804, after receiving descriptions of an improved French granulating machine, du Pont obtained a patent on a granulating machine that may or may not have been patterned after the French one; we have the description of his, but not of the French machine.[24] One of his powder agents, writing some time later about a patent granted a rival New York powder manufacturer, made this apt remark, "Mr. Rogers discovery is like many others, a European discovery brought to light here."[25]

Du Pont's success was in large measure due to his diligence in acquiring knowledge of the best methods in use in countries where black powder had been manufactured for over 500 years. Verification of this is found in one of his periodic reports to his father in which he stated, "In the midst of all my work I have neglected nothing that might facilitate or improve the manufacture."[26] The results of his efforts may be discerned in Secretary of the Treasury Gallatin's praise

[22] *Life of E. I. du Pont*, Vol. 5, pp. 253, 355, 359, 362-65; Vol. 6, pp. 158-59.

[23] *Ibid.*, Vol. 9, pp. 64-65.

[24] *Ibid.*, Vol. 7, pp. 31-35. This was patent No. 590, dated November 23, 1804.

[25] John Vaughan to E. I. du Pont, December 3, 1818, Old Stone Office Records, E. I. du Pont de Nemours and Company, E. M. H. L.

[26] *Life of E. I. du Pont*, Vol. 8, p. 28.

of the Brandywine factory in 1810 as "the most perfect establishment for making powder in America." [27]

4. *Papermaking*

Most cosmopolitan of the Brandywine manufacturers was Joshua Gilpin, a downstream neighbor of du Pont and member of the Philadelphia Gilpin merchant family. At the suggestion of Benjamin Franklin he and his uncle, Miers Fisher, had converted a Brandywine snuff mill into Delaware's first paper mill in 1787, and here made paper of many types for a variety of purposes for 50 years.[28] Franklin loaned them some French works on papermaking and received from Gilpin some of the first paper made for his use with a request for his opinion of its quality. Brissot de Warville was asked by Fisher during the first year of the mill's operations if he would assist by trying to secure some paper moulds and some experienced workmen from France, since Fisher and Gilpin had not been successful in obtaining men and moulds from England. De Warville praised the Gilpin-made paper as "equal to the finest made in France," both for writing and printing.[29]

With the mills in good running order under the supervision of his younger brother, Thomas Gilpin, who had come into the business in the 1790's, Joshua Gilpin took off on an extended "Grand Tour" of the the British Isles and the Continent for a six-year stay beginning in 1795. He made a subsequent, shorter visit from 1811 to 1815. Gilpin was a man of limitless curiosity who wanted to further his education beyond the limits of the grammar school he had attended. His library suggests a man of catholic tastes—a desire to be a "universal" man who knew something about everything—and one who sought the "polish" that came from extensive travel and contacts with interesting people and places. He privately published a volume of poetry and wrote an essay on a more mundane subject of internal improvements, the Chesapeake and Delaware Canal, a project promoted by the Gilpin family.

His practical side as merchant and manufacturer is displayed in some

[27] *American Watchman*, June 2, 1810.

[28] See Harold B. Hancock and Norman B. Wilkinson, "Thomas and Joshua Gilpin, Papermakers," in *The Paper Maker*, Vol. 27, No. 2 (1958); also by the same authors, "The Gilpins and their Endless Papermaking Machine," in *The Pennsylvania Magazine of History and Biography*, Vol. 81, No. 4 (October, 1957).

[29] Dard Hunter, *Papermaking in Pioneer America* (Philadelphia, 1952), pp. 83-84; Miers Fisher to Brissot de Warville, November 25, December 11, 1788, Letters to Brissot de Warville, Scioto and Ohio Land Co. Papers, New-York Historical Society; J. P. Brissot de Warville, *New Travels in the United States of America* (London, 1794), Vol. 1, pp. 362-63.

60 pocket-size memoranda books that he filled with notes and sketches made during his prolonged visits abroad. Possibly representing himself as more merchant than manufacturer, Gilpin seems to have had no difficulty, despite England's "closed door" policy toward foreigners ferreting out its manufacturing secrets, in gaining access to factories, in talking with owners, managers, and mechanics, and in recording what he learned; he made notes on processes, sketched installations and equipment, and commented on their relative merits. Connections with British merchants, his numerous Gilpin relatives in England, and letters of introduction may have gained for him access that would have been denied others less well-connected. The value of Gilpin's voluminous notes lies in the wealth of technical information they contain; in total, they offer a close look at industrial Britain at the end of the eighteenth century.[30]

What elements of British papermaking technology did Gilpin's diligent inquisitiveness allow him to "borrow" for adoption in his own mills? First, he acquired the services of a former paper mill owner, Lawrence Greatrake, who came to Delaware about 1801 to superintend the Gilpin mills. Another Englishman, Thomas Oakes, who had designed and erected mills, was engaged to enlarge and remodel the original Gilpin mill and erect new buildings. In English and Scotch paper mills Gilpin had observed the process of bleaching rags that were pulped for paper stock, and he became convinced that chlorine, rather than muriatic acid and alkalis, was the best chemical for bleaching. In 1804 a chlorine bleachery was installed in his mill, an innovation that marked him as the first American papermaker to use chlorine for taking the color out of rags. According to one authority on the American chemical industry, chlorine was not generally adopted by the American textile industry as a bleaching agent until the 1830's.[31]

The most significant "borrowing," one that revolutionized the American paper industry, was the papermaking machine that Thomas Gilpin constructed in 1816 based upon descriptions and drawings of English machines supplied him by his brother Joshua and by Greatrake. Joshua had familiarized himself with the development of the endless woven belt machine invented in France in 1798 by Louis Robert and subsequently perfected by Bryan Donkin and John Hall, who installed

[30] Joshua Gilpin's Journals are in the Pennsylvania State Archives, Harrisburg, Pa. Parts of his journals commenting on a dozen or more industries have been edited and published in the *Transactions* of the Newcomen Society, Vols. 32 and 33.

[31] Sidney M. Edelstein, "Origins of Chlorine Bleaching in America," *American Dyestuff Reporter*, Vol. 49, No. 8 (April 18, 1960), pp. 39-48; Williams Haynes, *The American Chemical Industry* (New York, 1954; 6 volumes), Vol. I, p. 140.

it in the Apsley paper mill of the Fourdrinier brothers in 1804. A second machine employing a wire-covered cylinder was invented by John Dickinson in 1809 and put into operation in Dickinson's Nash Mill. Gilpin and Greatrake visited both mills, and after talking with the owners and seeing the machines operate, realized their possibilities—"Whoever first gets one to work in America . . . will open a sure road to the fortunes of all concerned . . ." enthusiastically wrote Greatrake. Up to a point, Dickinson and the Fourdriniers were free with technical information, but certain essentials were not disclosed. To obtain these the Americans used various means, some proper, others suggesting bribery and "pirating," and got together enough data and drawings from which Thomas Gilpin, the mechanically talented member of the firm, could build his endless papermaking machine—a close resemblance to Dickinson's—which he patented in December, 1816, and put into production in February of the following year.[32]

In place of a single sheet of paper of limited size made by the slow hand method, Gilpin's new "wonderworking" machine turned out paper of smooth texture and excellent quality in a roll of any length, at greater speed, and with lower labor costs. It was estimated that the new machine, tended by two men and one boy, could produce as much as the old hand method did with 12 men and six boys, saving $6,000 to $12,000 yearly in wages.[33] It should also be noted that making paper in rolls of great length brought about subsequent changes in the printing trade and the publication of newspapers, periodicals, and books.

The Gilpins enjoyed the monopoly of the endless papermaking machine for about five years, but in 1822 John Ames, a Springfield, Massachusetts papermaker, inveigled construction details of the machine from a former employee, and about 1827 English machines of the Fourdrinier type were introduced into America.

Conclusion

The migration of industrial technique illustrated by these "borrowings" in the Brandywine mills is but one of numerous streams of influence flowing from Europe that shaped early American society. Our technological indebtedness has not, however, to the best of our

[32] "Richard A. Gilpin's Paper Making Book," Gilpin Collection, Historical Society of Pennsylvania. Scattered through the Gilpin Collection are memoranda with headings, "Dickinson's Machine," "The New Improvements in Paper Making," and "Ideas Relative to a Contemplated Improvement in the Paper Mill at Brandywine Generally," which bear upon the efforts to obtain information about the English machines.

[33] *American Watchman*, March 4, 1818.

knowledge, received the attention that has been given to the cultural transitions of law, religion, education, political philosophies, and the fine arts from the other side of the Atlantic. Possibly our preoccupation with our later inventiveness and eminence as the world's leading industrial nation has made it easy to overlook the first factory era when we were an apprentice nation, learning, imitating, and sometimes improving upon the machine technology that had been developing in western Europe since the mid-1700's.

If the industrial history of the Brandywine was typical, it follows that similar studies of the half dozen or more river valleys where the first factories appeared would provide a much more comprehensive picture of our technological debt to Europe. I believe we would conclude that Thomas Lea was wrong—the spirit, invention, and genius of Europe were far from decrepit; and Liancourt was right—Americans *were* adopting all the good inventions of Europeans by which they could improve their mills.

"Drive That Branch": Samuel Slater, the Power Loom, and the Writing of America's Textile History

JAMES L. CONRAD, JR.

In its broadest casting, Samuel Slater's story needs no lengthy introduction. He is best known for his efforts in 1790 at Pawtucket Village in Rhode Island, where he helped local mechanics and manufacturers construct America's first commercially successful waterpowered cotton-spinning machinery. Through his role in introducing machinery invented by Englishman Richard Arkwright as well as the managerial techniques to support it, including the factory system with its attendant child labor, he dramatically accelerated America's industrial beginnings. After 1790, Slater energetically participated in a number of highly profitable textile manufacturing ventures in four states. While these mills were not equal in size or capitalization to the much larger mills owned by the Boston Associates at Waltham and Lowell in Massachusetts, Slater's medium-sized holdings were successful for a long period. His mechanical and managerial accomplishments were reflected by his estate of over one million dollars at the time of his death in 1835.[1]

DR. CONRAD, professor of history at Nichols College, is currently writing a book-length manuscript on the early Rhode Island textile industry from 1789 to 1836. **This article is a revision of a paper presented at the 1988 SHOT meeting. The author is indebted to Laurence F. Gross, Patrick M. Malone, Thomas G. Smith, Faythe E. Turner, and the *Technology and Culture* referees for their helpful comments.**

[1]The best single work on Slater remains George S. White, *Memoir of Samuel Slater: The Father of American Manufactures* (Philadelphia, 1836; reprint, New York, 1967). Although numerous attempts have been made to portray Slater's life, no adequate biography exists. In fact, Slater offered little help in writing such a history. See his brief comments to the Rhode Island Historical Society in 1834 quoted in White, pp. 41–42. Furthermore, his sometimes prolific, generally sporadic, and almost always business-related correspondence tends to mask deeper feelings. For specific studies on Slater, see Barbara M. Tucker, *Samuel Slater and the Origins of the American Textile Industry, 1790–1860* (Ithaca, N.Y., 1984); "Samuel Slater, Francis Cabot Lowell, and the Beginnings of the Factory System in the United States," in *The Coming of Managerial Capitalism: A Casebook on the History of American Economic Institutions,* ed. Alfred D. Chandler, Jr., and Richard Tedlow (Homewood, Ill., 1985), pp. 140–69; Brendon Gilbane, "A Social History of Samuel Slater's Pawtucket, 1790–1830"

Within this general context, however, much has been written asserting that Slater's role could have been more productive. Historians generally have agreed that Slater should have provided stronger leadership in the Rhode Island textile industry's efforts to expand after the War of 1812. This interpretation sees Slater refusing to adopt power-loom weaving when it was first introduced in 1815, preferring instead to "put out" his yarn to handweavers.[2] His apparent reluctance to include a power-weaving "branch" in his manufacturing operations is viewed as a major weakness.[3] This failure to seize the moment left the Rhode Island textile industry without its longtime leader at a crucial point. As a consequence, industrial leadership was lost to Massachusetts mills patterned after Francis Cabot Lowell's Waltham factory, and Slater's story ends on a sour note.

Such a negative assessment of Slater's work with the power loom appears inconsistent with his earlier activities. How could the aggressive

---

(Ph.D. diss., Boston University, 1968); E. H. Cameron, *Samuel Slater: Father of American Manufactures* (Freeport, Maine, 1960); William R. Bagnall, *Samuel Slater and the Early Development of the Cotton Manufacture in the United States* (Middletown, Conn., 1890); D. H. Gilpatrick, "Samuel Slater, Father of American Manufacturers," *Proceedings of the South Carolina Historical Association* (1932): 23–34; Roger Burlingame, "Spinning Hero, Sam Slater," *North American Review* 246 (1938): 150–61; Frederick L. Lewton, "A Biography of Samuel Slater," unpublished manuscript in possession of Slater Mill, Pawtucket, Rhode Island, 1944; M. D. C. Crawford, *The Samuel Slater Story, 1768–1835* (Pawtucket, R.I., 1948); Arnold Welles, "Father of Our Factory System," *American Heritage* 9 (April 1958): 34–39, 90–92; Paul E. Rivard, *Samuel Slater: Father of American Manufactures* (Pawtucket, R.I., 1974); James L. Conrad, Jr., "The Evolution of Industrial Capitalism in Rhode Island, 1790–1836: Almy, the Browns, and the Slaters" (Ph.D. diss., University of Connecticut, 1973), and "The Making of a Hero: Samuel Slater and the Arkwright Frames," *Rhode Island History* 45 (1986): 1–13; Gary B. Kulik, "The Beginnings of the Industrial Revolution in America: Pawtucket, Rhode Island, 1672–1829" (Ph.D. diss., Brown University, 1980); Joseph Gustaitis, "Samuel Slater: Father of the American Industrial Revolution," *American History Illustrated* 24 (May 1989): 32–33. For comments on Slater's estate, see "Samuel Slater" (unsigned editorial), *Pawtucket Chronicle* (May 1, 1835); Leonard Bliss, Jr., *The History of Rehoboth* (Boston, 1836), p. 237.

[2]William R. Bagnall, *The Textile Industries of the United States* (Cambridge, Mass., 1893; reprint, New York, 1971), 1:162, 399, 547; James B. Hedges, *The Browns of Providence Plantations: The Nineteenth Century* (Providence, R.I., 1968), p. 183; C. Joseph Pusateri, *A History of American Business,* 2d ed. (Arlington Heights, Ill., 1988), p. 147; Cameron, p. 143; Tucker, p. 99; Peter Coleman, "Rhode Island Cotton Manufacturing: A Study in Economic Conservatism," *Rhode Island History* 23 (July 1964): 68; George Rogers Taylor, introduction to his *The Early Development of the American Cotton Textile Industry* (New York [J. & J. Harper edition], 1969), p. xii; and many others.

[3]The term "branch" was used by Almy and Brown, the Slaters, and others to denote a specific part of their manufacturing operation. For example, in 1827 Samuel Slater wrote to his son, John, "How many of the last looms have you gotten in motion—Do Drive that Branch." See Samuel Slater to John Slater (son), May 15, 1827, Slater Papers (hereafter cited as SPH), S. Slater & Sons, vol. 235, Baker Library, Harvard University Graduate School of Business Administration, Boston, Mass.

and confident Slater so committed to new machinery in 1790 and 1804 (the spinning mule) change, as these interpretations suggest? Did he really reject the power loom, or have errors been made in analyzing this phase of his career? My purpose here is twofold: first, to examine what has been written about the early American textile industry and determine how it has contributed to our understanding of Slater's approach to power-loom technology; and second, to revise where appropriate our understanding of Slater's post-1815 career. Contrary to the customary approach, I will argue that Samuel Slater aggressively supported and effectively utilized the power loom and other post-1815 technologies long before most of his contemporaries. He was an important force in the textile industry throughout his life. Indeed, a revised picture of Slater's last few decades, when viewed together with his earlier years in America, should be of interest to those now seeking models of successful industrial growth on a moderate rather than large scale.

The traditional understanding of Samuel Slater's use of the power loom had its beginnings in 19th-century histories of the cotton textile industry. These perceptions were strengthened over time by research based on extensive Slater business records and letters at the Harvard University Graduate School of Business Administration and at the Rhode Island Historical Society. The resultant scholarship, eventually consisting of over 150 years of accumulated assumptions and varying methodologies, has played a dominating role in our view of Slater.

Additional material, however, is now available which provides important new insights into the life of arguably America's earliest industrial innovator and entrepreneur. Central to this reassessment are previously unused letters and records of Samuel Slater, his brother John, and their sometime partners, William Almy and Obadiah Brown. Variously located at the Brown University Library, the Rhode Island Historical Society, the Historical Society of Pennsylvania in Philadelphia, and the Historical Manuscripts and Archives Department of the University of Connecticut, these documents significantly alter the picture of Slater that had become established before 1950.[4]

Samuel Slater's story first began to take form in an early biography of him written by George S. White and published in 1836, one year after the death of "the father of our manufacture of cotton." White, a

[4]See John Slater Papers (hereafter cited as JSP), Brown University Library, Providence, R.I. (part of this collection previously was located at the Rhode Island School of Design, Providence); Almy and Brown Papers (hereafter cited as A&BP), Rhode Island Historical Society (hereafter cited as RIHS), Providence; Cope Family Papers (hereafter cited as CFP), Historical Society of Pennsylvania, Philadelphia; Jewett City Cotton Manufacturing Company, Slater Company Papers (hereafter cited as SCPC), Historical Manuscripts and Archives Department, University of Connecticut, Storrs.

controversial Episcopalian minister whose English heritage drew him to Slater, ardently defended the Pawtucket machine builder and the factory system he introduced. Using selected Slater letters and comments from contemporaries, White presented the former English apprentice as a deliverer and portrayed his factory system as offering great social and moral benefits. Praising Slater, White asserted that he built the Arkwright machinery virtually alone, from memory, and repeated Slater's statement that he made the machinery "principally with his own hands." In the process, White embellished Slater's achievements, made errors of both omission and commission, and produced an incomplete biography. Nonetheless, few comparable studies have endured as long or have had as great an impact on subsequent historical writing as has White's *Memoir of Samuel Slater*.[5]

The first interpretations of Slater and the early textile industry built on the conceptual foundation introduced by White. Generally written by individuals with connections to the textile industry, these histories developed several themes. For instance, they saw individuals such as Slater and Lowell, who introduced the power loom and integrated factory at Waltham in 1815, as "informing souls" giving "direction and form" to the cotton textile industry through their machinery.[6] Whereas White had concentrated on Slater and Rhode Island, Nathan Appleton and Samuel Batchelder focused their industrial histories on Francis Cabot Lowell and on Waltham and Lowell in Massachusetts.[7] In 1858 Appleton, an early associate of Waltham loom builder Lowell, sought to promote a better image for a then-troubled industry. To develop pride in their earlier accomplishments, he focused on the power loom constructed at Waltham by Lowell in 1814 and the creation of an integrated mill designed to manufacture cotton into cloth within one building. In Appleton's opinion, "The power loom . . . changed the whole character of the manufacture."[8]

Several years later, a former colleague of Appleton's and a Lowell factory superintendent, Samuel Batchelder, published his industrial history. More detailed and analytical than Appleton, Batchelder introduced the idea that the early textile industry was composed of "two different systems or '*schools*' of manufacturing."[9] This perception of a

[5]For these quotations, see White, pp. 23, 42, 71, 74, 281.

[6]Nathan Appleton, quoted in Robert C. Winthrop, *Memoir of Nathan Appleton* (Boston, 1861), p. 24.

[7]Nathan Appleton, *Introduction of the Power Loom and the Origin of Lowell* (Lowell, Mass., 1858); Samuel Batchelder, *Introduction and Early Progress of the Cotton Manufacture in the United States* (Boston, 1863; reprint, Clifton, N.J., 1972).

[8]Appleton, p. 14.

[9]Batchelder, p. 73.

Massachusetts-Rhode Island dichotomy quickly was developed further by historians. Accordingly, the so-called Slater or Rhode Island system soon was understood to feature small mills run by individual proprietors, strong British ties, child and family labor, and diversified production reflecting a reluctance or inability to use the power loom. Its origins were attributed to Samuel Slater. In contrast, the Waltham system was characterized by large-scale factories with professional managers, large capitalization, joint-stock ownership, labor provided by young women living in company-owned boardinghouses, American-improved technology, standardized and inexpensive goods marketed through a single agent, and vertically integrated production units featuring the power loom. This "system" originated with Lowell, his mechanic Paul Moody, and their colleagues at Waltham.[10]

An immediate reaction to the Appleton and Batchelder emphases on contributions of Waltham and Lowell came from Rhode Island. Former mill owner Zachariah Allen claimed that Appleton had insulted Rhode Island manufacturers by portraying them as "lacking sagacity to appreciate the advantage of improved machinery." To counter Appleton's claims, the Rhode Island Society for the Encouragement of Domestic Industry (RISEDI) immediately asked its subcommittee on manufactures for a report on "the introduction of the Power Loom and other machines for the manufacture of cotton in this state." Over the next five years, the society published approximately twenty-five letters and reports written by Rhode Islanders who had worked on or around the first Rhode Island power looms. Although relying on fading memories and sometimes secondhand information, these writers nonetheless offered the best descriptions of early power-loom building in Rhode Island. Allen and the RISEDI writers directly challenged Appleton's claims that Lowell was the first to use the "water" loom in America and the first to put all manufacturing processes under one roof. Rhode Island mechanics and manufacturers, according to Allen, had achieved these goals before Lowell arrived at Waltham, although not on the scale later reached by Lowell's Boston Manufacturing Company.[11]

[10]For a more detailed description of the Rhode Island system, see Peter J. Coleman, *The Transformation of Rhode Island, 1790–1860* (Providence, R.I., 1969), pp. 71–107. For the Massachusetts system, see George R. Taylor, *The Transportation Revolution, 1815–1860* (New York, 1951), pp. 231–32; George S. Gibb, *The Saco-Lowell Shops: Textile Machinery Building in New England, 1813–1949* (Cambridge, Mass., 1950), pp. 33–39, 58–62.

[11]For these quotations and references, see Zachariah Allen, "Cotton Manufacture in America," unpublished manuscript, n.d., Zachariah Allen Papers (hereafter cited as Allen Papers), RIHS; RISEDI, *Transactions for the Year 1861* (1862), pp. 30–31. Letters from Rhode Islanders were published in RISEDI, *Transactions for the Year 1861* (1862), pp. 76–125, and *Transactions for the Year 1864* (1865), pp. 59–84. Allen believed that the mill of John Franklin in New Providence had achieved "arrangements of all processes for

These early narrative histories and recollections contained few specific references to Slater's use of the power loom. Appleton never mentioned him. Batchelder did acknowledge Slater's contributions in setting up America's first Arkwright spinning machinery in 1790, but he was less certain about Slater's employment of the power loom. On this point Batchelder wrote that in 1815 Samuel Slater and his brother John had the chance to support Scottish-born power-loom builder William Gilmore's (sometimes Gilmour) efforts to construct power looms at the relatively new mill of Almy, Obadiah Brown, and the two Slaters in Smithfield, Rhode Island. This was an important opportunity since Gilmore's power-weaving machinery eventually gained wide acceptance in America. Rhode Islanders, however, saw business prospects in 1815 and 1816 as "discouraging," according to Batchelder, and he concluded that Slater and his Providence partners, Almy and Obadiah Brown, were not interested in additional expansion.[12] For his information on Slater, Batchelder relied on White's Slater biography.

Even the Rhode Island "recollections" scarcely mentioned Slater. Reported only was Gilmore's short stay at the Smithfield mill managed by John Slater before his eventual success at the nearby Lyman and Coventry mills. Gilmore's first American-built water looms and support machinery became fully operational in March or April 1818.[13] Since the RISEDI writers and Allen were countering Appleton's claim that the first successful American power loom was the work of Lowell at Waltham, what Slater may have done after 1818 was unimportant and was not discussed. Furthermore, this was a Rhode Island–Massachusetts battle over a legitimate claim to an industrial milestone, and Slater was not Rhode Island–born.

Greater acknowledgment eventually did go Slater's way in 1890 when industrial America celebrated its Cotton Centennial and applauded his

---

the conversion of cotton into cloth" in 1813 and therefore was the first in America to develop an integrated approach to cloth manufacturing. See Allen, p. 30.

[12]Batchelder, p. 70.

[13]This date is sometimes given as 1817. Rhode Island mechanic Perez Peck recalled that the operation of the first waterpowered looms occurred in 1817. See Perez Peck to Elisha Dyer, RISEDI, *1864,* p. 83. Batchelder also believed that power-loom weaving began in 1817 since Rhode Islanders gave money to Gilmore for use of his loom on May 31, 1817, thereby suggesting that the loom was then available. See Batchelder (n. 7 above), p. 102. The 1818 date, however, is provided by Job Manchester who worked at the Lyman mill. Manchester stated that he was sure of the year and he undoubtedly was aware of the claim for 1817. See Job Manchester, RISEDI, *1864,* p. 70. Possibly, the loom had been completed in 1817, but the dresser and warper may have required another year. Most now accept the 1818 date. See Julia Bonham, "Cotton Textile Technology in America: Three Centuries of Evolutionary Change" (Ph.D. diss., Brown University, 1979), p. 131. Other Rhode Islanders were vague about the timing. For example, see H. L. Lyman to Dyer, RISEDI, *1861,* p. 77; Gideon C. Smith to Dyer, RISEDI, *1861,* p. 91.

work with the Arkwright spinning frames.[14] During this celebration two significant publications appeared which focused on the early textile industry. Written by William R. Bagnall, one concerned Slater and the other the textile industry itself.[15] Both remain important.

Bagnall, a former minister and manufacturer, repeated the heroic interpretation of Slater and the Arkwright frames so much a part of the 1890 celebration.[16] At the same time he also commented on what he believed was Slater's failure to use the power loom. Bagnall wrote that loom builder Gilmore had been supported at Smithfield by John Slater but he in turn was "overruled by his older and more conservative brother, Samuel." The older Slater's decision to reject the power loom, as Bagnall saw it, was due both to his preference for cotton spinning and to the poor postwar business climate. Bagnall labeled Samuel Slater a "spinner" or specialist in spinning yarn for manufacturers and for domestic use, an explanation that seemed to justify Slater's failure to introduce power weaving. Leaving little to question, however, Bagnall disapprovingly referred to Slater as an "entrepreneurial conservative" and concluded that this attitude had cost his mills the "honor of introducing into Rhode Island a new departure in cotton manufacture."[17]

For his information, Bagnall, like Batchelder and others before him, relied on White. In a surprisingly important passage in his *Memoir of Samuel Slater,* White had written that when loom builder Gilmore arrived in Boston, "John Slater, Esq. invited him to Smithfield, Rhode Island . . . but Mr. Slater could not prevail on the whole of his partners to engage him in the trial." Bagnall assumed that White's "Mr. Slater" was John Slater and concluded that Samuel Slater was among "the whole of his partners" and therefore opposed to the loom.[18] This interpretation then

[14]Refer to Massena Goodrich, "Pawtucket and the Slater Centennial," *New England Magazine* 3 (October 1890): 138–56; George Rich, "The Cotton Industry of New England," *New England Magazine* 3 (October 1890): 167–91; "Pawtucket Cotton Centenary," *Harper's Weekly* 34 (October 1890): 771. Also see Massena Goodrich, *Historical Sketch of the Town of Pawtucket* (Pawtucket, R.I., 1876).

[15]Bagnall, *Samuel Slater* (n. 1 above), and *Textile Industries* (n. 2 above).

[16]For an analysis of the heroic interpretation, see Conrad, "The Making of a Hero" (n. 1 above).

[17]For quotations, see Bagnall, *Samuel Slater,* p. 62, and *Textile Industries,* pp. 547, 162, 399. Bagnall's conclusion undoubtedly was prompted by Slater's great skill as a "spinner" and his continued interest in this "branch." As late as 1825 Slater appears to have promoted an "Improved Vertical Spinner" apparently of his own design. See "Slater's Improved Vertical Spinner," *New England Farmer* (June 3, 1825).

[18]For quotations, see White (n. 1 above), p. 389; and Bagnall, *Textile Industries,* pp. 62, 399. James Montgomery also relied heavily on White in his 1840 book on cotton manufacturing in the United States and Great Britain, but did not mention Slater. See James Montgomery, *A Practical Detail of the Cotton Manufacture of the United States of America*

became the conceptual linchpin for the traditional view of Slater's approach to power-loom technology.

Such a reading of White has to be questioned. White used "Mr. Slater" only when referring to Samuel Slater, the subject of his biography. Tall, taciturn, and patrician-like, Samuel Slater—and no one else, including John Slater—was "Mr. Slater." When rereading White in this light, Samuel Slater, with his brother, John, is shown supporting the power loom. Unfortunately for their mills, the Slater brothers failed to convince partners Almy and Obadiah Brown of its value. White understood Samuel Slater's position but simply did not state it clearly.[19] And, since White made only this one reference to Slater and the power loom, it became a damning one. Consequently, Samuel Slater emerged from the 1890s with two strong although inaccurate images: one focused on heroic accomplishment based on his introduction of the Arkwright spinning frames, the other suggested hesitation and indecision because of his failure to use the power loom. Both views became accepted.

By the turn of the century, professionally trained scholars began to employ new perspectives and methodologies when writing about American business history. In the 1920s and 1930s, historians Caroline Ware and Arthur Cole moved beyond earlier narrative and descriptive histories by constructing growth and developmental models of the early textile industry. Ware, for example, divided the industry into the "old form" (the Rhode Island system) and the "new form" (the Waltham system).[20] In the process they supported the customary account of Slater's rejection of the power loom. Indeed, it became a basis for the models themselves. Ware's research found Slater first using the power loom in 1823; Cole settled on 1825 although he probably was referring to power looms used in making woolen goods.[21] Ware's date was

and the State of the Cotton Manufacture of that Country Contrasted and Compared with that of Great Britain (Glasgow, 1840; reprint, New York, 1969), p. 154.

[19]The Bagnall interpretation of White's passage was questioned first in Conrad, "Evolution of Industrial Capitalism" (n. 1 above), p. 339. Kulik, "The Beginnings of the Industrial Revolution in America" (n. 1 above), p. 320, agreed with this analysis. White's brief discussion of the power loom did contain hints regarding his position. For example, White was extremely positive about the power loom's potential, noting that "handlooms were immediately superseded" when the power loom appeared. Had Samuel Slater rejected power weaving, White would have had to explain further—and he offered no explanation. For these brief references, see White, pp. 388–90.

[20]Caroline F. Ware, *The Early New England Cotton Manufacture: A Study in Industrial Beginnings* (1931; reprint, New York, 1966), p. 60.

[21]Ibid., p. 75; Arthur H. Cole, *The American Wool Manufacture* (Cambridge, Mass., 1926; reprint, New York, 1969), 1:125. Alfred Chandler selected 1828 as the date "Samuel Slater finally accepted the Lowell challenge." See Alfred D. Chandler, Jr., ed., introduction to *The New American State Papers: Manufactures* (Wilmington, Del., 1972), 1:17. Ware's 1823 date,

approximately nine years after the power loom appeared in Waltham and five years after Gilmore's power-weaving machinery operated at the Lyman mill in Rhode Island. Ware concluded that Slater was "loath to abandon the method he had employed for thirty years . . . [he was] the most tenacious supporter of the old form."[22]

Leadership for this increased scholarly interest in business history soon became centered at the Research Center in Entrepreneurial History located at Harvard University. Between 1948 and 1958, scholars such as Cole, Leland Jenks, Thomas Cochran, Fritz Redlich, Joseph Schumpeter, William Miller, John Sawyer, Hugh Aitken, and David Landes, along with others, met often to discuss the role of the entrepreneur. Under their direction, the study of business history gradually shifted from narrative and descriptive approaches featuring unique individuals and single firms to the performances of larger groups in greater conceptual frameworks. In the process, Alfred D. Chandler, Jr., emerged to lead colleagues in the examination of the role played by managerial skills in the rise of the "modern" American business enterprise.[23] As a result, the Waltham system was seen as a prototype of the first modern factory and thus was elevated to the highest level of organizational superiority.[24]

These historians found no reason to doubt earlier interpretations regarding Slater's failure to use the power loom. Furthermore, since Slater and his firms could not be placed within the context of large-scale manufacturing operations assumed to be driving America's industrial growth, he was gradually relegated to a lesser role.[25] N. S. B. Gras and his colleague, Henrietta M. Larson, agreed that Slater's mills were later than others in adopting power weaving, a conclusion that appeared in a casebook used at the Harvard Business School in 1939.[26] A more recent

however, generally has been accepted by historians. For instance, see David J. Jeremy, *Transatlantic Industrial Revolution: The Diffusion of Textile Technologies between Britain and America, 1790–1830s* (Cambridge, Mass., 1981), p. 103; Jonathan Prude, *The Coming of Industrial Order: Town and Factory Life in Rural Massachusetts, 1810–1860* (Cambridge, 1983), p. 123.

[22]Ware, *The Early New England Cotton Manufacture,* p. 74.

[23]Hugh G. J. Aitken, "Entrepreneurial Research: The History of an Intellectual Innovation," in *Explorations in Enterprise,* ed. Hugh G. J. Aitken (Cambridge, Mass., 1965), pp. 3–19. For the major publications of Alfred D. Chandler, Jr., establishing this approach, see *Strategy and Structure: Chapters in the History of the Industrial Enterprise* (Cambridge, Mass., 1962), *The Visible Hand: The Managerial Revolution in American Business* (Cambridge, Mass., 1977), and *Scale and Scope: The Dynamics of Industrial Capitalism* (Cambridge, Mass., 1990).

[24]Chandler and Tedlow, eds. (n. 1 above), pp. 140–41; Chandler, ed., introduction to *The New American State Papers* (n. 21 above).

[25]For example, see Chandler, *Visible Hand,* pp. 57–60, 67.

[26]N. S. B. Gras and Henrietta M. Larson, "Samuel Slater and the American Textile Industry, 1789–1835," in *Casebook in American Business History* (New York, 1939), p. 227. Gras, however, did describe Samuel Slater as an important example of industrial

edition of this casebook series now focusing on managerial capitalism and edited by Chandler and Richard S. Tedlow concludes similarly that Slater stood "firm against new invention . . . [partly] to concentrate on the part of cotton manufacture which he knew best: spinning."[27]

Influenced by this work, scholars from across the historical profession, including historians of technology, held to the customary view of Samuel Slater and the power loom. With each additional acceptance, this position became a firmer part of the historical fabric. Some understood Slater as "a prime example" of "ingrained traditionalism," and as illustrative of the "technological lag" evidenced by the Rhode Island cotton textile industry. More specifically, H. J. Habakkuk saw Slater as typically English in his conservatism and concluded he never used the power loom since spinning "alone was enough." Historian Hannah Josephson wrote that he "had the caution and limitations of a mechanic with one skill" and that his "lack of enterprise cost him the leadership of the industry." And, in a recent study, Barbara M. Tucker found Slater "bound" by traditional or preindustrial values that resulted in his failure to fully adopt the power loom. In contrast, Tucker described Lowell as forward-looking with an "innovative technology."[28]

Beginning in the 1960s, however, traditional views of the textile industry experienced direct challenges. Scholars loosely grouped as "new" social historians began to react to what they understood was the increasing inability of Americans to understand their social structure. By focusing on the process of social change and the contributions of less visible Americans, they began to rewrite America's history, including the traditional view of the early textile industry. As a result, long-held concepts such as the vision of a heroic Slater, the Slater-centric nature of the Rhode Island textile industry, the existence of the Massachusetts–Rhode Island dichotomy, and the dominant place given to the Waltham system all have been questioned. Given greater roles in the industry's past are Rhode Island artisans, mill workers, handloom weavers, women mill workers at Lowell, Slater's Providence partners, and even the

---

capitalism in spite of his delay in adopting the power loom. Refer to N. S. B. Gras, *Business and Capitalism; an Introduction to Business History* (1939; reprint, New York, 1971), p. 217.

[27]Chandler and Tedlow, eds. (n. 1 above), p. 157.

[28]For these quotations and other references, refer to Pusateri (n. 2 above), p. 147; H. J. Habakkuk, *American and British Technology in the Nineteenth Century* (Cambridge, 1962), p. 114; Coleman, "Rhode Island Cotton Manufacturing" (n. 2 above), p. 72; Hannah Josephson, *The Golden Threads: New England Mill Girls and Magnates* (New York, 1949), p. 15; David J. Jeremy, "Innovation in American Textile Technology during the Nineteenth Century," *Technology and Culture* 14 (1973): 44; W. Paul Strassman, *Risk and Technological Innovation: American Manufacturing Methods during the Nineteenth Century* (Ithaca, N.Y., 1959), pp. 101–2, 111–12; Tucker, *Samuel Slater* (n. 1 above), pp. 90, 99, 101, 123.

machinery in place when Slater arrived.[29] Certainly the image of an industry featuring the Massachusetts–Rhode Island dichotomy does not exist as it once did. Historians such as Philip Scranton, Anthony F. C. Wallace, Cynthia J. Shelton, and Jonathan Prude, among others, have demonstrated that other places and different cultural settings have contributed significantly to the broader mosaic that is the early cotton textile industry.[30]

This revisionist approach can be applied to Slater as well. Historian James Hedges was right when, after supporting the traditional view of Slater and the power loom, he questioned whether or not "the whole story" was fully understood.[31] While large Massachusetts textile-manufacturing firms had spoken through Appleton, Batchelder, and the Harvard-based historians, and the Rhode Island mill people had told their stories through the RISEDI *Transactions,* no one had spoken for

[29]Revisionist studies concerning various aspects of the early cotton textile industry include Gary Kulik, "Patterns of Resistance to Industrial Capitalism: Pawtucket and the Strike of 1824," in *American Workingclass Culture,* ed. Milton Cantor (Westport, Conn., 1979), pp. 209–39; Gary J. Kornblith, " 'Cementing the Mechanic Interest': Origins of the Providence Association of Mechanics and Manufacturers," *Journal of the Early Republic* 8 (1988): 355–87; Barbara Tucker, "The Family and Industrial Discipline in Ante-bellum New England," *Labor History* 21 (1979–80): 55–74; Gail Fowler Mohanty, "Experimentation in Textile Technology, 1788–1790, and Its Impact on Handloom Weaving and Weavers in Rhode Island," *Technology and Culture* 29 (1988): 1–31, and "Putting Up with Putting Out: Power Loom Diffusion and Outwork for Rhode Island Mills, 1821–1829," *Journal of the Early Republic* 9 (1989): 191–216; James L. Conrad, Jr., "Entrepreneurial Objectives, Organizational Design, Technology, and the Cotton Manufactory of Almy and Brown, 1789–1797," *Business and Economic History* 13 (1984): 7–19, and "The Making of a Hero" (n. 1 above), pp. 3–13; Paul E. Rivard, "Textile Experiments in Rhode Island, 1788–1789," *Rhode Island History* 33 (1974): 35–45; Thomas Dublin, *Women at Work: The Transformation of Work and Community in Lowell, Massachusetts, 1826–1860* (Baltimore, 1974); Jonathan Prude, "The Social System of Early New England Textile Mills: A Case Study, 1812–1840," in *The New England Working Class and the New Labor History,* ed. Herbert G. Gutman and Daniel H. Bell (Urbana, Ill., 1987), pp. 90–127; David J. Jeremy, "British Textile Technology Transmission to the United States: The Philadelphia Region Experience, 1770–1820," *Business History Review* 47 (1973): 24–52; Carroll W. Pursell, Jr., "Thomas Digges and William Pearce: An Example of the Transit of Technology," *William and Mary Quarterly* 21 (1964): 551–60.

[30]Philip Scranton, *Proprietary Capitalism: The Textile Manufacture at Philadelphia, 1800–1885* (Cambridge, 1983); Anthony F. C. Wallace, *Rockdale: The Growth of an American Village in the Early Industrial Revolution* (New York, 1972); Cynthia J. Shelton, *The Mills of Manayunk: Industrialization and Social Conflict in the Philadelphia Region, 1787–1837* (Baltimore, 1986); Prude, *The Coming of Industrial Order* (n. 21 above). Also see Gary Kulik, Roger Parks, and Theodore Penn, eds., *The New England Mill Village, 1790–1860* (Cambridge, Mass., 1982); Richard M. Candee, "Early New England Mill Towns of the Piscataqua River Valley," in *The Company Town: Architecture and Society in Early Industrial Life,* ed. John S. Garner (New York, 1992), pp. 111–38; among others.

[31]Hedges (n. 2 above), pp. 182–83.

Slater except White—and he had been misunderstood. Without an accurate picture of Samuel Slater, the early industry's broader landscape is incomplete. Individuals such as Slater do count. Furthermore, at the present time, with American entrepreneurial skills and capabilities under question at home and abroad, his story provides interesting and important insights into the use of technology and its relationship to entrepreneurial success in other than large-scale manufacturing systems.

Any rewriting of the Slater story must involve far more than his reaction to the power loom. Many false ideas about Slater's activities have evolved to cloud his image and make assessments difficult. For one, Slater's role and status in his early partnerships with Providence manufacturers Almy, Smith Brown, and Obadiah Brown have been greatly exaggerated. As important as Slater and his machinery may have been for America's early industrial course and for its historians, he did not have full control of the mills he supervised, at least between 1790 and 1812. He made recommendations to his Providence partners, Almy and Obadiah Brown, regarding the operation of their Pawtucket and Smithfield mills—and they made the decisions.

In return for mechanical expertise and mill management skills, Almy and the Browns gave Slater a one-third share in the Almy, Brown, and Slater mill built at Pawtucket in 1793 and a one-fourth share of the larger Smithfield mill constructed in 1806. John Slater, Samuel's brother, was a partner in the Smithfield mill along with Almy and Obadiah Brown. While profits were divided according to shares, all decisions regarding financial matters were made by the partnership of Almy and Brown located in Providence. The Providence partners supplied the capital for these manufacturing ventures, controlled all purchases, marketed all products, and received a percentage from all transactions. As unfair as this may seem—and it contributed to friction among the partners and certainly was one reason the Slaters started their own mills—this was the way the Providence Quaker manufacturers ran their operation.[32]

The industrial preeminence of Almy and Brown had been hard-earned and should not be slighted. During the twenty-year period which began in 1789, William Almy, Moses Brown, and Smith Brown at first, and then Obadiah Brown, had been active participants in the industry's emergence. They successfully dealt with English competition, deficient American technical knowledge, American overproduction, scarce specie, inadequate transportation systems, and uncertainties brought on by the Embargo of 1807. By 1809, Almy and Obadiah Brown, with the Slaters and with other partners in Warwick, Rhode Island, controlled

[32]See draft of Smithfield Agreement, May 26, 1806, A&BP, RIHS.

approximately 30 percent of the operating spindles within 30 miles of Providence, a total that was equal to 20 percent of the operating spindles in the United States. Perhaps understandably, and in spite of impressive financial success, they had begun to show the wearing effects from two decades of industrial pioneering.[33]

War in 1812 dramatically added to their burdens with its external threats and internal commercial complications. Goods had to be protected from "Cruisers in the [Long Island] Sound" and a possible British invasion of Rhode Island, costly land transportation was now necessary to reach the lucrative New York–Baltimore-Philadelphia market area, sharply increasing domestic competition required effective responses, and a near collapse of the national currency and banking system had to be endured. Almy and Obadiah Brown were understandably upset. As a result of increased competition locally and nationally, they found mills "*scattered all over the country*" making it difficult to sell yarns and cloth for cash. On occasion they complained of having to accept paper currency "no better than Oak leaves."[34]

Their lowest point occurred in 1814 when a general increase in circulating banknotes coupled with the national government's policy of deficit financing led to severe inflation and then a general suspension of specie payments. As a result, banknotes from outside New England were unacceptable in Providence, and specie was unavailable due to demand. Almy and Brown rightly feared the loss of their cotton cloth and yarn valued at approximately $250,000 then in the hands of agents south of New Haven. The partners complained that "to look forward seems nothing but darkness and confusion." They even discussed closing their operation although they also admitted that their sales in 1812 and 1813 had never been better. The anticipation of peace by 1814 should have helped, but it only brought back painful recollections of the chaos caused by British imports during the 1790s and just added to their anguish.[35]

[33]See "A List of cotton manufacturers Within Thirty Miles of Providence, Nov. 14, 1809," Carter-Danforth Papers, RIHS; *American State Papers: Documents, Legislative and Executive of the Congress of the United States: Finance* (Washington, D.C., 1832–59), 1:433. For related correspondence, refer to A&BP and Moses Brown Correspondence, 1790–1812, RIHS. For Almy and Brown's early organization, see James L. Conrad, Jr., "Entrepreneurial Objectives" (n. 29 above). Also see Caroline Ware, "The Effects of the American Embargo, 1807–1809," *Quarterly Journal of Economics* 40 (1926): 672–88.

[34]For quotations, see Almy and Brown to Elijah Waring, April 21, 1813, Letterbook 7, A&BP; Almy and Brown to Daniel Waldo, June 19, 1813, Letterbook 8, A&BP; Almy & Brown to Daniel Cooledge, September 1, 1809, Letterbook 7, A&BP.

[35]For quotation and references, see Almy & Brown to Elijah Waring, October 13, 1814, Letterbook 8, A&BP; Moses Brown to William Wilson, April 24, 1814, Letterbook 8, A&BP; Almy & Brown to M. Morgan & Co., January 6, 1814, Letterbook 8, A&BP.

These insecurities understandably affected their ability to function within the old partnership structure. Moreover, this uncertainty appeared at a crucial time as America prepared to add to its growing collection of industrial machinery. In 1815, Samuel Slater complained that "my Friends A & B [Almy and Brown] . . . are at times rather more alarmed than perhaps would be for the best," and he was right. Others noticed the differences between the Slaters and the Browns and Almy. One business contact later stated to the Slaters that he knew "from observation your views and theirs [Almy and Brown] were somewhat dissimilar in regard to the extension of the business." This undoubtedly applied to the development of power weaving. Unlike his partners, Samuel Slater realized that some sacrifices had to be made. "After then," as he optimistically put it, "business will be more regular."[36]

In fact, while Almy and Brown agonized over their future roles, the Slater brothers moved forward aggressively even before the war began. First Samuel Slater sold his shares in S. Slater and Company in Rehoboth, Massachusetts, a mill he had begun in 1799 with his in-laws, the Wilkinsons. He then built two medium-sized mills in Oxford in south-central Massachusetts between 1812 and 1816. With partners who had worked for him in Pawtucket, he constructed the Slater and Tiffany cotton-spinning mill and the Oxford Dye House for dyeing and finishing yarn and cloth. In both cases, Slater obtained additional support by borrowing from Almy and Brown.[37] At the same time, his brother acquired a major interest in a cotton-spinning mill in West Boylston, Massachusetts, appropriately named "John Slater and Company." He also borrowed from their Providence partners. Considering the collateral provided by the Slaters' shares in the Pawtucket and Smithfield mills, Almy and Brown took little risk.

As the Slaters expanded their mill holdings into Massachusetts, a small group of talented Rhode Islanders began to design and build power looms. By 1813, nearly two years before Lowell's first power loom was fully operational at Waltham, at least six Rhode Island mechanics had constructed power looms of different designs literally from the ground up. As early as October 21, 1812, John Pitman, a local minister and mill owner, watched "two weaving looms going by water" at Henry Franklin's Union Factory in Providence.[38] These probably were early

[36]For quotations, see Samuel Slater to Jeremiah Brown, July 31, 1815, CFP; E. B. Robinson to Saml & John Slater, August 21, 1818, A&BP; Samuel Slater to Jeremiah Brown, March 27, 1815, CFP.

[37]Conrad, "Evolution of Industrial Capitalism" (n. 1 above), pp. 230–33.

[38]Entry, October 21, 1812, John Pitman Diary, RIHS. Later John Waterman, the former manager of Franklin's mill, recalled "several" power looms at work there in 1813. These looms required that a full-time mechanic be in attendance, a consideration that probably

vertical looms built and patented that year by John Thorp. Three months later, the ubiquitous Pitman saw Elijah Ormsbee's looms operating at the Rutenburg Mills in Olneyville on the outskirts of Providence.[39] Other Rhode Island loom builders included Samuel Bydensburg, Silus Shepard, Thomas R. Williams, Job Manchester, and a Mr. Ingraham of Pawtucket.[40] In each case, however, problems caused by defective machinery, poor or nonexistent preparatory machines, excessive expense due to the need for additional mechanics, inadequate waterpower, or a general lack of capital all worked to delay permanent acceptance of their looms.

During this six years or so of initial experimentation, the power loom's potential must have been discussed by the Slater brothers. They surely understood that this loom greatly reduced operating costs and produced better-quality standard cloth goods. Probably, these discussions were known to John Slater's partner at West Boylston, Boston merchant Robert Rogerson. It was Rogerson who brought Gilmore to the Almy, Brown, and Slaters' Smithfield mill in October 1815. He must have been reasonably certain that Gilmore would be welcomed by the Slaters. Gilmore reportedly stated that the Slaters encouraged him to build power looms at Smithfield, and this makes sense given their previous commitment to mechanization.[41]

There also must have been a sense of urgency in their work since the Boston Manufacturing Company at Waltham had commenced power weaving in 1815. The somewhat crude power loom introduced at Waltham by Lowell was similar to Englishman Robert Miller's wiper loom patented in 1796. In the Waltham-built loom, wipers or cams provided a uniform motion for the operation of loom parts including the movement which opened the shed for the shuttle's passage. This use of cams or eccentric wheels, however, achieved only moderate success in

---

resulted in their later replacement by Gilmore looms. See "John Waterman's Statement," July 10, 1861, Allen Papers, p. 30.

[39]Entry, January 8, 1813, John Pitman Diary, RIHS. Samuel Greene believed that this was the first attempt at power-loom weaving in Rhode Island. Refer to Greene to Allen, June 14, 1861, Allen Papers.

[40]Refer to Job Manchester, RISEDI, *1864* (n. 11 above), pp. 63–69; also Samuel Greene to Zachariah Allen, June 14, 1861, Allen Papers. Also see Appleton (n. 7 above), p. 9; Z. Allen, credit, May 6, 1813, to Samuel Blydensburg, Allen Papers; Lyman to Dyer, RISEDI, *1861* (n. 11 above), pp. 76–77. For a more detailed discussion of early Rhode Island loom builders, see Gail Fowler Mohanty, " 'All Other Inventions Were Thrown into the Shade': The Power Loom in Rhode Island, 1810–1830," in *Working in the Blackstone River Valley: Exploring the Heritage of Industrialization,* ed. Douglas M. Reynolds and Marjory Meyers (Woonsocket, R.I., 1990), pp. 77–88; Kulik, "The Beginnings of the Industrial Revolution in America" (n. 1 above), pp. 313–19.

[41]Lyman to Dyer, RISEDI, *1861* (n. 11 above), p. 76.

weaving coarse cloth. Problems arose, according to historian Richard L. Hills, as cams were not fully satisfactory in striking or beating up the weft or filling and the shuttle could be caught in the shed when the reed advanced too soon. This led to warp breakages and costly machine stoppages.[42]

In contrast, Gilmore's loom, the so-called scotch or crank loom, relied on a crank-and-lever mechanism based on William Horrocks's variable-speed batten motion patented in Great Britain in 1813 (see fig. 1). Horrocks's motion gave a quicker and sharper stroke to the weft with a faster withdrawal of the reed, thereby allowing the shuttle more time to move through the shed. This in turn permitted the use of larger bobbins, reduced warp breakage, lessened downtime, and ultimately made it possible to add more filling per inch. Furthermore, the American-built, iron-framed Horrocks loom could be operated at higher speeds resulting in a greater number of beats per minute than the lighter, wood-framed Lowell loom. Horrocks claimed that his loom produced stronger and more uniform goods "better wrought in all respects" while able to utilize yarns of a higher count. David J. Jeremy referred to Horrocks's improvement over the wiper as a "major incremental invention in power loom technology."[43]

On Gilmore's arrival at Smithfield in 1815, he immediately went to work building his looms at the Almy, Brown, and Slaters' mill. Samuel Slater's support for this project cannot be doubted. Gilmore's work on the power looms apparently had progressed beyond the planning stages when the elder Slater wrote to his brother in November 1816, "Are you making any progress with water looms at Smithfield, I do think we ought to get some underway as soon as it is practicable."[44]

[42]Richard L. Hills, *Power in the Industrial Revolution* (Manchester, 1970), pp. 224–25. I am indebted to Gail Fowler Mohanty for her explanation of certain aspects of power-loom operation. For an 1808 view of the wiper and crank looms, see John Duncan, *Practical and Descriptive Essay on the Art of Weaving* (Glasgow, 1808), pp. 272–76, pl. XIV. See also Jeremy, *Transatlantic Industrial Revolution* (n. 21 above), pp. 62–65, 99.

[43]Quotations are from William Horrocks, Great Britain Patent no. 3,725, July 31, 1813 (Queen's Printing Office, 1854), p. 4; Jeremy, *Transatlantic Industrial Revolution* (n. 21 above), pp. 62–65. Also refer to "Weaving," in *The Cyclopaedia: or Universal Dictionary of Arts, Sciences, and Literature,* ed. Abraham Rees (London, n.d.), 38, pt. 1, n.p. Also see Allen, "Cotton Manufacture in America," p. 24, Allen Papers; Batchelder (n. 7 above), p. 61; William R. Bagnall, "Sketches of Manufacturing Establishments in New York City and of Textile Establishments in Eastern States," ed. Victor S. Clark, unpublished typescript, microfilm ed. (North Andover, Mass., 1977), 3:1997.

[44]S. Slater to J. Slater, November 4, 1816, JSP. This letter was introduced in Conrad, "Evolution of Industrial Capitalism" (n. 1 above), pp. 337–40, and reviewed in Kulik, "The Beginnings of the Industrial Revolution in America"(n. 1 above), p. 320. These two studies, however, did not provide evidence of Slater's early use of the power loom, merely an indication of initial interest.

Fig. 1.—Power-loom weaving at the mill of Swainson, Birley, and Co. near Preston, England, in the 1830s, featuring female weavers, a weaving-room superintendent, and the 1813 version of William Horrocks's power loom, first patented in 1803. A similar engraving appears in George S. White's 1836 biography of Samuel Slater. (Edward Baines, Jr., *History of the Cotton Manufacture in Great Britain* [London, 1835], facing p. 239.)

Significant barriers, however, stood in their way. Economic instability and the general state of the textile industry in 1815 did not help. Appleton and Batchelder were correct in this respect. In a general sense, overbuilding, fear of imports, and inadequate tariff protection combined to hurt and, in some cases, destroy Rhode Island mills.[45] Then too, the Slaters' partners, Almy and Brown, had stated as early as 1808 that they wanted to concentrate on the manufacture of yarn into cloth "by others rather than ourselves."[46] This inclination, coupled with their underlying concern about general economic and industrial conditions, meant Almy and Brown support could not be counted on for several years.

Moreover, Gilmore's crank loom, as good as it ultimately turned out to be, was not an immediate success. In spite of the Slater brothers' early support, Gilmore had to be shifted to other tasks at their Smithfield mill. According to one source, Gilmore complained that he was being kept on only "to build a hydrostatic pump to water the meadows."[47] Possibly, Gilmore simply failed to produce successful power-weaving machinery, and the Slaters' already hesitant partners assumed that his looms were as inefficient as those developed earlier by Rhode Islanders.[48]

Subsequent events suggest that this possible assessment was not altogether wrong. When Gilmore left Smithfield for the nearby Lyman Manufacturing Company, his machinery was found to have defects. Although he was a talented machinist, he had no practical weaving experience. Some Rhode Islanders were exceedingly critical of his earliest efforts. One observer reported that the first Gilmore looms were "too light and otherwise faulty." Another recalled that Gilmore's "warper [a machine which placed yarn side-by-side in a warp and then wound the yarn on a beam] worked badly, [his] dresser [a machine for sizing the warp to strengthen it before weaving] worse, and [the] loom would not run at all."[49]

According to Manchester, who was there, power-weaving machinery built in both the Lyman and the Coventry mills by Gilmore wove its first

[45]Editorial, *Manufacturers' and Farmers' Journal* (December 3, 1821).

[46]Almy & Brown to John Wintringham, June 23, 1808, A&BP.

[47]Lyman to Dyer, RISEDI, *1861* (n. 11 above), p. 76.

[48]Gary Kulik concluded that Almy and Brown decided to remain with what they understood best. See Kulik, "The Beginnings of the Industrial Revolution in America" (n. 1 above), p. 320. This assessment, however, does not address the broad dimensions of their dilemma.

[49]For quotations, see Job Manchester, RISEDI, *1864* (n. 11 above), p. 70; Lyman to Dyer, RISEDI, *1861* (n. 11 above), pp. 76, 78. Also see "Manufacture of Cotton" (unsigned article), *Pawtucket Chronicle* (May 29, 1830); Jeremy, *Transatlantic Industrial Revolution* (n. 21 above), pp. 287, 294.

cloth in March or April 1818, more than two years after Gilmore's arrival at Smithfield. Even then the dresser still did not work properly. In the final stages, Gilmore received help from Pawtucket mechanic and inventor David Wilkinson, who was Samuel Slater's brother-in-law. Even with Wilkinson's assistance and the efforts of other qualified Rhode Islanders, an "intemperate Englishman" had to correct the dresser's last lingering problems in 1818.[50] For his assistance and ten dollars, Wilkinson received rights to the Gilmore loom and began to assemble power-loom machinery as early as December 1817.[51]

In the meantime, Gilmore returned to the Almy, Brown, and Slaters' mill in Smithfield and immediately resumed building power-weaving machinery there.[52] His continuing presence, along with the availability of Wilkinson-built loom frames, allowed the Smithfield mill to quickly catch up, if indeed it had ever been behind. Apparently even Almy and Obadiah Brown were convinced. This development, however, has not been noted by historians. By mid-May 1818, no more than one month or so after power-loom weaving was said to have started at the Lyman and Coventry mills, the Smithfield water looms wove marketable cloth. On May 22, 1818, the Smithfield mill shipped its Philadelphia agents twenty-three pieces of brown shirting, proudly noting on the invoice that they had sent "Water Loom cloth stout & good & the first that has been wove." This appears to be the earliest documentable date for the commercial weaving of American cotton goods on the Gilmore loom. The initial shipment totaled 1,956 yards and included 725 yards of "water loom shirting of a Superior Quality." During the next year and a half, approximately 80 percent of Smithfield cloth sent to the Philadelphia market appears to have been manufactured on "water" looms.[53]

[50]Allen (n. 11 above), p. 40. By 1818, still another promising loom had been introduced into Rhode Island by David Fales, although little seems to be known about it. Refer to Peleg Wilbur, RISEDI, *1861* (n. 11 above), p. 99; Job Manchester, RISEDI, *1864* (n. 11 above), p. 78. According to an 1830 notation in the Providence Steam Cotton Manufacturing Company records, Fales continued to sell his loom, at least into the 1830s. See "Memorandum, October 26, 1830, 8 prs of loom sides of his [Fales] pattern," Slater Companies, Providence Steam Manufacturing Company, vol. 18, SPH. For a better understanding of Fales, see Mohanty, " 'All Other Inventions' " (n. 40 above), pp. 84–85. Also see John W. Lozier, "Taunton and Mason: Cotton Machinery and Locomotive Manufacture in Taunton, Massachusetts, 1811–1861" (Ph.D. diss., Ohio State University, 1975), pp. 29–31.

[51]Notice of David Wilkinson and Samuel Greene, *Rhode Island American* (December 12, 1817); Allen (n. 11 above), p. 40.

[52]Smith to Dyer, RISEDI, *1861* (n. 11 above), p. 91.

[53]For quotations, see Invoice, May 22, 1818, Almy, Brown & Slaters to J. & M. Brown, Box 39f, CFP; Almy, Brown & Slaters to I. & M. Brown, May 22, 1818, CFP. Totals for Smithfield shipments in 1818 and 1819 were determined by examining each invoice that referred to shirting, sheeting, and "Water Loom" cloth priced at about thirty cents a yard or roughly

Encouraged by their power-loom success at Smithfield, the Slaters began to install power-loom machinery in their Massachusetts mills. One year after water looms were introduced at Smithfield, John Slater and Company in West Boylston announced it had "water loom shirtings" to sell as well as "shirtings similar to the Waltham bleach'd shirtings." Samuel Slater's efforts in Oxford, however, were not as successful, although his intentions were obvious. In October 1819, he told his Philadelphia agent that he hoped the health of John Tyson, his partner at the Oxford Dye House, "would be restored so as to get some water looms in motion . . . and . . . that he [Tyson] shall lose no time in getting them in motion so as to make *more* [emphasis mine] white goods." Slater's letter suggests that water looms may have operated at the Oxford Dye House prior to October 1819. And prospects for Slater's other Oxford mill were even less promising. In response to a question about getting water looms started, Slater's Oxford Factory agent wrote, "I fear the time is far distant when that desirable event will take place."[54] Quite probably the Oxford Factory experienced this "event" in 1821.[55]

The existence of power looms at West Boylston and Smithfield is easily verified. According to the 1820 *Census of Manufactures,* J. Slater and

---

the same price as Waltham manufactured cloth. See Invoices, October-November 1819, Almy, Brown & Slaters to J. & M. Brown, Box 39f, CFP. Also see Entries, Day Book no. 4, May 20, 1818, to June 24, 1818, Almy, Brown and Slaters mill, Slatersville, Rhode Island, microfilm copy, Old Sturbridge Village Research Library, Sturbridge, Mass. For Waltham prices, see Jeremy, *Transatlantic Industrial Revolution* (n. 21 above), p. 185.

[54]For quotations, see R. Rogerson to J. & M. Brown & M. D. Lewis, May 8, 1818, June 1, 1818, CFP; Samuel Slater to J. & M. Brown & M. D. Lewis, October 29, 1819, CFP; A. W. Porter to J. & M. Brown & M. D. Lewis, December 6, 1819, box 39f, CFP. John Tyson's correspondence in 1817 confirmed that the Oxford Dye House was weaving cotton goods although he probably was then referring to goods woven on handlooms. See Slater & Tyson to J. M. Brown, November 20, 1817, CFP.

[55]There is additional scattered evidence to support a beginning date of 1821. On December 1, 1820, Tyson reported from the Oxford Dye House that a "Mr. Burney has commenced warping." See Tyson to Slater, December 1, 1820, H. Nelson Slater Papers (hereafter cited as NSP), Slater Mill Historic Site, Pawtucket, R.I. Also a "debit" on the Oxford mill accounts notes a payment of $769.37 on "Pitcher & Gay's bill of 8 looms & Interest Settled on the 18th of March, 1823." Important here is the reference to "interest." Since six months' credit generally was extended and much more frequently allowed to good customers, it is quite probable that one year, possibly two, had elapsed between delivery and payment for the looms. Consequently, installation could have been in 1821, possibly even earlier. Refer to Oxford Mill Account, Slater: Almy & Brown, 1793–1833, 4:300, Slater Companies, SPH. Entries in the records of the Providence Iron Foundry, makers of Gilmore loom frames, report that a large number of castings (although not identified as "loom castings") were purchased by Slater's Oxford mills between 1819 and 1824, providing additional indications of the power loom's presence there in 1819 or 1820. Refer to Ledger, 1817–1823, Providence Iron Foundry, vols. 3, 5, Slater Companies, SPH.

Company is listed as operating twenty-four power looms while the Almy, Brown, and Slaters' Smithfield mill, according to the census taker, had seventy looms at work.[56] Furthermore, Gilmore continued to work at Smithfield, at least through 1820 when he introduced a faster version of his power loom. Prospective customers were informed that he could be reached at "Messrs Almy, Brown & Slater's factory in Smithfield, R.I."[57] His continued presence must have improved the Smithfield power-weaving operation and contributed to the Slaters' increasing knowledge of the technology. Three years later, according to another source, the Smithfield mill had 116 operating power looms.[58]

Unfortunately, the existence of textile machinery in other Slater-owned mills simply cannot be documented. The 1820 *Census of Manufactures* for the Oxford-Dudley area of Massachusetts does not list local mills separately or itemize their power-weaving machinery. However, if Samuel Slater's Oxford mill operated merely one-half the number of looms reported by J. Slater and Company (in 1817, the Oxford Factory had more spindles than J. Slater and Company, suggesting it was a larger mill), the Slater brothers in West Boylston, Oxford, and Smithfield commanded possibly as many as 110 power looms or nearly 15 percent of the reported total number of power looms in American cotton mills in 1820. While this number did not equal the 175 power looms of the Boston Manufacturing Company at Waltham, it nonetheless testifies to a significant Slater commitment to power weaving.[59]

Samuel Slater's support for power weaving prior to 1820 was emphatically demonstrated in other ways as well. With David Wilkinson, Benjamin and Charles Dyer, and Salmon Fobes, Slater began the Providence Iron Foundry in December 1817. Since one of the foundry's primary functions was to make cast-iron power-loom frames (the Wilkinson-built Gilmore loom became the first iron-framed loom in America), its role in the diffusion of Gilmore power-loom technology cannot be overemphasized. Organized when Gilmore agreed to allow Wilkinson to build and

[56] *Census of Manufactures, 1820,* National Archives, Washington, D.C., record group 29, M-279, roll 2: Massachusetts, p. 73, Rhode Island, pp. 4, 8. John Slater officially reported forty power looms at the Smithfield mill, whereas census taker Joseph Mann noted that seventy looms were in operation. It would seem that Mann had little reason to lie.

[57] William Gilmour's notice of "Patent Power Looms, on a new Construction," *Manufacturers' and Farmers' Journal* (April 17, 1820).

[58] Robert Grieve and John P. Fernald, *The Cotton Centennial, 1790–1890* (Providence, R.I., 1891), p. 35. In the opinion of Grieve and Fernald, the Gilmore loom put Rhode Island manufacturers "a decade ahead of their contemporaries."

[59] Refer to *Census of Manufactures, 1820,* Massachusetts, p. 33. For a comparison of loom-spindle ratios between Massachusetts and Rhode Island mills, see Mohanty, "Putting Up with Putting Out" (n. 29 above), p. 209.

sell his power loom, the foundry's opening clearly was timed to take advantage of an anticipated demand in loom sales.[60]

Providence Foundry records show that Samuel Slater was fully involved. He owned one-quarter of the foundry, chaired the organization's first meetings, and his name was given to customers seeking foundry services including the purchase of loom castings.[61] This was hardly the action of someone who opposed power-loom utilization. Between 1818 and 1820, nine mills purchased "loom castings" from the Providence Iron Foundry, while as many as forty others bought unidentified "castings."[62] Prospective loom users could either purchase loom castings from the foundry and then finish the looms at the mill site or buy assembled looms from David Wilkinson and Company in Pawtucket. Wilkinson quickly became the major buyer of foundry loom castings, followed by the Almy, Brown, and Slaters' Smithfield mill. With his partner, Samuel Greene, Wilkinson sold more than one hundred looms throughout New England and as far south and west as Georgia, Louisiana, and Pittsburgh.[63]

In spite of the Slaters' efforts, many conditions continued to work against the rapid adoption of power weaving in Rhode Island even after 1820. Although the Gilmore power loom was an inexpensive machine, the overall cost of introducing power weaving was a deterrent to prospective loom users with little investment capital. While the loom itself cost only from seventy to ninety dollars, additional expenses could be large. This involved the construction of separate weave shops or additions to present structures and the expansion of waterpower facilities including the gearing necessary to drive the machinery, as well as the building or acquisition of additional machinery such as dressers,

[60]Daybook, December 20, 1817, Providence Iron Foundry, 2:25, Slater Companies, SPH; Notice of Providence Iron Foundry, *Rhode Island American* (May 26, 1818); Bagnall, *Textile Industries* (n. 2 above), p. 549; "Samuel Greene's History of the Power Loom in Rhode Island, June 17, 1861," Allen Papers; Job Manchester, RISEDI, *1864* (n. 11 above), p. 70. Wilkinson's role in the Providence Iron Foundry frequently is mentioned, while Slater's generally is ignored. For example, see Jeremy, *Transatlantic Industrial Revolution* (n. 21 above), p. 103.

[61]Ledger, 1819–23, Slater: Almy & Brown, vol. 6, Slater Companies, SPH; Proceedings, April 30, 1819, June 18, 1819, Providence Iron Foundry, vol. 1, Slater Companies, SPH; Record of Deeds, Providence City Archives, January 31, 1818, 41:210, and March 1, 1819, 42:333–38, Providence City Hall, Providence, R.I. Customers seeking foundry services or products were advised to go either to the foundry on the west side of Providence or to "B. & C. Dyer & Co. [also referred to as the Providence Dyeing, Bleaching, and Calendering Company], or to Samuel Slater, or to David Wilkinson." See Notice of Providence Iron Foundry, *Rhode Island American* (May 26, 1818).

[62]Ledger, 1817–23, vol. 5, Providence Iron Foundry, Slater Companies, SPH.

[63]Mohanty, " 'All Other Inventions' " (n. 40 above), p. 83; Jeremy, *Transatlantic Industrial Revolution* (n. 21 above), p. 103.

warpers, and other support machines. For example, in 1823 the warper and dresser alone were sold by David Wilkinson and Company and its Pawtucket competitor, Pitcher and Gay, for $650.[64] Undoubtedly, a limited number of looms were placed in existing structures, but operations of thirty or more power looms generally required costly additions or completely new facilities. Then too, some Rhode Islanders, especially those in the Pawtucket area along the Blackstone River, may not have had enough waterpower to operate additional machinery including water looms.[65] The fact that Gilmore's power-loom system cost less to install than comparable patent-controlled Waltham machinery is almost irrelevant in view of difficult financial conditions and general attitudes in Rhode Island.

As for Samuel Slater, the assumption usually made is that he had the capital to expand virtually at will. This conclusion also does not square with the facts. Although Slater's partnership status in five textile mills and the Providence Iron Foundry correctly suggests increasing wealth and industrial influence, he did not have surplus capital available for investment. Only a few years before, both Samuel and John Slater had used their credit with Almy and Brown to start their Massachusetts mills. To obtain additional capital after 1818, Slater had to collect long-standing balances owed to him by Almy and Brown as well as his partners in his Oxford mills. In 1819 he received $5,000 as his share of the profits in the Almy, Brown, and Slater Pawtucket mill for the period 1803–18. Two years later, he requested and received $10,000 from the Smithfield mill account. Slater's chances of obtaining additional distributions from the Pawtucket and Smithfield mills became even more difficult in 1822 with the death of Obadiah Brown and the fiery destruction of the Smithfield mill. As a result, the partnership of Almy and Brown sold some of its assets to satisfy Brown's $120,000 legacy and had to borrow money to rebuild at Smithfield.[66]

Beyond the problems of Slater and his partners, depressed conditions continued to limit Rhode Island power weaving. Overextended opera-

[64]S. Slater to J. Slater, October 21, 1823, JSP; David Wilkinson to John Slater, November 5, 1823, JSP; Advertisement for water looms placed by Adams and Foster, *Manufacturers' and Farmers' Journal* (February 27, 1823); Jeremy, *Transatlantic Industrial Revolution,* p. 209.

[65]Kulik, "The Beginnings of the Industrial Revolution in America" (n. 1 above), p. 340.

[66]Daybook Entries, December 1, 1821, June 12, 1823, and other dates, Slater: Almy & Brown, vols. 4, 6, Slater Companies, SPH; Agreement, Samuel Slater, William Almy, Obadiah Brown, February 19, 1819, A&BP; Almy & Brown to John Slater, August 4, 1821, JSP. According to Moses Brown, Obadiah's death resulted in "Legacies which will require more than $120,000 to discharge." Refer to Moses Brown to Mary Bowers, October 30, 1822, Moses Brown Papers (hereafter cited as MBP), RIHS. Also see Brown to Gould Brown, February 11, 1826, MBP; "Account of Payments, Legacy of Obadiah Brown," August 13, 1824, Samuel Austin Papers in New England Yearly Meeting of Friends Archives at RIHS.

tions were hurt by a series of economic crises occurring between 1816 and 1820. During the Panic of 1819, for example, the Almy, Brown, and Slaters' Smithfield mill cut the number of operating breaker cards from twenty-six to thirteen, and the partners agreed not to enlarge the mill. Times were bad. Moses Brown explained that businesses in Rhode Island "made money plenty" during the war, but, with the bank crisis in 1814, they began to experience a "dullness" that continued into the 1820s. By 1820, many Rhode Island spindles were stopped; the agent of Almy and Brown's spinning mill in Warwick, Rhode Island, reported that it was "still."[67]

This combination of events made Rhode Island textile manufacturers and potential investors reluctant to invest further, thereby providing the Slaters and others with little encouragement. Job Manchester, the mechanic and loom builder who helped to start the first looms at the Coventry mill in 1818, believed that once Rhode Islanders realized that power-loom weaving was possible, "they seemed to pause and take breath before attempting to make further advances." The opinionated Zachariah Allen went further. He blamed Rhode Island difficulties on the huge capital expenditures of the Boston Manufacturing Company at Waltham, the sudden shift from war to peace, and an "indifference" on the part of some Rhode Islanders because of "extraordinary profits . . . from the exclusion of foreign fabrics [during the Embargo and war]." This negativism sharply contrasted with the enthusiasm growing in Waltham as well as with Samuel Slater's measured optimism.[68]

Not only did Slater experience opposition to expansion from his Providence partners, he also encountered it in at least one of his Massachusetts partners. In 1822, Slater had gone into partnership with weaver Edward Howard at Dudley. Howard was one of the foremost weavers of woolen goods in America. In itself, the partnership with Howard illustrates Slater's ongoing efforts to diversify his manufacturing operation. Immediately, Slater and Howard purchased power looms from pioneer loom makers William H. Howard of Worcester, Massachusetts, and local Delano Pierce.[69] Such aggressiveness led Cole to call

[67]Almy, Brown & Slaters to Almy & Brown, June 8, 1819, A&BP; Moses Brown, "Ruff Essay about the Dullness of Business," n.d., Samuel Austin Papers, RIHS; *Census of Manufactures, 1820* Rhode Island (n. 56 above), p. 100.

[68]For quotations, see Job Manchester, RISEDI, *1864* (n. 11 above), p. 75; Allen (n. 11 above), p. 30. In 1816 Nathan Appleton wrote to his brother that "many factories are now stopping and they grandly consider themselves as lordly saving themselves. Ours at Waltham . . . [is] to increase its capital from 100 to 200,000." N. Appleton to Sam. Appleton, January 20, 1816, Nathan Appleton Papers, 1815–25, Massachusetts Historical Society, Boston; Slater to Jeremiah Brown, March 27, 1815, CFP.

[69]Charles G. Washburn, "Worcester—Manufacturing and Mechanics Industries," in *History of Worcester County, Massachusetts* (Philadelphia, 1879), 2:1605–16; George F.

Slater a "pioneer in the allied wool manufacture" and Elizabeth Hitz to refer to the Slater and Howard factory as "a fine illustration of a factory with a very up-to-date technology."[70] Unfortunately Slater and Howard looms remained unused as Howard was either unwilling or unable to give full attention to power-loom weaving, preferring to weave fine fabrics on handlooms.

When the Slater brothers finally had the opportunity to work together without interfering or reluctant partners, they developed a model cotton textile mill that featured power weaving. In 1823 they purchased the Jewett City Cotton Manufacturing Company located in northeastern Connecticut, 15 miles south of the Oxford Factory, for the bargain price of $17,000.[71] The Slaters' first order of business at Jewett City was to add power looms. Clearly this had been planned in advance. Within six months, looms and dressers were on the way to Jewett City from Pawtucket, and gearing capable of running forty looms was being constructed at the mill site. In less than one year, the small Jewett City mill was manufacturing 3,538½ yards of cloth in one week; 95 percent was woven on the twenty newly installed water looms.[72]

Far from threatening Slater's approach to manufacturing, the new machinery presented him with an unequaled opportunity to analyze and control production in his mills. This was his strength. Full mechanization of the weaving process allowed the close monitoring of mill and worker outputs. In December 1824, the Jewett City agent proudly informed him that their thirty-six operating looms had averaged 21⅔ yards a day. Rising to his calculating best, Slater always pushed harder. He later wrote from Dudley, "I hope the 40 looms there [at Jewett City] will shortly come up to my 32 looms here in quantity [of woven goods produced]." In 1825 Slater pointed with pride to the fact that "several of my weavers averaged 40 yards per loom every day last week."

Daniels, *History of the Town of Oxford, Massachusetts* (Oxford, Mass., 1892), p. 198. Also see Agreement with George Hall, formerly of Ireland, & Finisher, and Samuel Slater and Edward Howard, April 22, 1823, Slater & Howard Papers, vol. 22, Slater Companies, SPH.

[70]Cole (n. 21 above), facing p. 224; Elizabeth Hitz, "A Technical and Business Revolution: American Woolens to 1832" (Ph.D. diss., New York University, 1978), pp. 314, 345.

[71]Minutes of Directors' Meetings, April 17, 1823, SCPC. Also see Bagnall, *Textile Industries* (n. 2 above), 1:595.

[72]For quotations and for the Jewett City mill's first year in operation, see Samuel Slater to E. W. Fletcher and Fletcher to Slater, November 1823 to May 1824, SCPC; J. & S. Slater to S. Slater, June 17, 1824, SCPC. Also refer to correspondence, S. & J. Slater to Samuel Slater and to John Slater, 1823–24, SCPC; Samuel Slater to J. Slater (brother), October 21, 1823, JSP; David Wilkinson & Co. to Jno. Slater, November 5, 1823, JSP; Entry, January 1, 1824, Slater: Almy & Brown, vol. 4, Slater Companies, SPH; Jewett City Mill to John Slater, July 1, 1824, SCPC.

Eventually, the power loom's consistency made it possible to rate workers' performances and to assign "black marks for defects in weaving" for those who did not meet his standards. The manufacture of inexpensive standard cloth had become central to Slater's expanding manufacturing system. In 1827 his advice to his son regarding power-loom weaving was emphatic: "Do Drive that Branch." This was essentially the same directive he had given to his brother at Smithfield eleven years before.[73]

In essence, a careful reexamination of Samuel Slater's activities suggests his image must be redrawn. Vague references to a conservative mentality are misleading if not completely wrong. Slater's lifelong commitment to machine technology and his vision of an efficient and productive manufacturing society mandated his use of the power loom. He was not an inventor or innovator in the sense of an Arkwright or a Wilkinson, nor did he have the metalworking abilities of many Rhode Island artisans. Instead, he installed machinery already tested. By 1824 his operation was impressively diversified. The Smithfield, Dudley-Oxford, and Jewett City mills specialized in plain cotton goods with some fancy goods woven by hand, Slater and Howard's mill in Dudley specialized in handwoven woolen goods, and the Pawtucket mill continued to produce high-quality cotton yarns and threads. Since a satisfactory power loom for fancy goods would not be fully perfected until the 1830s, he had no choice but to put out some yarns to handloom weavers located beyond his mill compounds. In 1828 he added a steam-powered mill to his system, and this made him unique in his industry.

Perhaps the most unfortunate result of the conservative approach to Slater has been the tendency of historians to overlook his activities after 1815. As a consequence, his role in applying steam power to textile manufacturing generally has been either misunderstood or underestimated.[74] Within the context of Slater's experience in America, the use of steam for power was just another indication of his acceptance and

[73]For quotations, see Jewett City Mill to Samuel Slater, December 2, 1824, SCPC; Samuel Slater to Ezra Fletcher, February 14, 1826, SCPC; Samuel Slater to John Slater (son), July 28, 1823, S. Slater & Sons, vol. 235, Slater Companies, SPH; Samuel Slater to John Slater (son), March 22, 1825, S. Slater & Sons, vol. 235, Slater Companies, SPH. A record of "black marks against weavers" was initiated on February 19, 1831. Refer to Steam Cotton Mfg. Company records, vol. 10, 1824–31, Slater Companies, SPH. Samuel Slater to John Slater (son), May 15, 1827, S. Slater & Sons, vol. 235, Slater Companies, SPH; S. Slater to J. Slater, November 4, 1816, JSP.

[74]Peter Coleman is one of the few to credit Slater with "pioneering" the use of steam power in the textile industry. See Coleman, *The Transformation of Rhode Island* (n. 10 above), p. 108. Others spend little time on this part of his career. Caroline Ware, for example, states only that "Slater tried to use steam power in 1828." See Ware, *The Early New*

utilization of new technology.[75] Contemporary Zachariah Allen saw the new mill as "an experiment of the practicality of employing *steam power.*"[76]

Although the Providence Steam Cotton Mill was only the second steam-powered cotton mill in Rhode Island (the first was at Newport), it was the largest and most effective. Operational first in December 1828, within a year the steam cotton mill produced 3,000–4,000 yards of "unsurpassed" cloth per week.[77] Slater's function as a partner in the Providence Steam Cotton Manufacturing Company organized in 1827 as an extension of the Providence Iron Foundry became even more important in 1832 when his steam mill partners failed and he became the mill's "sole proprietor."[78] He was then 62. In 1835, at the time of his death, its 183 looms were weaving 15,000 yards weekly.[79] In the process, the steam mill contributed immensely to Providence's vigorous involvement in building steam engines and its rise as a manufacturing city.

Of course Slater can be criticized. His operations were not perfect. His reliance on the partnership form of organization with family management, his desire to keep manufacturing units relatively small, and his reluctance to accept outside financing imposed limitations on future expansion. He was highly critical of organizations financed, as he put it, by "those, who, themselves engaged in other pursuits, have invested the nett [*sic*] profits of their businesses in manufacturing, and left the latter to the superintendence of others."[80] And, as Elizabeth Hitz and Barbara Tucker have pointed out, his administrative practices were badly in need

England Cotton Manufacturer (n. 20 above), p. 82. Also refer to Tucker, *Samuel Slater* (n. 1 above), p. 96; Cameron (n. 1 above), p. 147; Bagnall, *Samuel Slater* (n. 1 above), pp. 67–69.

[75]For a brief discussion regarding the role of Slater's steam cotton mill, see Carroll W. Pursell, Jr., *Early Stationary Steam Engines in America: A Study in the Migration of a Technology* (Washington, D.C., 1969), p. 84.

[76]Zachariah Allen, *Recent Improvements in the Useful Arts in Europe, and in the United States of America* (Providence, R.I., 1829), p. 352, note.

[77]John Clark to J. & M. Brown & M. D. Lewis, December 10, 1828, Steam Cotton Manufacturing Co., 1827–32, vol. 14, Slater Companies, SPH; "Fine Cotton" (unsigned editorial), *Rhode Island American and Providence Gazette* (June 5, 1829).

[78][Louis McLane, secretary of the treasury], *Documents Relative to Manufactures in the United States Collected and Transmitted to the House of Representatives,* Executive Document no. 308, 22d Cong., 1st sess. (1833; reprint, New York, 1969), 1:951 (hereafter cited as the McLane Report). The Providence Iron Foundry became the Providence Steam Manufacturing Company on October 11, 1827. See Providence Iron Foundry Minutes, October 11, 1827, vol. 1, Slater Companies, SPH. From that point on, the minutes of the two companies were combined since the same partners were involved. Refer to minutes of meeting, April 17, 1829, Steam Cotton Manufacturing Company, vol. 11, Slater Companies, SPH.

[79]"Description of Steam Mill Estate," S. Slater & Sons to William Whitney, December 19, 1835, NSP.

[80]McLane Report, p. 929.

of updating by the mid-1820s.[81] Yet he did not entirely reject the corporate approach. In 1827, with his partners in the Providence Iron Foundry, he requested its incorporation only to be rejected by a nervous Rhode Island legislature overly cautious about granting too much power to manufacturing organizations.[82] Essentially, his approach reflected a dynamic middle ground between the large-scale Waltham-Lowell mills and the small, individually owned and operated mills found throughout the country. Machine technology was utilized aggressively by Slater, but within the self-established limitations of his medium-sized manufacturing operation.

Those who knew and watched Samuel Slater supply the best descriptions of the man. One unidentified contemporary writer sensitively touched on Slater's contribution to Rhode Island manufacturing, noting that the steam mill went "further to establish the feasibility of manufacturing by steam in New England than all the theories ever broached."[83] This is precisely the point. Slater was not concerned with theory. He was a practical manufacturer who worked with machinery and people on the factory floor. This view is emphasized by perhaps the most perceptive observer of all, William Anthony, an artisan and later a mill owner who was in Pawtucket in 1790 when the first Arkwright machines were constructed and then followed Slater thereafter. Anthony referred to Slater as "a manufacturer who could both build and use machinery."[84] The essence of this statement should not be lost among the clutter of Slater myths. Samuel Slater understood the economic advantages of new machinery, including the power loom and the steam engine, as well if not better than anyone else of his era. His attitude toward machine technology was always positive and always consistent. He never wavered in this respect.

[81]Hitz (n. 70 above), p. 345; Tucker, *Samuel Slater* (n. 1 above), pp. 189–93.

[82]Coleman, *The Transformation of Rhode Island* (n. 10 above), p. 111.

[83]"Steam vs. Water" (unsigned editorial), *Manufacturers' and Farmers' Journal* (May 26, 1836). The Providence Steam Mill was compared favorably to English mills by Englishman J. S. Buckingham, who visited Providence in 1838. See J. S. Buckingham, *America: Historical, Statistic, and Descriptive* (New York, 1841), 2:439.

[84]William Anthony, quoted in *Manufacturers' and Farmers' Journal* (October 11, 1827); "Cotton Manufacturers," *Worcester County Republican* (March 11, 1929).

Raw Materials Supply and Technological Change in the American Charcoal Iron Industry

RICHARD H. SCHALLENBERG AND DAVID A. AULT

In the early 19th century, charcoal iron smelting was carried out in iron "plantations"—rural, small-scale operations which employed a primitive, age-old technology. Such crude techniques resulted in an enormous waste of the raw materials, particularly the charcoal fuel. The name "plantation" derives from the fact that these enterprises were often nearly self-sufficient communities, producing not only iron but also most of their own food, and were therefore as much agricultural as industrial operations. The quality of the iron produced at these plantations was usually quite good, but its price, due to the inefficiency of the processes used, was high.

Two historians of the American iron industry—Peter Temin and Louis Hunter—have claimed that these characteristics of the charcoal iron plantation were dictated by the nature of the fuel used.[1] Temin and Hunter maintain that charcoal blast furnaces had to be widely dispersed in rural areas in order to be near the source of their fuel. Both claim that wood could not be carried far because of high transport costs, and that charcoal could not be transported far because it was too brittle. The furnaces themselves had to be kept small, or else their consumption of charcoal would have rapidly deforested the woods in their immediate area. As a result of these factors, the charcoal iron plant was forced to retain the small, primitive, agricultural pattern of the iron plantation.

Temin in particular has linked these two aspects of iron production—namely, fuel and technology—in such a way as to suggest that the all-important transition from charcoal to coal smelting was a prerequisite for any improvements in smelting technology: "In this time [i.e., the 1840s] the blast furnace began the transition from medieval technology and organization to more modern forms of pro-

At the time of the publication of this article Dr. Schallenberg was in the History Department and Dr. Ault in the Computer Science Department at Virginia Polytechnic Institute and State University. Dr. Schallenberg is now deceased.

[1] Peter Temin, *Iron and Steel in Nineteenth Century America: An Economic Inquiry* (Cambridge, Mass., 1964), pp. 62–76; Louis C. Hunter, "Factors in the Early Pittsburgh Iron Industry," in *Facts and Factors in Economic History* (New York, 1967), pp. 433–34.

duction. Charcoal was replaced by mineral fuel, and the semi-feudal iron plantation was replaced by the urban establishment and the company town."[2]

Using price and production statistics, Temin and Hunter both demonstrate that the *fuel* cost for producing a ton of iron at a charcoal iron plantation was far greater than the fuel cost at a contemporary, urban coke furnace plant.[3] During the second half of the 19th century, therefore, the charcoal ironmaster became caught in an impasse. His antique technology could not compete economically with that of the coal smelters, and yet he was forced by the nature of his fuel to retain that old technology. Trapped in this situation, he was driven out of business in the latter half of the century.

In a previous paper we challenged this intimate linkage between fuel and technology by demonstrating that many late 19th-century American charcoal stacks were almost identical technologically with coke or anthracite coal furnaces.[4] In the present paper we extend this criticism by showing how the adoption of new smelting and charcoaling technologies greatly reduced wood consumption, thereby making this older fuel into a far less costly factor of production than it had been at the typical antebellum iron plantation.[5]

By analyzing technological innovation in charcoal iron plants in different parts of the United States, we intend to prove that problems inherent in ore supply, and not wood supply, were the chief limiting factors in large-scale improvement and enlargement of postbellum charcoal furnaces. Our regional analysis will show that those areas of the country which possessed large quantities of cheap, rich ore expanded their charcoal iron production in the late 19th century, whereas those areas having little ore rapidly declined as charcoal iron producers, regardless of timber supplies. Although we do not deny that the problems inherent in the use of wood as a blast furnace fuel did contribute to the decline of charcoal smelting vis-à-vis coke and coal smelting, our analysis will demonstrate that past economic interpretations based solely upon the differences in fuels are misleading as an explanation of the overall decline of the late 19th-century American charcoal iron industry.

[2]Temin, p. 82.

[3]Ibid., pp. 62–76; Hunter, p. 435.

[4]Richard H. Schallenberg, "Evolution, Adaptation and Survival: The Very Slow Death of the American Charcoal Iron Industry," *Annals of Science* 32 (1975): 341–58.

[5]For the design of the computer program used to analyze data for this paper, see David A. Ault and Richard H. Schallenberg, "Charcoal Iron Industry Analysis: Data Preparation and Computer Design," Virginia Polytechnic Institute and State University Technical Report no. CS 75020-R (September 1975).

The Economics of Innovation

During the first half of the 19th century, the chief charcoal iron producing states were Pennsylvania, Ohio, New York, Virginia, Connecticut, Maryland, Missouri, Tennessee, and Kentucky. The furnaces in these regions were constructed according to age-old patterns: short and squat, with low blast pressure and temperature. The waste of raw materials was great and the output of iron small.[6] Almost all of the charcoal furnace plants in these regions retained the antiquated operating conditions and equipment after the mid-19th century.

This lack of innovation in the older charcoal iron producing states stood in marked contrast to the activities of coal smelters, who, with the quickening tempo of industrialization after the 1830s, began to restructure totally the technology of American iron smelting. In only three areas of the country did charcoal ironmasters adopt the innovations pioneered by the coal smelters: northern Alabama–western Tennessee–northwestern Georgia, Michigan–northeastern Wisconsin, and Missouri.

Furnaces of the iron plantation pattern were built between 30 and 35 feet in height. In the 1840s the height of new or remodeled furnaces was often raised to 40–45 feet, and, after the Civil War, up to as much as 65 feet (see table 1). All antebellum furnaces were built of masonry in the traditional squat, truncated-pyramid shape. A substantial number of charcoal stacks used hot blast before the Civil War, although both the blast pressure and temperature of these furnaces were low compared with later practice.

The charcoal was made by age-old techniques which burned much of the wood to ash. At no antebellum charcoal furnace plant was any attempt made to condense by-product wood chemicals produced by the charcoaling process. Furnaces were "open topped," so that unconsumed carbon monoxide and heat escaped from the top of the furnace into the atmosphere instead of being utilized in the plant. Given such methods, it is probable that, for every cord of wood which actually contributed to the reduction of ore in the furnace, another two or even three cords were completely wasted.

Charcoal furnaces built in the Alabama, Michigan, and Missouri regions after the Civil War were usually between 55 and 65 feet in height, with very hot, high temperature blast fed into the stack through a set of multiple, water-cooled tuyeres. Older furnaces usually had only a single tuyere. The increased height of the new fur-

[6]For a description of the iron plantation type of iron works, see James M. Ranson, *Vanishing Ironworks of the Ramapos* (New Brunswick, N.J., 1966).

TABLE 1

AVERAGE HEIGHTS OF OPERATING CHARCOAL BLAST FURNACES, 1859–1908

	Mass.	Conn.	N.Y.	Maine	Pa.	Md.	Va.	Ky.	Tenn.	Ga.	Ohio	Ala.	Mich.	Mo.	Wis.
1859	35/6	29/13	36/30	42/11	37/123	32/19	33/26	35/36	34/44	29/7	36/37	31/3	37/6	36/6	30/1
1876	35/5	33/10	35/16	38/1	30/40	37/15	33/24	41/18	35/13	38/8	37/36	46/12	44/30	40/10	...
1886	32/4	34/9	41/9	50/11	33/25	37/14	35/23	38/3	44/9	50/2	36/17	54/10	49/25	55/4	48/10
1898	32/3	34/5	32/3	...	36/4	42/2	33/1	...	56/3	65/1	49/4	55/4	55/4	54/9	60/1
1908	33/3	36/3	32/2	...	39/4	36/1	42/2	35/1	46/3	...	41/6	56/5	60/11	60/1	60/1

SOURCE.—American Iron & Steel Association, *Directory of the Iron and Steel Works in the United States and Canada;* John Peter Lesley, *The Iron Manufacturer's Guide to the Furnaces, Forges and Rolling Mills of the United States* (New York, 1859).

NOTE.—The number to the left of the slash is the average height of all operating charcoal furnaces in that state; the number to the right of the slash is the number of all furnaces operating for which height data were available.

naces saved charcoal, since the carbon monoxide reducing agent produced by the partial combustion of the charcoal had more contact time with the iron oxide. The hot blast, higher pressures, and multiple tuyeres also saved on charcoal consumption by speeding up reaction rates. The higher blast pressures of postbellum advanced stacks were produced by replacing water wheels with steam engines and by using cast iron, machined blast cylinders instead of the leather bellows or wood-stave cylinders of the iron plantation.

Unfortunately, quantitative data on blast temperatures and pressures used at charcoal furnaces are too scanty to draw precise conclusions. However, these data indicate that temperatures at modern stacks were much higher than the 300°–400° F which was common at those antebellum charcoal furnaces which used hot blast at all.[7] Furthermore, we can derive some idea of comparative blast temperature and pressure conditions between furnaces in different states by the indirect process of comparing furnace heights and widths. Data on these parameters are fortunately very complete.

Older charcoal furnaces of the iron plantation type had squat internal dimensions, with a narrow top diameter and a strong outward batter to the walls down to the bosh, so that the bosh diameter was quite large compared to the top. Indeed, the total height of such furnaces might be only three to four times the bosh width. The reason for this shape was that the descending reaction mass in these furnaces was fairly nonfluid and sluggish due to the low temperatures and pressures of the blast. The widely flaring walls and large diameter bosh tended to prevent the sluggish reaction mass from clogging or "hanging up" in the furnace.[8] Thus the height-to-width ratios (width here being bosh width) of iron plantation type furnaces were relatively small.

More modern charcoal furnaces, by adopting the higher temperature blasts and pressures pioneered by the coal smelters, could be built with narrower internal dimensions and with more vertical walls. The narrower cross section at the bosh, coupled with the use of multiple tuyeres around the perimeter of the hearth, created more uni-

[7]E.g., the following blast temperatures were used at these advanced Alabama and Michigan charcoal stacks: Alabama—Gadsden furnace: 900° F (1883); Michigan —Jackson Iron Company furnace: 725° F (1883), Lake Huron furnace: 900° F (1887), Martel furnace: 1,700° F (1886), Antrim furnace: 950° F (1886) (*Journal of the United States Association of Charcoal Ironworkers* (hereinafter: *Charcoal Ironworkers Journal*) 3 [1883]: 85, 87–88; ibid., 7 [1887]: 208; ibid., 6 [1886]: 61; Woodward Iron Company, *Alabama Blast Furnaces* [Woodward, Ala., 1948], p. 72).

[8]A "hung-up" furnace was not only a nuisance, but a danger as well, since the mass of material, if suspended high up in the furnace, could let go and come crashing down with explosive force.

form and fluid conditions throughout the cross section of the reaction mass at this point, thus producing a more homogenous pig iron.

Figure 1 shows schematic drawings of two furnaces typical of the old and new patterns of charcoal blast furnace construction. The Howard furnace has the characteristic bottle-shaped internal lines and external pyramid shape of antique stacks. The outward slope of the interior walls from top to bosh is over 1 in 6 for Howard, as compared to only 1 in 18 for Gladstone. The Gladstone furnace's weight is supported by a cast-iron shell, borne upon iron columns below the bosh. The furnace's walls from bosh to hearth thus could be kept fairly thin, permitting easy access at the hearth by a large number of tuyeres—five on this stack. The massive masonry pile of the Howard furnace allowed room for only two tuyeres. Although the Gladstone is somewhat smaller than most early 20th-century coke blast furnaces, its overall shape, both internally and externally, as well as the tuyere arrangement and auxiliary apparatus (stoves, bells, blowing cylinders, etc.), is identical to standard coke furnace technology of the turn of the century.[9]

Figure 2 is a graph of the height-to-width ratios of all operating charcoal blast furnaces in several states from 1820 onward. The higher curves for Michigan and Alabama are indicators of the use of more modern and powerful hot blast stoves and blast machinery at the average furnace in these states. In addition to these innovations, table 1 reveals that charcoal ironmasters in Alabama, Michigan, Wisconsin, and Missouri were also increasing the absolute height of their furnaces during the same period.[10]

All of the above-mentioned factors contributed to the reduction of charcoal consumption per ton of iron smelted, thus improving the competitive position of innovative charcoal stacks versus coal stacks in the later 19th century. Given this technological differential between charcoal furnaces of the iron plantation type and those of the Michigan, Alabama, and Missouri regions—a differential which became increasingly greater as the end of the century approached—any economic analysis of the decline of the charcoal iron industry which is based upon technical constants derived from the iron plantation model must therefore be reexamined to determine precisely how inappropriate it is for the innovative parts of the industry. Since previ-

[9]J. E. Johnson, Jr., *Blast-Furnace Construction in America* (New York, 1917); see esp. Warwick thin-lined, p. 265, and Port Henry, p. 267.

[10]American Iron & Steel Association, *Directory of the Iron and Steel Works of the United States and Canada* (hereafter cited as *Directory*); John Peter Lesley, *The Iron Manufacturer's Guide to the Furnaces, Forges and Rolling Mills of the United States* (New York, 1859).

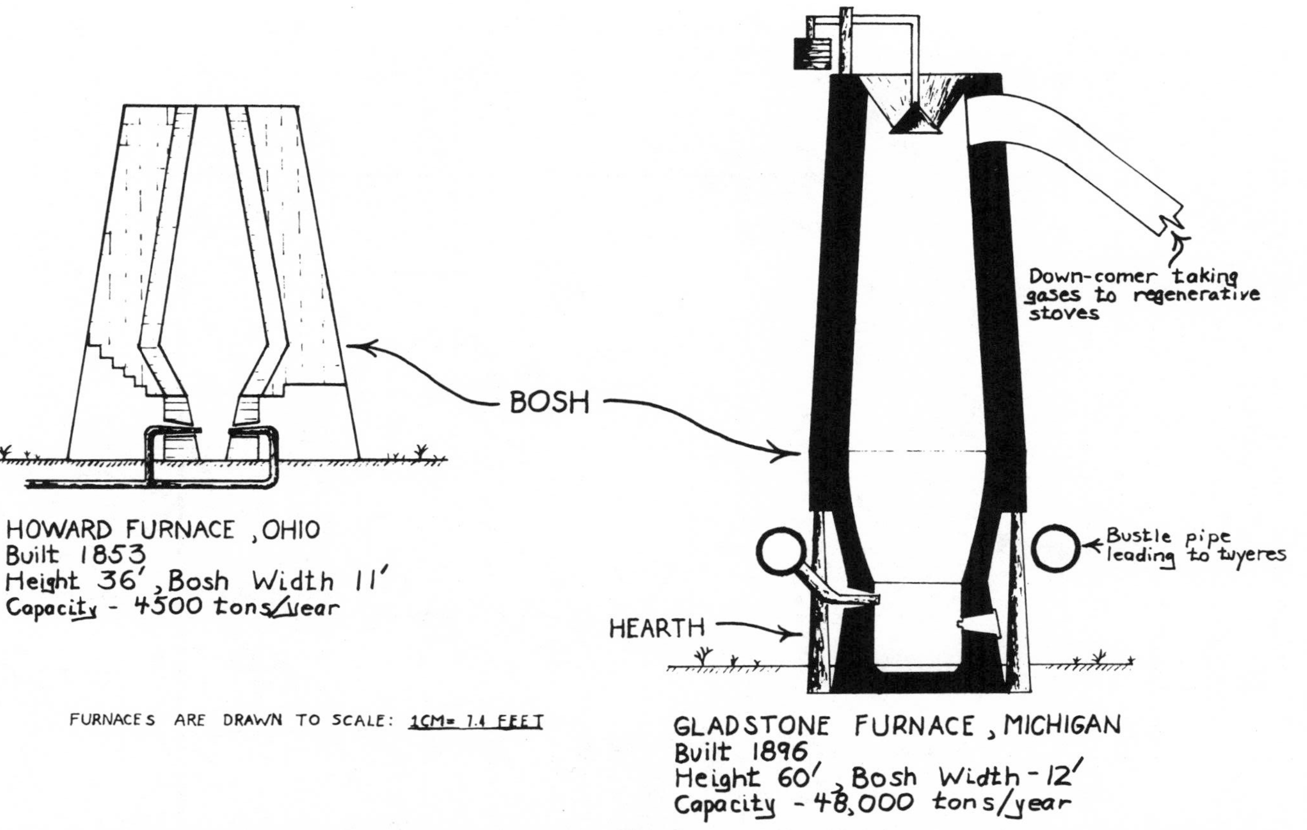

FIG. 1.—Schematic sections of Howard and Gladstone charcoal iron blast furnaces. (Simplified drawings by author from originals in "Charcoal Iron Industries of Hanging Rock Region," *Charcoal Ironworkers Journal* 2[1881]: 314, and "Most Modern of Charcoal Furnaces," *Iron Trade Review* [January 2, 1896], pp. 32–36.)

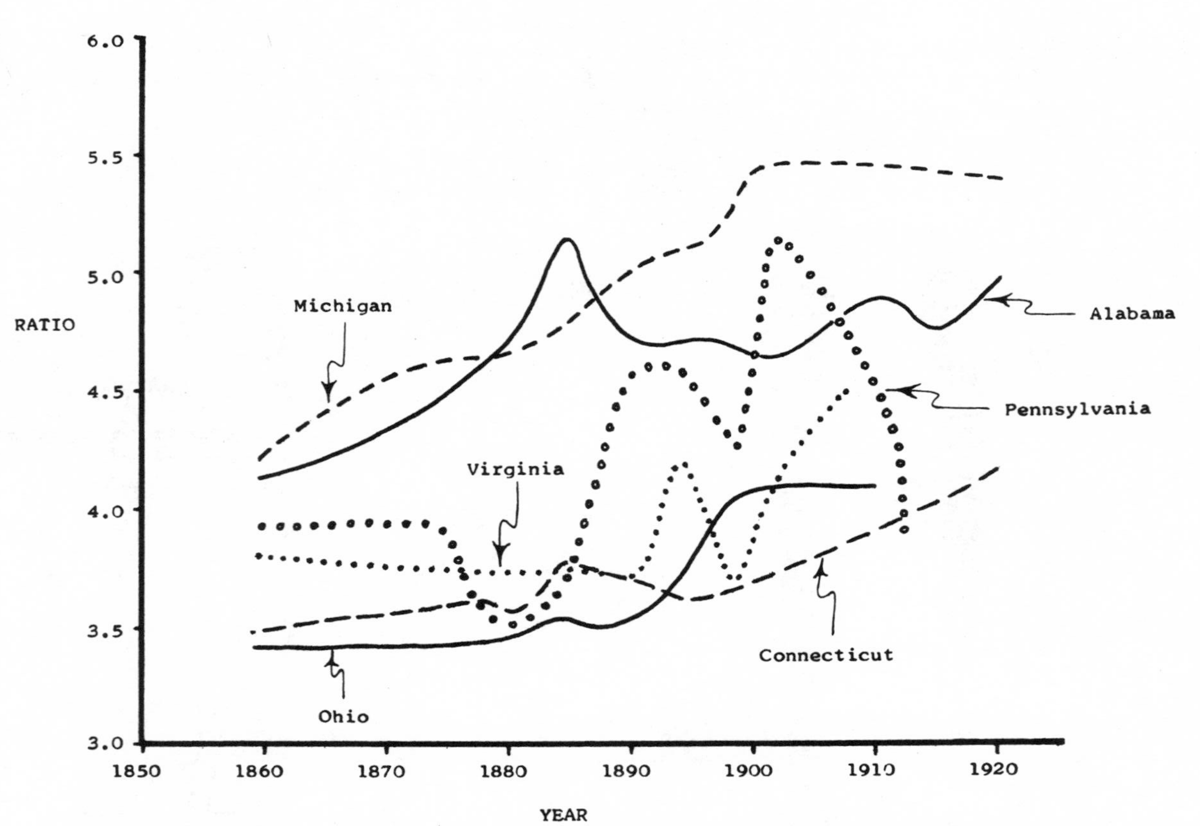

FIG. 2.—Height/bosh width ratios of operating charcoal blast furnaces, 1859–1920, by state. (From American Iron & Steel Association, *Directory of the Iron and Steel Works of the United States and Canada,* and John Peter Lesley, *The Iron Manufacturer's Guide to the Furnaces, Forges & Rolling Mills of the United States* [New York, 1859].

ous analyses have used such technical constants to study the charcoal-versus-coal price differential, we shall restrict our reexamination to this argument.

In the immediate antebellum period the average height of a Pennsylvania charcoal furnace was 37 feet, which was fairly typical of the national average of 35.5 feet.[11] Alfred Chandler says that the amount of charcoal needed to smelt a ton of pig iron at such Pennsylvania stacks was 240–150 bushels per ton; we will call this ratio the furnace's efficiency. Temin gives 180 as the efficiency of Pennsylvania iron plantations, and, for the country as a whole, Hunter gives between 250 and 150 bushels per ton for furnaces in the pre–Civil War period.[12]

If we compare these efficiency figures with those of postbellum charcoal stacks which adopted the kinds of innovations mentioned above, we find a very dramatic improvement. Although reliable data on furnace efficiencies are available for only a fraction of postbellum charcoal furnaces, table 2 is representative of conditions at the average Michigan, Alabama, Wisconsin, and Missouri stacks of the late 19th–early 20th centuries.

Although efficiency figures even for such advanced stacks as those included in this table may vary widely, it is clear that the increased height of many postbellum charcoal furnaces, coupled with the use of higher pressure blowing apparatus and hotter blast stoves, could produce a very marked saving in wood costs.

However, the above discussion by no means settles the question of rising wood costs. A popular historical assumption which has been made about the decline of the charcoal iron industry is that wood prices *must* have risen as the 19th century progressed, thus worsening the competitive situation between charcoal and coal irons.[13] The argument is that, as the charcoal furnace consumed the wood in its surrounding area and then had to import it from further afield, and as population growth created competition from lumbering interests, the price of a cord of timber rose. The higher cost of wood meant higher charcoal prices, and thus higher iron prices. Stated in these

[11] Lesley.

[12] Alfred D. Chandler, "Anathracite Coal and the Beginnings of the Industrial Revolution in the United States," *Business History Review* 46 (1972): 162; Temin, p. 65; Louis C. Hunter, "Influence of the Market upon Technique in the Iron Industry in Western Pennsylvania up to 1860," *Journal of Economic and Business History* (February 1929), pp. 262–63.

[13] See, e.g., Hunter, "Influence," pp. 261–62; Hunter, "Factors," p. 435; Woodham W. Cauley, "A Study of the Accounting Records of the Shelby Iron Company" (master's thesis, University of Alabama, 1949), p. 26.

TABLE 2

BUSHELS OF CHARCOAL NEEDED TO SMELT A TON OF CHARCOAL IRON AT ADVANCED FURNACES

State and Furnace	Date of Data	Height	Efficiency (Bushels/Ton)
Michigan:			
Antrim	1887	48.0	81.5
Bangor	1881	43.0	100.0
Deer Lake	1881	49.0	107.0
Elk Rapids	1882	48.0	98.0
Hamtramck	1883	53.0	93.0
Manistique	1926	58.6	87.5
Martel	1882	53.0	80.0
Newberry	1926	60.0	103.0
Spring Lake	1886	46.0	84.0
Wells	1912	60.0	100.0
Wisconsin:			
Hinkle	1890	60.0	82.0
Hinkle	1896	60.0	72.9
Hinkle	1926	60.0	84.0
Iron Mountain	1883	40.0	88.0
Alabama:			
Shelby	1881	60.0	120.0
Tecumseh	1881	60.0	107.0
Woodstock	1881	50.0	114.0
Missouri:			
Midland	1882	55.0	84.0

SOURCE.—*Journal of the Iron and Steel Institute* 34 (1880): 707; ibid., 50 (1896): 349; *Iron Age* (October 16, 1890), p. 630; *Charcoal Ironworkers Journal* 2 (1881): 24, 126, 134, 169; ibid., 3 (1882): 18, 59; ibid., 7 (1887): 95–96, 208.

terms, the argument is logical and makes sense. However, it does not stand up to empirical evidence.

If wood prices did go up as a result of these developments, then we should expect to find evidence for a greater increase in prices in those areas where the developments were most apparent. In other words, consumption of wood should have progressed further in the older charcoal producing regions of the east, therefore creating higher wood prices there than in newer areas, such as Michigan, Wisconsin, and Alabama, which began large-scale smelting only after the Civil War.

Wood prices also should have been lower in these latter regions, since they were less populous than those of the east, and competition for timber should not have been as great. We therefore tried to find evidence for higher prices for wood in such regions as New England, Pennsylvania, and New York, as contrasted with the newer smelting areas. The available price data, although too scanty to be conclusive, provide little evidence for any price differential unfavorable to the older regions.

Hunter indicates that in the period 1828–60 wood cost 5¢–10¢ per cord in Pennsylvania. The Richmond Iron Works in Massachusetts paid 50¢–80¢ per cord in the mid-1880s. In Oregon the Oswego furnace paid 90¢–100¢ per cord in 1881. A cord cost $1.25–$1.75 for the Spring Lake, Michigan, stack in 1886. The Antrim, Michigan, furnace paid $1.30 a cord in 1887.[14]

Another assumption which has been made by almost everyone who has written on the American charcoal iron industry is that charcoal iron furnaces were forced, by the nature of the fuel that they used, to locate their smelting operations in very close proximity to the source of that fuel. Thus Temin says, "Charcoal was produced by exploiting extensive tracts of woodland and could not be transported with facility; as late as 1874, it ordinarily was transported only 2–5 miles. The extensive nature of the production of charcoal and the difficulty of transporting it dictated a rural, isolated location for the the blast furnaces using charcoal. The nature of the location dictated in its turn the use of the plantation system. As a result, iron plantations were the only sources of pig iron in the 18th century and they lasted as long as charcoal was used as a fuel."[15]

The evidence does not support this statement. Although many charcoal iron blast furnaces were located in rural areas, we find no evidence to support the contention that rural locations were essential for the successful operation of these stacks. If anything, the opposite situation was true: isolated, rural settings appear to have limited the growth potential of charcoal stacks, with more advanced stacks located nearer urban centers.

Proximity to the source of wood does not seem to have dictated the location of the charcoal furnace. On the contrary, the two chief factors in the location of charcoal furnaces were proximity to large-scale ore deposits and proximity to a means of transportation. Figure 3 reveals that charcoal furnaces in 1859 were built only in the immediate proximity of ore beds, and, except in some areas of western Virginia, along canals and navigable rivers.[16] Figures 4 and 5 indicate

[14]Hunter, "Influence," pp. 241–81; *Charcoal Iron Workers Journal* 7 (1887): 51; ibid., 2 (1881): 289; ibid., 6 (1886): 95; ibid., 7 (1887): 209.

[15]Temin, p. 83.

[16]Lesley. On the maps presented in this paper, charcoal blast furnaces from a few states such as Texas, Utah, Colorado, Washington, and Oregon have been omitted, since there was rarely more than a single stack operating in any of these states. Furnaces are located on the maps to no greater accuracy than the limits of county boundaries. Arabic numbers inside circles indicate the number of furnaces operating in each state which we were unable to locate by county.

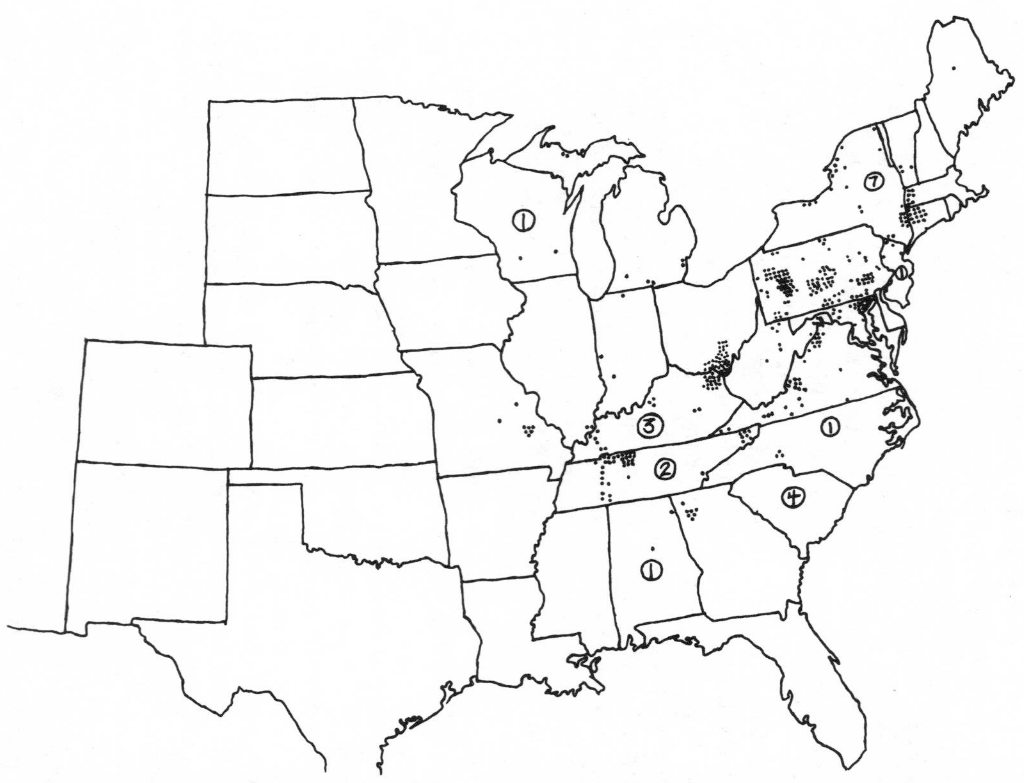

FIG. 3.—Location of operating charcoal blast furnaces in 1859; circled numbers indicate furnaces known to be in operation but whose location could not be determined. (From Lesley, *Guide* [see fig. 2].

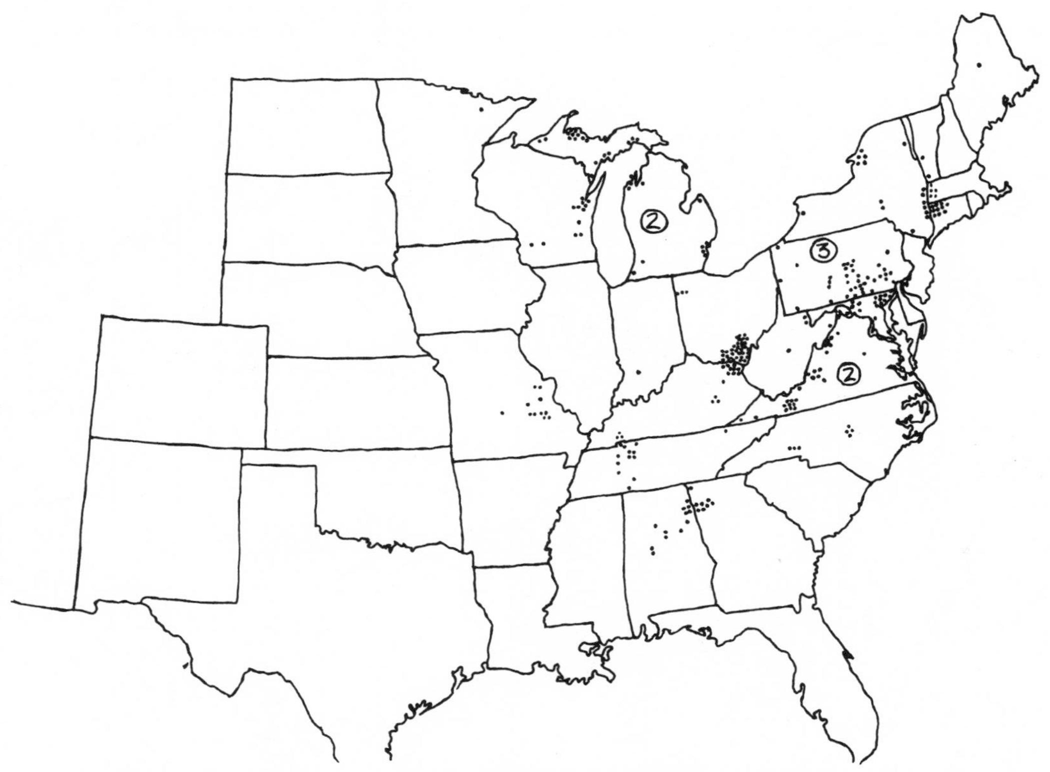

Fig. 4.—Location of operating charcoal blast furnaces in 1876; circled numbers indicate furnaces known to be in operation but whose location could not be determined. (From *Directory*, 1876 ed. [see fig. 2].

FIG. 5.—Location of operating charcoal blast furnaces in 1890; circled number indicates furnace known to be in operation but whose location could not be determined. (From *Directory*, 1890 ed. [see fig. 2].

that this pattern was retained after the Civil War, except that in this period furnaces began to follow the lines of railroads rather than canals and rivers.[17]

An economic study of the postbellum Pennsylvania and Connecticut charcoal iron industries by John Tyler indicates that the largest and most successful plants in these states deliberately were built very close to their ore supplies and had their charcoal transported in from a considerable distance.[18] During the postbellum period, in those regions where railroad facilities were available, it seems to have been no great economic or technological problem to ship charcoal a hundred miles or more.[19]

A close examination of figures 3, 4, and 5 tends to confirm Tyler's thesis. The large number of stacks in the neighborhood of Baltimore shows that immediate proximity to backwoods timberlands was not essential for the successful operation of such furnaces. As figures 4 and 5 show, these Baltimore-area stacks were among the most long-lived of all eastern charcoal furnaces. Not only did these furnaces survive for a long time, but their average output was among the highest of all eastern furnaces. Large charcoal furnaces located in or near urban centers were not an unusual phenomenon after the Civil War.

The postbellum increase in the number of Michigan charcoal stacks was almost as great in the south of the state in and around Detroit as it was in the remote upper peninsula. Not only was the number of furnaces being built around Detroit as great as in the north, but the capacity of these furnaces was equally great—in 1894, when the average design capacity of all Michigan stacks was 20,000 tons per year, the five furnaces in the Detroit–Wayne County region had an almost equal average capacity of 16,000 tons.[20] The common factor about all the Michigan furnaces was not that they were all dispersed in the middle of vast forests, but that they could all be easily supplied with cheap, rich ore because of their location on the shores of the Great Lakes. The wood for these stacks was often brought from considerable distances in specially built railroad boxcars.[21]

[17]*Directory,* 1876 and 1890 eds.

[18]The Barnum-Richardson Co. in Connecticut had its charcoal shipped 110 miles by railroad from Vermont. In the 1880s, the Isabella furnace in Pennsylvania had its fuel shipped 300 miles. This practice was not engaged in because there was no local wood, but because the total cost for charcoal delivered at the furnace was lower this way (John D. Tyler, "The Charcoal Iron Industry in Decline, 1855–1925" [master's thesis, University of Delaware, 1967], pp. 24–29).

[19]Ibid., p. 24.

[20]*Directory,* 1894 ed.

[21]*Charcoal Ironworkers Journal* 6 (1885): 116; "Most Modern of Charcoal Furnaces," *Iron Trade Review* (January 2, 1896), pp. 32–36.

Therefore, it is untrue that charcoal furnaces had to be built close to their fuel supplies. This is particularly true in the case of technologically advanced, high-output stacks, which could afford large capital investment in timber lands and transport systems. Although long distance shipment of wood or charcoal was obviously not a preferred method of operation, it was equally not the great handicap which some have supposed.[22]

However, the fact that groups of large charcoal furnaces were built near or in urban centers in the postbellum period, and that they continued to operate for considerable spans of time, would lead one to expect a rise in timber prices as these large-consumption operations deforested the timberlands in their vicinity. Quantitative data on this issue are again very scanty and any arguments pro and con can only be tentative. However, we do possess some very complete data from the largest charcoal furnaces ever to operate in the South which tend to refute the idea that massive consumption of timber must have driven prices up.

In 1849 the Shelby Iron Company built a 29-foot high charcoal furnace in Shelby County, Alabama—not far from Birmingham.[23] This furnace plant was of the traditional iron plantation design and had a capacity typical of such plants—3,000 tons per year. A second furnace was built in 1863 to meet war needs. This had a 56-foot height and a capacity of 6,000–7,000 tons a year. This was the first blast furnace to operate in Alabama with waste gas recovery equipment.

In 1868, the original stack having been abandoned, a new furnace was built with a height of 60 feet and a 14-foot bosh.[24] This was a very large and advanced furnace for such an early period. It had seven tuyeres, an iron shell with integral water cooling, and produced 8,500 tons per year. In 1888 the 1863 stack was replaced by a twin of the 1868 furnace. By 1890 these two stacks had been improved so that they then produced 25,000 tons per year between them, and by 1894 this production rate had been increased to 40,000 tons.[25]

In 1847, when the Shelby furnaces were just beginning operation,

[22]See, e.g., William T. Hogan, *Economic History of the Iron and Steel Industry in the United States* (Lexington, Mass., 1971), pp. 19–22; Hunter, "Factors," p. 431; Temin, p. 83.

[23]For the early history of the Shelby works, see Joyce Jackson, "History of the Shelby Iron Co., 1862–68" (master's thesis, University of Alabama, 1948); Robert H. McKenzie, "Horace Ware: Alabama Iron Pioneer," *Alabama Review* 26 (1973): 157–72; Frank E. Vandiver, "The Shelby Iron Company in the Civil War," *Alabama Review* 1 (1948): 12–26, 111–27, 203–17; James F. Doster, "The Shelby Iron Works Collection in the University of Alabama Library," *Business History Review* 26 (1952): 214–17. See also n. 13 above.

[24]Woodward Iron Co. (n. 7 above), p. 126.

[25]*Directory*, 1890 and 1894 eds.

the area around the furnace plant was relatively unpopulated and timber was plentiful. In that year, furnace account sheets show that $10 worth of charcoal was required to smelt a ton of pig iron.[26] During the course of the next sixty years, these furnaces, as well as the other large furnaces of northeastern Alabama (see figs. 4 and 5) must have done as much to deforest timberlands in their area as almost any other furnaces in the nation. Moreover, the mushrooming growth of the nearby Birmingham area in the decades before the turn of the century created severe competition from lumbering interests for wood.[27] Nevertheless, in the years 1910 and 1911, furnace account sheets reveal that it cost $9.32 and $8.29, respectively, in charcoal costs to smelt a ton of iron.[28]

However, it is hypothetically possible for late 19th-century ironmasters at large furnaces to have paid continually *higher* prices for their wood, and yet have benefited from continually *lower* charcoal costs. The explanation of this phenomenon demonstrates the final, but most significant error in the use of technical constants derived from iron plantation operations in economic analyses of the overall postbellum decline of the American charcoal iron industry. Namely, it fails to take into account the enormous advances in charcoal-making technology during the second half of the 19th century.

Iron plantation charcoaling techniques required no capital investment but were rather expensive from the point of view of labor costs. The wood was carbonized on the forest floor under a cover of mud and wet leaves. Heat was supplied by allowing a portion of the wood to burn completely. Accordingly, the collier required a considerable degree of craft skill to produce a maximum output of good charcoal per cord of wood, particularly under adverse weather conditions such as high wind. As the pile of wood was being "coaled," it had to be continually tended. If the covering of wet material dried out, or if holes developed in it, or if the wind blew part of it away, the entire pile could burst into flames. On the other hand, some holes had to be left in the cover to allow some oxygen in, and if these filled up the pile could cool and go out. Uniform coaling conditions within the pile were hard to maintain, particularly as the pile shrank and changed shape with the progress of the operation. The skill required to make good "open pit" charcoal made the collier the best paid worker at an iron plantation. The maximum output of this age-old process was 35–38 bushels of charcoal per cord of wood burned.[29]

[26]Woodward Iron Co., p. 122.

[27]Cauley, p. 26.

[28]Ibid., p. 281.

[29]For data on charcoaling techniques, see, e.g., *Charcoal Ironworkers Journal* 2 (1882): 210–14, 313; ibid., 3 (1882): 26; ibid., 6 (1885): 51, 61; ibid., 7 (1886): 115.

Shortly before the Civil War the first charcoal kilns were introduced. These became popular throughout the industry after the war. The kiln was a permanent shell built of masonry, brick, or sheet iron. It had little vents which could be opened or closed by the collier to control carefully the rate of carbonization. The possibility of the wood being accidentally converted to ash was thus minimized. However, the wood was still coaled by the heat produced from its own partial combustion. Although a few kilns had off-gas collection pipes for production of wood chemical distillates, this practice was rare. The maximum production of kilns was between 45 and 50 bushels of charcoal per cord. Figure 6 shows a typical masonry kiln of the late 19th century. This partially collapsed kiln stands just outside Marquette, Michigan, and was used by the Carp River Furnace from circa 1874 to 1907.

In the 1870s and 1880s, as the sophistication of some charcoal iron plants approached that of Pittsburgh coal iron furnaces, the retort

FIG. 6.—Carp River furnace masonry charcoal kiln, Marquette, Michigan. (Author's reconstructed rendering of partially destroyed remains at Marquette.)

system of producing charcoal became popular. Whereas kilns were usually inexpensive, homemade affairs, a wood retort plant was an expensive, complicated, contracted piece of work. Retort plants were constructed according to the same pattern as present-day by-product coking ovens—the wood was carbonized by external heat (usually off-gases from the furnace or waste heat from the steam engine boilers) and no air was allowed into the retort. The methanol, tar, and other volatile materials driven off were passed into an integral chemical plant for condensation and separation.

Retort systems were designed by professional consulting engineers and chemists—in fact, the profession of chemical engineering has one of its roots in this technology. The maximum output of retort plants was 60–65 bushels of charcoal per cord or wood. Moreover, in addition to the large savings in wood, the retort technology also provided the charcoal plant with very substantial chemical by-product sales. One example will demonstrate how important such sales could be in offsetting the high cost of wood as a smelting fuel. In 1911, the Newberry furnace in Michigan was producing, on the basis of our calculations, approximately $9.70 worth of pig iron, $5.50 worth of methanol, and $3.50 worth of acetate of lime for every cord of wood carbonized.[30]

Although the initial capital cost of erecting a retort plant was high, once established these highly mechanized operations reduced labor costs, since a battery of retorts could be tended by a few, relatively unskilled men. From what was said previously about the costs of charcoal at the Shelby works, it is significant to note that in 1882 this furnace erected one of the most advanced retort and wood chemical plants of the late 19th century.[31] Figure 7 shows a sixteen-retort plant of the Mathieu pattern (a fairly sophisticated design), set up for chemical recovery. This plant was erected at Luther, Michigan, in 1884.

The above discussion has shown that technical constants derived from the structure and practice of the iron plantation are not appropriate for an analysis of the decline of the charcoal iron industry. The following illustrative example demonstrates how far off calculations based on such constants can be.

Temin says that at the average antebellum iron plantation an acre of forest yielded 30 cords of wood, each cord yielding 40 bushels of charcoal and each ton of pig iron smelted requiring 180 bushels of charcoal in the furnace. The average furnace produced about 1,000

[30]Data based on private communication to R. Schallenberg by Philip Hamilton, last manager of the Newberry Plant; price data from Williams Haynes, *American Chemical Industry*, vol. 1 (New York, 1954), indices.

[31]*Charcoal Ironworkers Journal* 3 (1882): 4–6, 209.

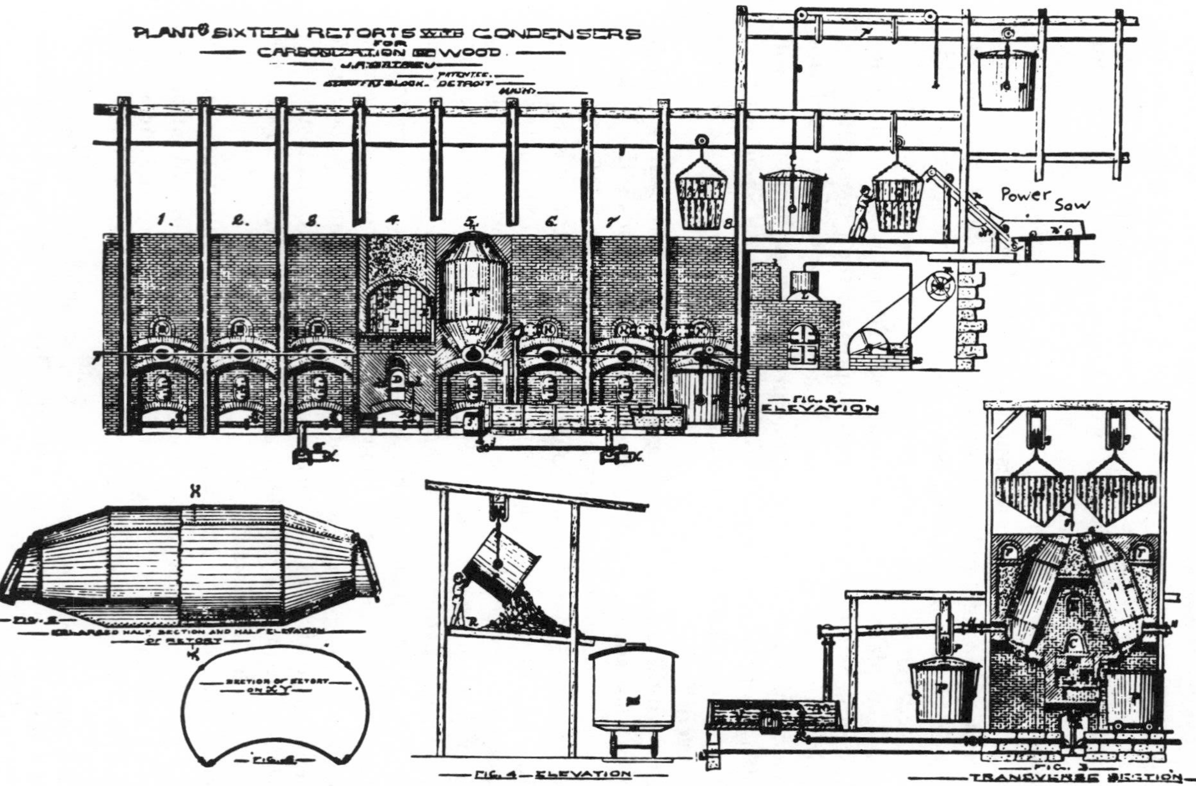

FIG. 7.—By-product wood carbonization retort plant erected at Luther, Michigan, 1884. (From *Charcoal Ironworkers Journal* 5 [1884]: 10.)

tons per year.[32] These figures, although not completely in agreement with our calculations, may be taken as fairly accurate for the 1840s and 1850s.[33] On this basis, an average furnace would have required 150 acres of woodland per year.

In 1887 the Antrim Iron Company in Mancelona, Michigan, reported that it was producing a ton of pig iron with only 81.5 bushels of charcoal—less than half of Temin's antebellum figure.[34] The contemporary Michigan charcoal production rate was 43 bushels of charcoal from a cord of wood.[35] We will assume that the woodlands used by the Antrim furnace had a density of 60 cords per acre, which would be a minimal figure for first-growth Michigan timberland.

On the basis of these figures, such a furnace required about 537 acres per year—three and one-half times as much woodland acreage for seventeen times as much iron as Temin's antebellum iron plantation, or about 22 tons per acre. The contemporary Oswego furnace in Oregon was even better off vis-à-vis forest use, consuming an acre of woodland for every 33 tons of iron smelted, or requiring only one-fifth of the forest acreage per ton of the antebellum iron plantation![36]

The Opportunity for Innovation

Given the great advantages which charcoal iron plants could have obtained from investment in innovation, it remains to be explained why charcoal ironmasters in so few areas of the country took advantage of these innovations. The crucial factor in restricting innovation was the problem of obtaining large quantities of rich, low-priced iron ore. This is not to suggest that the more intuitively important problem of timber supply did not also have a powerful effect on the economics and development of the post-1840 charcoal iron industry. However, regional patterns of technological innovation indicate that technical advance took place only in those regions where rich, low-priced iron ore was available in large quantities.

After the middle of the 19th century, charcoal iron began to come into competition with coal-smelted iron. We have demonstrated that

[32]Temin, pp. 65, 83.

[33]See Schallenberg, p. 345, n. 14. Also, 30 cords per acre can only be taken as accurate for second-growth woodland (see "The Charcoal Iron Industries of the Hanging Rock Region," *Charcoal Ironworkers Journal* 2 [1881]: 313).

[34]*Charcoal Ironworkers Journal* 7 (1887): 208–9.

[35]See Schallenberg, p. 349, n. 24.

[36]E. W. Crichton, "The Oswego Furnace," *Charcoal Ironworkers Journal* 2 (1881): 288–89.

charcoal iron produced in advanced furnaces could be economically competitive with coal irons.[37] Such advanced stacks relied upon large-output operations to provide the capital and incentive to invest in innovation. For example, a considerable amount of wood had to be carbonized in order to justify the large capital investment involved in setting up a retort wood chemical recovery plant. The new, regenerative hot blast stoves, which provided higher temperature blast, and the steam engines and cast-iron, machined blast cylinders, which provided higher pressure blasts, were also very expensive investments. Low-output furnaces could not afford them.

Antebellum charcoal furnaces in all parts of the country had low output—average design capacity varied between 1,000 to 3,000 tons per year (see table 3). The iron plantation model of organization and scale fitted these furnace plants well. Most of them were built in clusters around those relatively small-scale ore deposits upon which the American iron industry depended before the Civil War—for example, the Cornwall field in Connecticut, the southern Ohio deposits, and the deposits in western Virginia (see fig. 3).

After the Civil War, the number of charcoal furnaces operating in the United States fell rapidly. It is clear from table 4 that, although the industry in Pennsylvania, Connecticut, New York, Maryland, Ohio, and a number of other eastern states declined quickly in the postbellum period, the industry in states such as Michigan, Wisconsin, and Alabama increased rather dramatically.[38] Charcoal smelting operations in these latter areas did not begin to decline until the 1890s, whereas operations in the older regions began their decline at least thirty years earlier.

We have explained elsewhere the overall expansion of the industry up to 1890 and its subsequent decline in terms of demand factors —until the 1890s the demand for charcoal iron continually grew, but subsequently declined as charcoal iron became technologically obsolescent.[39] However, the regional differences in patterns of expansion and decline—as indicated in the preceding paragraph—are explainable only in terms of changing patterns of ore supply which began just before the Civil War.

The growth of the Michigan-Wisconsin and Alabama-Tennessee-Georgia charcoal iron industries was made possible by their location near the new, national ore fields. The Great Lakes fields and the Alabama field were not opened up to large-scale exploitation until the

[37] Schallenberg, p. 354–55.

[38] *Directory*.

[39] Schallenberg.

TABLE 3

AVERAGE DESIGN CAPACITY OF CHARCOAL FURNACES, 1859–1925 (Tons/Year × 10^3)

	Conn.*	N.Y.	Pa.	Md.	Va.†	Tenn.	Ga.	Ohio‡	Ala.	Mich.	Mo.	Wis.
1859	1.7	2.1	1.5	2.1	1.3	1.7	1.0	3.0	0.9	3.1	3.0	1.3
1876	3.3	3.6	2.0	2.9	3.5	4.2	4.2	4.2	6.5	7.7	8.4	4.8
1880	3.8	3.8	2.3	3.2	3.1	2.9	3.4	4.1	5.6	10.0	9.8	6.1
1884	3.9	4.8	3.7	4.5	5.4	4.7	5.4	4.2	8.8	13.4	10.8	7.8
1886	4.0	6.4	3.3	4.5	3.4	9.1	4.3	4.5	8.3	14.6	14.3	9.0
1890	4.4	6.9	3.9	6.9	3.5	11.3	8.3	3.8	13.1	19.2	16.3	16.6
1894	4.2	5.8	3.6	6.9	5.2	15.3	12.3	4.0	13.9	19.5	14.3	18.8
1898	4.7	5.0	2.6	5.5	2.3	13.7	15.5	3.6	15.4	25.1	20.0	45.0
1901	...	22.5	3.6	...	...	...	17.0	...	1.8	26.0	...	...
1904	...	5.0	3.2	6.0	...	...	16.0	...	6.9	32.0	36.7	45.0
1908	5.0	1.9	3.0	6.0	8.4	8.3	...	18.1	17.3	32.3	...	...
1912	5.0	...	...	...	...	...	...	...	...	38.0	...	...
1916	5.0	5.0	2.4	6.0	14.0	5.0	...	3.5	20.3	31.9	20.0	45.0
1925	...	5.0	...	...	...	4.8	...	...	22.7	30.6	45.0	...

SOURCE.—See table 1.
*Data for Connecticut and Massachusetts are very similar.
†Includes West Virginia.
‡Data for Ohio and Kentucky are very similar.

TABLE 4

NUMBER OF OPERATING CHARCOAL BLAST FURNACES, 1859–1930

	Mass.	Conn.	N.Y.	Maine	Pa.	Md.	N.C.	S.C.	Va.*	Ky.	Tenn.	Ga.	Ohio	Ala.	Mich.	Mo.	Wis.
1859	7	14	32	1	126	19	4	4	38	30	44	7	38	3	7	8	3
1876	5	10	15	1	40	15	7	0	33	17	13	8	36	12	30	11	12
1880	5	10	16	1	38	15	7	0	34	17	20	8	34	10	26	10	10
1884	4	9	15	1	31	14	6	0	35	13	9	6	21	12	27	9	11
1886	4	9	9	1	25	14	2	0	26	3	9	2	17	10	25	4	10
1890	4	8	10	1	16	9	2	0	19	3	9	3	13	14	25	3	6
1894	4	7	6	0	14	6	0	0	8	3	9	3	10	15	20	2	6
1898	3	5	3	0	4	2	0	0	1	0	3	1	5	4	9	0	1
1901	1	2	3	0	4	1	0	0	2	0	1	2	7	3	9	1	1
1904	2	3	4	0	5	1	0	0	4	0	3	3	7	6	10	1	1
1908	3	2	2	0	4	1	0	0	4	1	2	0	5	5	11	1	1
1912	2	3	0	0	4	1	0	0	0	1	1	1	1	4	11	1	1
1916	2	2	1	0	4	1	0	0	2	1	1	2	1	3	11	1	2
1920	1	3	1	0	4	1	0	0	0	0	6	2	0	2	9	3	1
1925	0	0	1	0	0	0	0	0	0	0	2	0	0	3	9	2	1
1930	0	0	0	0	0	0	0	0	0	0	2	0	0	0	5	0	1

SOURCE.—See table 1.
*Includes West Virginia.

1860s.[40] Before the opening up of the Marquette range in the late 1850s, the Michigan charcoal iron industry consisted of no more than a half-dozen furnaces of the iron plantation type.[41] Exploitation of the rich Lake ores permitted this industry to expand and invest in technological innovation. Average design capacity of Michigan stacks climbed from 3,000 tons per year in 1859 to a maximum of 38,000 tons per year in 1912; a thirteenfold increase in output per stack.[42]

The charcoal smelting industry in Alabama began taking advantage of the rich northern Alabama ore fields at the same time that the Michigan–western Wisconsin furnace masters began exploiting the Lake ores.[43] The Alabama charcoal blast furnaces never attained the same size or degree of technological sophistication as that of the Michigan-Wisconsin stacks, but the rate of change in the structure of the Alabama industry after the Civil War was even greater than in the north.[44] Average design capacity for Alabama charcoal furnaces was the lowest in the nation in 1859—900 tons per year.[45] By 1908 this had increased almost twentyfold to 17,300 tons, and, by 1925, had risen to its all-time high of 22,700 tons per furnace.[46]

The charcoal iron industry in the contiguous states of Tennessee and Georgia was also able to take advantage of the Alabama ores. Although table 4 indicates a drop in the number of operating furnaces in Tennessee from 1859 to 1894 of forty-four to nine, the average production capacity of the furnaces which remained increased by almost a factor of ten—from 1,700 tons to 15,300 tons per year.[47] Total Tennessee production of charcoal iron rose from 34,000 tons in 1872 to a maximum of 61,400 tons in 1893 (see table 5). The advanced stacks which were responsible for this production were located near the Tennessee River and could be economically supplied with Alabama ore.

[40] For reasons for the metallurgical and economic superiority of Lake Superior ores, see Kenneth Warren, *The American Steel Industry, 1850–1870: A Geographical Interpretation* (Oxford, 1973), pp. 115–17.

[41] Lesley.

[42] Ibid; *Directory,* 1912 ed.

[43] Hogan, p. 20.

[44] E.g., although many Alabama furnaces had fairly large hot blast stoves, none of these were of the regenerative pattern, as were many of the Michigan and Wisconsin charcoal stacks (see Woodward Iron Co.). In addition, the collection of wood chemicals, except for a few plants, seems to have been absent in the South (ibid., and Edward H. French and James R. Withrow, "The Hardwood Distillation Industry of America," *Ohio State University Bulletin* 19, no. 7 [1914]: 30).

[45] Lesley.

[46] *Directory,* 1908 and 1925 eds.

[47] Ibid., 1894 ed.; Lesley.

TABLE 5

PRODUCTION OF CHARCOAL PIG IRON 1866–1910 (Net Tons/Year × 10^3)

	Mass.	Conn.	N.Y.	Maine	Pa.	Md.	Va.	Ky.	Tenn.	Ga.	Ohio	Ala.	Mich.	Mo.	Wis.
1866	...	...	...	...	57.8	...	...	...	...	...	87.9	...	35.4	...	...
1867	...	...	...	...	60.2	...	...	...	...	...	89.5	...	55.7	...	...
1872	12.8	22.7	19.8	...	45.0	29.0	21.4	39.7	34.1	2.9	95.6	12.5	86.8	45.6	27.8
1973	15.7	26.9	29.3	0.8	45.9	30.3	22.5	42.2	34.5	7.5	100.5	22.3	113.9	39.5	38.9
1874	17.8	14.5	28.3	1.7	40.9	25.0	23.5	36.6	37.2	4.3	92.8	32.9	128.9	49.1	28.9
1875	10.0	10.9	11.5	2.0	34.5	...	15.4	22.3	18.0	3.8	61.9	25.1	101.8	39.8	26.5
1876	5.0	10.2	8.1	...	23.1	13.9	7.4	17.2	10.1	0.5	48.9	23.3	82.5	24.1	26.3
1877	2.9	14.4	16.6	1.9	29.6	17.9	6.2	17.0	11.2	4.0	42.1	24.8	75.2	28.6	22.2
1878	1.4	15.9	15.8	1.2	29.4	17.8	6.3	16.9	11.2	2.5	33.5	23.9	70.9	16.9	27.5
1879	5.0	16.8	18.1	1.2	35.9	19.7	7.7	12.7	7.6	4.1	43.4	31.9	101.5	17.8	31.4
1880	9.9	22.6	27.8	3.6	43.4	33.1	14.0	21.2	16.7	7.3	69.2	37.7	154.4	15.8	42.9
1881	12.4	28.5	30.5	4.4	51.9	27.6	39.0	16.8	19.0	13.4	66.2	44.2	187.0	43.2	47.7
1882	10.3	24.3	30.7	4.1	49.9	28.3	26.1	17.1	37.6	15.6	58.7	55.5	210.2	54.3	55.3
1883	10.5	19.9	25.7	4.4	38.3	29.8	16.9	13.5	35.0	13.0	40.5	57.4	173.2	34.1	55.8
1884	4.9	14.2	23.5	...	23.2	15.1	14.8	7.9	18.8	9.6	24.9	59.5	172.8	31.6	25.8
1885	0.9	17.5	14.7	0.4	12.1	10.4	12.6	4.7	31.2	5.8	18.0	77.6	143.1	21.8	19.6
1886	8.1	19.4	14.4	5.1	16.7	7.8	6.0	6.3	28.7	0.5	16.4	82.1	190.7	20.2	28.5
1887	11.1	21.7	26.5	4.4	11.9	15.5	9.5	6.2	46.2	...	18.5	95.2	213.5	40.7	47.5
1888	13.2	21.6	19.9	5.6	15.2	15.7	7.4	5.1	51.9	0.3	21.9	94.1	213.3	28.3	69.8
1889	7.8	24.1	19.1	5.2	15.9	16.3	8.9	6.8	50.2	2.3	22.5	110.4	214.4	32.7	80.5
1890	5.5	22.5	17.9	1.2	18.3	16.8	7.0	3.7	54.0	6.6	26.2	110.4	258.5	33.9	95.0
1891	10.0	24.4	11.6	...	11.8	10.3	3.6	3.8	51.2	17.6	22.1	87.3	238.7	16.1	105.4
1892	8.8	19.2	18.3	...	13.2	10.9	1.9	3.6	56.2	11.1	21.2	89.0	206.5	31.5	91.9
1893	8.8	14.0	7.2	...	5.2	5.2	0.6	2.4	61.4	13.8	20.5	75.3	131.6	17.5	57.6
1894	0.2	8.3	8.6	...	5.3	3.7	...	...	7.7	14.1	14.9	40.4	106.6	7.3	25.8
1895	5.3	6.3	5.8	...	5.0	...	...	...	21.5	15.0	11.8	21.1	102 1	1.8	51.4

SOURCE.—American Iron & Steel Association; *Bulletin of the AISA; Charcoal Ironworkers Journal; Iron Age.*

TABLE 5 *(Continued)*

	Mass.	Conn.	N.Y.	Maine	Pa.	Md.	Va.	Ky.	Tenn.	Ga.	Ohio	Ala.	Mich.	Mo.	Wis.
1896	2.1	11.4	5.8	...	3.0	5.4	1.6	...	30.4	16.0	12.8	33.4	167.4	10.5	46.4
1897	3.7	9.3	6.0	...	2.2	5.2	0.2	...	27.3	19.2	8.7	16.7	148.5	12.9	18.9
1898	4.1	7.1	7.4	...	3.6	2.4	...	...	19.4	15.3	7.1	41.1	165.3	10.5	...
1899	2.8	11.3	7.9	...	4.1	——1.9——		...	32.5	...	7.3	46.7	150.0	3.6	30.0
1900	3.7	11.4	8.8	...	3.8	——6.6——		——3.5——		25.6	8.6	64.5	175.0	6.0	32.0
1901	3.8	9.4	25.3	...	5.4	——5.7——		——3.2——		30.6	11.3	59.4	190.8	——55.4——	
1902	3.8	13.6	38.3	...	4.7	——4.9——		...	7.1	35.5	12.1	67.8	173.8	——62.4——	
1903	3.7	16.2	36.3	...	4.6	——6.5——		...	17.0	46.8	10.6	73.1	274.1	——67.6——	
1904	3.5	9.9	33.5	...	2.9	——5.9——		...	9.2	27.6	1.1	34.2	192.1	——58.0——	
1905		——18.9——		...	3.7	——2.4——			——24.5——		4.7	28.6	235.9	——76.6——	
1906		——22.6——		...	3.0	——5.5——			——30.2——		6.2	28.9	315.2	——73.4——	
1907		——21.4——		...	2.6	——1.7——			——22.9——		2.7	39.3	330.3	——68.9——	
1908		——16.2——		...	2.8	——3.7——			——21.8——		2.8	26.7	160.9	——44.5——	
1909		——23.1——		...	3.0	——6.3——			——16.5——		...	37.6	259.5	——75.0——	
1910		——18.6——		...	4.6	——1.8——			——10.5——		1.2	39.9	292.1	——72.8——	

TABLE 6

REPRESENTATIVE SOUTHERN FURNACES, 1894

State and Furnace	Height	Bosh	Design Capacity (Tons/Year)
Tennessee:			
Aetna	55.0	11.0	18,000
Cumberland	60.0	11.0	15,000
LaGrange	65.0	12.0	18,000
Mannie no. 1	60.0	12.0	36,000*
Mannie no. 2	60.0	12.0	
Napier	60.0	12.0	18,000
Warner no. 1	55.0	11.0	36,000*
Warner no. 2	45.0	9.0	
Alabama:			
Attalla	55.0	11.0	18,000
Clifton no. 1	55.0	12.5	30,000*
Clifton no. 2	56.0	13.5	
Decatur	60.0	12.0	18,000
Ironaton no. 1	55.0	12.0	33,000*
Ironaton no. 2	60.0	14.0	
Montgomery	60.0	12.0	18,000
Piedmont	60.0	12.0	No Data
Georgia:			
Rome	65.0	12.0	15,500

*Indicates total production of a double furnace.
SOURCE.—*Directory* (see table 1).

The most technologically advanced of postbellum southern charcoal furnaces—the Shelby stacks—have already been described. A fairly large number of Alabama-Tennessee-Georgia area charcoal furnaces approached the Shelby works in size, output, and sophistication. Table 6 is representative of the best southern furnaces in 1894.

Missouri was the third section of the country to possess an advanced, albeit small, charcoal iron industry after the Civil War. Tables 3 and 4 indicate that, although the design capacity of Missouri furnaces was equal to those of Michigan, the number of such stacks was considerably less. By 1890, when Michigan, Alabama, Tennessee, and Wisconsin still had twenty-five, fourteen, nine, and six operating furnaces, respectively, Missouri had only three.[48] However, fourteen years earlier the relative status of Missouri was closer to that of the other advanced charcoal smelting states—in 1876 Michigan, Albama, Tennessee, Wisconsin, and Missouri had thirty, twelve, thirteen, twelve, and eleven operating stacks, respectively.[49]

[48]*Directory*, 1890 ed.
[49]Ibid., 1876 ed.

This comparison most clearly demonstrates the effect of ore supply on the charcoal iron industry. The furnaces of Missouri depended for their survival upon two rich, but comparatively limited iron ore deposits—Pilot Knob and Iron Mountain.[50] These fields are located near each other in the southeastern part of the state. The opening up of these two fields coincided with that of the Marquette field, and during the early postbellum period it looked as though these fields might actually contribute more to total American iron production than the Marquette deposit.[51]

The output of Pilot Knob and Iron Mountain grew after the Civil War—which accounts for the fairly large number of charcoal furnaces in the state in the early postbellum period. However, these fields already began to decline in terms of absolute output in the 1880s, at the same time that the Marquette field itself was being dwarfed by the exploitation of newer Great Lakes ore deposits.[52] The decline in the Missouri charcoal iron industry reflects the decline of these local fields.

Figures 3, 4, and 5 also demonstrate the dependence of the Missouri charcoal iron industry on these local ore reserves. The clusters of these furnaces, located in the southeastern part of the state near the ore deposits, show that the most crucial factor in the location of charcoal blast furnaces was proximity to iron ore, not wood. Instead of being dispersed so as not to compete for timber reserves, the postbellum Missouri furnaces were all clustered around the ore fields. Concentration around iron ore sources, therefore, and not dispersal, would appear to have been more important for the survival of the larger-scale charcoal iron industry.

As figures 3, 4, and 5 and tables 3 and 4 demonstrate, the charcoal iron industry in every other section of the country except those discussed above went into a decline beginning in the 1860s. Not only did the number of furnaces in operation in these states fall, but among those furnaces which remained in operation, innovation and increase in production were minimal or nonexistent. These factors were common to the industry in northern Tennessee, southern Ohio, northern Kentucky, western and eastern Pennsylvania, Maryland, western Virginia, eastern New York, and western Massachusetts, and Connecticut. The only other common factor about the industry in all these states was lack of proximity to the rich, extensive ore fields opened up beginning in the 1860s.

[50]"Where Our Iron Ore Comes From," *Charcoal Ironworkers Journal* 8 (1887): 194.
[51]Warren, p. 44.
[52]"Where Our Iron Ore Comes From" (n. 50 above).

As a result, the cost per ton of iron produced in these furnaces was too high for the ironmasters to sell their product for anything but specialty purposes. Furnace capacity remained low, and ironmasters concentrated on producing quality rather than quantity metal. Accordingly, they had neither the money nor the incentive to invest in the kind of technological innovation which Alabama and Michigan ironmasters were engaging in. Such low capacity furnaces could continue to survive in regions where a specialty market still existed for their high-priced metal. This is why the eastern Pennsylvania, Maryland, Connecticut, and Massachusetts charcoal industry retained a considerable degree of vitality after the Civil War at the same time that the western Pennsylvania furnaces were disappearing very rapidly. The eastern furnaces were linked economically to the large east coast specialty manufacturing establishment.[53]

Conclusions

Historians of the American iron industry have not been mistaken in stressing the importance of the higher price of charcoal versus coal as one of the contributing factors to the relative decline of the charcoal smelting industry after the middle of the last century. However, it has been the purpose of this paper to demonstrate that the economic and technological problems of wood as a factor of production have been treated too simplistically by past historians, and that the economics of ore supply had at least as much to do with the decline of the charcoal iron industry as did problems of wood supply.

In their analyses of the charcoal iron industry, these historians have chosen a *static* model—the iron plantation—from which to derive the technical constants they used. In periods of very rapid industrial change, such as the period under examination in this paper, such static models should be regarded with suspicion. The theory that iron plantation technology was dictated by problems inherent in wood supply, and that the economics of such a smelting technology led to the overall decline of the charcoal industry after the middle of the 19th century, places undue emphasis on fuel problems. It is true that the static model provides a fairly accurate image of blast furnace plants in most areas of the nation even in the decades after the Civil War. However, this lack of innovation was a constraint produced by problems of ore supply, and not wood.

[53]In 1890 there were twenty-four railroad-car wheel foundries in the New York–New England area, as well as a large number of specialty metal working shops in the New England region which depended upon the unique metallurgical properties of charcoal iron—e.g., the gun shops of the Connecticut Valley (*Directory*, 1890 ed.). Maryland also possessed car wheel foundries which needed a great deal of charcoal iron.

Wherever charcoal ironmasters could take advantage of supplies of cheap, rich ore to produce large amounts or iron which were reasonably competitive with coal irons, they did so. The almost unlimited amounts of such ore which ironmasters in Michigan, Wisconsin, Alabama, and Tennessee had available encouraged them to build increasingly larger furnaces. These stacks permitted the use of charcoal-saving innovations, which, because of their expense, were not feasible for small-scale smelters. In order to make charcoal iron which could compete with coal irons it was necessary for the charcoal ironmaster to do *both* of these things simultaneously—that is, use the new ores and adopt the innovations. Ironmasters located in regions where such ores were not easily obtainable could not hope to compete with coke irons, and therefore their iron of necessity remained a specialty metal. Trapped in such a geographically limited situation, their adoption of technological innovations was minimal.

Bursting Boilers and the Federal Power

JOHN G. BURKE

I

When the United States Food and Drug Administration removes thousands of tins of tuna from supermarket shelves to prevent possible food poisoning, when the Civil Aeronautics Board restricts the speed of certain jets until modifications are completed, or when the Interstate Commerce Commission institutes safety checks of interstate motor carriers, the federal government is expressing its power to regulate dangerous processes or products in interstate commerce. Although particular interests may take issue with a regulatory agency about restrictions placed upon certain products or seek to alleviate what they consider to be unjust directives, few citizens would argue that government regulation of this type constitutes a serious invasion of private property rights.[1]

Though federal regulatory agencies may contribute to the general welfare, they are not expressly sanctioned by any provisions of the U.S. Constitution. In fact, their genesis was due to a marked change in the attitude of many early nineteenth-century Americans who insisted that the federal government exercise its power in a positive way in an area that was non-existent when the Constitution was enacted. At the time, commercial, manufacturing, and business interests were willing to seek the aid of government in such matters as patent rights, land grants, or protective tariffs, but they opposed any action that might smack of governmental interference or control of their internal affairs. The government might act benevolently but never restrictively.

The innovation responsible for the changed attitude toward government regulation was the steam engine. The introduction of steam power

Now deceased, PROFESSOR BURKE was in the Department of History at the University of California, Los Angeles, at the time this article was published.

1 See, e.g., *Report* on Practices and Procedures of Governmental Control, Sept. 18, 1944 (House of Representatives document 678, ser. 10873 [Washington: 78th Congress, 2d session]), p. 3, where it is stated: "Regulation, seen through modern eyes is not a violent departure from the ways of business to which the nation is both habituated and strongly attached . . . . regulation . . . enjoys, as a system, in large measure the confidence and approval of the parties concerned."

was transforming American culture, and while Thoreau despised the belching locomotives that fouled his nest at Walden, the majority of Americans were delighted with the improved modes of transportation and the other benefits accompanying the expanding use of steam. However, while Americans rejoiced over this awesome power that was harnessed in the service of man, tragic events that were apparently concomitant to its use alarmed them—the growing frequency of disastrous boiler explosions, primarily in marine service. At the time, there was not even a governmental agency that could institute a proper investigation of the accidents. Legal definitions of the responsibility or negligence of manufacturers or owners of potentially dangerous equipment were in an embryonic state. The belief existed that the enlightened self-interest of an entrepreneur sufficed to guarantee the public safety. This theory militated against the enactment of any legislation restricting the actions of the manufacturers or users of steam equipment.

Although the Constitution empowered Congress to regulate interstate commerce, there was still some disagreement about the extent of this power even after the decision in *Gibbons* v. *Ogden*, which ruled that the only limitations on this power were those prescribed in the Constitution. In the early years of the republic, Congress passed legislation under the commerce clause designating ports of entry for customs collections, requiring sailing licenses, and specifying procedures for filing cargo manifests. The intent of additional legislation in this area, other than to provide for these normal concomitants of trade, was to promote commerce by building roads, dredging canals, erecting lighthouses, and improving harbors. Congress limited its power under the commerce clause until the toll of death and destruction wrought by bursting steamboat boilers mounted, and some positive regulations concerning the application of steam power seemed necessary. Thomas Jefferson's recommendation that we should have "a wise and frugal Government, which shall restrain men from injuring one another, shall leave them otherwise free to regulate their own pursuits of industry and improvement" took on a new meaning.[2]

Although several historians have noted the steamboat explosions and the resulting federal regulations, the wider significance of the explosions as an important factor in altering the premises concerning the role of government vis à vis private enterprise has been slighted.[3] Further, there has been no analysis of the role of the informed public in this matter.

[2] Thomas Jefferson, "Inaugural Address," *Journal of the Executive Proceedings of the U.S. Senate*, I (Washington, 1828), 393.

[3] The most authoritative work is Louis C. Hunter, *Steamboats on the Western Rivers* (Cambridge, Mass., 1949), pp. 122–33, 271–304, 520–46.

The scientific and technically knowledgeable members of society were —in the absence of a vested interest—from the outset firmly committed to the necessity of federal intervention and regulation. They conducted investigations of the accidents; they proposed detailed legislation which they believed would prevent the disasters. For more than a generation, however, successive Congresses hesitated to take forceful action, weighing the admitted danger to the public safety against the unwanted alternative, the regulation of private enterprise.

The regulatory power of the federal government, then, was not expanded in any authoritarian manner. Rather, it evolved in response to novel conditions emanating from the new machine age, which was clearly seen by that community whose educations or careers encompassed the new technology. In eventually reacting to this danger, Congress passed the first positive regulatory legislation and created the first agency empowered to supervise and direct the internal affairs of a sector of private enterprise in detail. Further, certain congressmen used this precedent later in efforts to protect the public in other areas, notably in proposing legislation that in time created the Interstate Commerce Commission. Marine boiler explosions, then, provoked a crisis in the safe application of steam power, which led to a marked change in American political attitudes. The change, however, was not abrupt but evolved between 1816 and 1852.

II

Throughout most of the eighteenth century, steam engines worked on the atmospheric principle. Steam was piped to the engine cylinder at atmospheric pressure, and a jet of cold water introduced into the cylinder at the top of the stroke created a partial vacuum in the cylinder. The atmospheric pressure on the exterior of the piston caused the power stroke. The central problem in boiler construction, then, was to prevent leakage. Consequently, most eighteenth-century boilers were little more than large wood, copper, or cast-iron containers placed over a hearth and encased with firebrick. In the late eighteenth century, Watt's utilization of the expansive force of steam compelled more careful boiler design. Using a separate condenser in conjunction with steam pressure, Watt operated his engines at about 7 p.s.i. above that of the atmosphere. Riveted wrought-iron boilers were introduced, and safety valves were employed to discharge steam if the boiler pressure exceeded the designed working pressure.

Oliver Evans in the United States and Richard Trevithick in England introduced the relatively high-pressure non-condensing steam engine almost simultaneously at the turn of the nineteenth century. This development led to the vast extension in the use of steam power. The high-pressure engines competed in efficiency with the low-pressure type, while their compactness made them more suitable for land and water vehicular transport. But, simultaneously, the scope of the problem faced even by Watt was increased, that is, the construction of boilers that would safely contain the dangerous expansive force of steam. Evans thoroughly respected the potential destructive force of steam. He relied chiefly on safety valves with ample relieving capacity but encouraged sound boiler design by publishing the first formula for computing the thickness of wrought iron to be used in boilers of various diameters carrying different working pressures.[4]

Despite Evans' prudence, hindsight makes it clear that the rash of boiler explosions from 1816 onward was almost inevitable. Evans' design rules were not heeded. Shell thickness and diameter depended upon available material, which was often of inferior quality.[5] In fabrication, no provision was made for the weakening of the shell occasioned by the rivet holes. The danger inherent in the employment of wrought-iron shells with cast-iron heads affixed because of the different coefficients of expansion was not recognized, and the design of internal stays was often inadequate. The openings in the safety valves were not properly proportioned to give sufficient relieving capacity. Gauge cocks and floats intended to ensure adequate water levels were inaccurate and subject to malfunction by fouling with sediment or rust.

In addition, there were also problems connected with boiler operation and maintenance.[6] The rolling and pitching of steamboats caused

[4] Greville and Dorothy Bathe, *Oliver Evans* (Philadelphia, 1935), pp. 151, 253. Also, see Walter F. Johnson, "On the Strength of Cylindrical Steam Boilers," *Journal of the Franklin Institute* (hereinafter cited as "JFI"), X, N.S. (1832), 149. Evans' formula reveals that he considered that a safe design tensile strength for good quality wrought iron was about 42,000 p.s.i. and that a factor of safety of 10 should be used to arrive at a safe shell thickness.

[5] For reports of defective design and poor quality material see: Charles F. Partington, *An Historical and Descriptive Account of the Steam Engine* (London, 1822), p. 85; Committee on Steamboats *Report*, May 18, 1832 (House of Representatives document 478, ser. 228 [Washington: 22d Congress, 1st session]), pp. 44, 170 (hereinafter cited as "Doc. 478"). Also, *JFI*, VI, N.S. (1830), 44–51; VIII, N.S. (1831), 382; IX, N.S. (1832), 28, 100, 363; X, N.S. (1832), 226–32; XVII, N.S. (1836), 298–302; XX, N.S. (1837), 100, 103.

[6] For operating difficulties see: Partington, *op. cit.*, p. 118; *JFI*, V, N.S. (1830), 402; VI, N.S. (1830), 9; VIII, N.S. (1831), 277, 289–92; VIII, N.S. (1831), 309, 313, 382; IX, N.S. (1832), 20–22. Also, Secretary of the Treasury, *Report on Steam Engines*, Dec. 13, 1838 (House of Representatives document 21, ser. 345 [Washington: 25th Congress, 3d session]), p. 3 (hereinafter cited as "Doc. 21"). The whole number of steam engines in the United States in 1838 was estimated at 3,010: 800

alternate expansion and contraction of the internal flues as they were covered and uncovered by the water, a condition that contributed to their weakening. The boiler feedwater for steamboats was pumped directly from the surroundings without treatment or filtration, which accelerated corrosion of the shell and fittings. The sediment was frequently allowed to accumulate, thus requiring a hotter fire to develop the required steam pressure, which led, in turn, to a rapid weakening of the shell. Feed pumps were shut down at intermediate stops without damping the fires, which aggravated the danger of low water and excessive steam pressure. With the rapid increase in the number of steam engines, there was a concomitant shortage of competent engineers who understood the necessary safety precautions. Sometimes masters employed mere stokers who had only a rudimentary grasp of the operation of steam equipment. Increased competition also led to attempts to gain prestige by arriving first at the destination. The usual practice during a race was to overload or tie down the safety valve, so that excessive steam pressure would not be relieved.

III

The first major boiler disasters occurred on steamboats, and, in fact, the majority of explosions throughout the first half of the nineteenth century took place on board ship.[7] By mid-1817, four explosions had taken five lives in the eastern waters, and twenty-five people had been killed in three accidents on the Ohio and Mississippi rivers.[8] The city council of Philadelphia appears to have been the first legislative body in the United States to take cognizance of the disasters and attempt an investigation. A joint committee was appointed to determine the causes of the accidents and recommend measures that would prevent similar occurrences on steamboats serving Philadelphia. The question was referred to a group of practical engineers who recommended that all boilers should be subjected to an initial hydraulic proof test at twice the intended working pressure and additional monthly proof tests to be conducted by a competent inspector. Also, appreciating the fact that marine engineers were known to overload the safety valve levers, they advocated placing the valve in a locked box. The report of the joint

---

on steamboats, 350 in locomotives, and 1,860 in manufacturing establishments. The majority of these engines were put into service after 1830. The term "practical engineer" was reserved for a designer or builder of engines, while engine-room operatives were called "engineers." The complaints about the incompetence of the latter are very frequent in the literature.

[7] *Doc. 21*, p. 3.

[8] Bathe, *op. cit.*, p. 250.

committee incorporated these recommendations, but it stated that the subject of regulation was outside the competence of municipalities. Any municipal enactment would be inadequate for complete regulation. The matter was referred, therefore, to the state legislature, and there it rested.[9]

Similar studies were being undertaken abroad. In England, a fatal explosion aboard a steamboat near Norwich prompted Parliament to constitute a Select Committee in May 1817 to investigate the conditions surrounding the design, construction, and operation of steam boilers. In its report, the committee noted its aversion to the enactment of any legislation but stated that where the public safety might be endangered by ignorance, avarice, or inattention, it was the duty of Parliament to interpose. Precedents for legislation included laws covering the construction of party walls in buildings, the qualification of physicians, and the regulation of stage coaches. The committee recommended that passenger-carrying steam vessels should be registered, that boiler construction and testing should be supervised, and that two safety valves should be employed with severe penalties for tampering with the weights.[10]

No legislation followed this report, nor were any laws enacted after subsequent reports on the same subject in 1831, 1839, and 1843.[11] The attitude of the British steamboat owners and boiler manufacturers was summarized in a statement that the prominent manufacturer, Sir John Rennie, made to the Select Committee in 1843. There should be, he said, no impediments in the application of steam power. Coroners' juries made such complete investigations of boiler explosions that no respectable manufacturer would risk his reputation in constructing a defective boiler. Constant examination of boilers, he argued, would cause serious inconvenience and would give no guarantee that the public safety would be assured. Admittedly, it would be desirable for steam equipment to be perfect, but with so many varied boiler and engine designs, it would be next to impossible to agree on methods of examination. Besides, he concluded, there were really few accidents.[12]

In this latter remark, Sir John was partially correct. In England, from 1817 to 1839, only 77 deaths resulted from twenty-three explosions.[13] This record was relatively unblemished compared to the slaughter in the United States, where in 1838 alone, 496 lives had been lost as

[9] *Ibid.*, p. 255; *JFI*, VIII, N.S. (1831), 235–43.

[10] Parliamentary Sessional Papers, *Report* (1817), VI, 223.

[11] *Ibid.* (1831), VIII, 1; (1839), XLVII, 1; (1843), IX, 1.

[12] *Ibid.* (1842), IX, 383–84.

[13] *Ibid.* (1839), XLVII, 10.

a result of fourteen explosions.[14] The continued use of low-pressure engines by the British; the fact that by 1836 the total number of U.S. steamboats—approximately 750—was greater than the total afloat in all of Europe; and the fact that the average tonnage of U.S. steamboats was twice that of British vessels, implying the use of larger engines and boilers and more numerous passengers, accounted for the large difference in the casualty figures.[15]

In France, the reaction to the boiler hazard was entirely different than in Great Britain and the United States. Acting under the authority of Napoleonic legislation, the government issued a Royal Ordinance on October 29, 1823 relative to stationary and marine steam engines and boilers.[16] A committee of engineers of mines and civil engineers prepared the regulations, but the scientific talent of such men as Arago, Dulong, and Biot was enlisted to prepare accurate steam tables.[17] By 1830, amendments resulted in the establishment of a comprehensive boiler code. It incorporated stress values for iron and copper and design formulas for these materials. It required the use of hemispherical heads on all boilers operating above 7 p.s.i. and the employment of two safety valves, one of which was enclosed in a locked grating. Boiler shells had to be fabricated with fusible metal plates made of a lead-tin-bismuth alloy and covered with a cast-iron grating to prevent swelling when close to the fusing point. Boilers had to be tested initially at three times the designed working pressure and yearly thereafter. The French engineers of mines and government civil engineers were given detailed instructions on the conduct of the tests and were empowered to remove any apparently defective boiler from service. The proprietors of steamboats or factories employing boilers were liable to criminal prosecution for evasion of the regulations, and the entire hierarchy of French officialdom was enjoined to report any infractions.[18]

Proper statistics proving that this code had a salutary effect in the prevention of boiler explosions are not available. It is certain that some

[14] The number of explosions and the loss of life occasioned thereby, listed throughout this paper, were obtained by a comparison and tabulation of the figures listed in *Doc. 21*, pp. 399–403, and in the Commissioner of Patents, *Report*, Dec. 30, 1848 (Senate document 18, ser. 529 [Washington: 30th Congress, 2d session]), pp. 36–48 (hereinafter cited as "Doc. 18").

[15] Department of the Interior, Census Office, *10th Census* (Washington, 1883), IV, 6–7; *JFI*, IX, N.S. (1832), 350.

[16] *Archives Parlementaires* (Paris, 1864), III, ser. 2, 732; *JFI*, VII, N.S. (1831), 272.

[17] *JFI*, X, N.S. (1832), 106; *Doc. 478*, p. 145.

[18] *JFI*, VII, N.S. (1831), 272, 323, 399; VIII, N.S. (1831), 32; X, N.S. (1832), 105, 181.

explosions occurred despite the tight regulations. Arago, writing in 1830, reported that a fatal explosion on the "Rhone" resulted from the tampering with a safety valve and pointed out that fusion of the fusible metal plates could be prevented by directing a stream of water on them.[19] Undoubtedly, in some instances the laws were evaded, but Thomas P. Haldeman, an experienced Cincinnati steamboat captain said in 1848 that the code had been effective. He wrote: "Since those laws were enforced we have scarcely heard of an explosion in that country. . . . What a misfortune our government did not follow the example of France twenty years ago."[20] Significantly, both Belgium and Holland promulgated boiler laws that were in all essentials duplicates of the French regulations.[21]

IV

From 1818 to 1824 in the United States, the casualty figures in boiler disasters rose, about forty-seven lives being lost in fifteen explosions. In May 1824 the "Aetna," built in 1816 to Evans' specifications, burst one of her three wrought-iron boilers in New York harbor, killing about thirteen persons and causing many injuries. Some experts attributed the accident to a stoppage of feedwater due to incrustations in the inlet pipes, while others believed that the rupture in the shell had started from an old fracture in a riveted joint.[22] The accident had two consequences. Because the majority of steamboats plying New York waters operated at relatively low pressures with copper boilers, the public became convinced that wrought-iron boilers were unsafe. This prejudice forced New York boat builders who were gradually recognizing the superiority of wrought iron to revert to the use of copper even in high-pressure boilers. Some owners recognized the danger of this step, but the outcry was too insistent. One is reported to have said: "We have concluded therefore to give them [the public] a copper boiler, the strongest of its class, and have made up our minds that they have a perfect right to be scalded by copper boilers if they insist upon it."[23] His forecast was correct, for within the next decade, the explosion of copper boilers employing moderate steam pressures became common in eastern waters.[24]

19 *Ibid.*, V, N.S. (1830), 399, 411.

20 *Doc. 18,* p. 180.

21 Parliamentary Sessional Papers, *Report* (1839), XLVII, 180.

22 Bathe, *op. cit.*, p. 237; *JFI,* II (1826), 147.

23 *Doc. 21,* p. 425.

24 *Ibid.*, pp. 105, 424; *JFI,* XIII, N.S. (1834), 55, 126, 289.

The second consequence of the "Aetna" disaster was that it caught the attention of Congress. A resolution was introduced in the House of Representatives in May 1824 calling for an inquiry into the expediency of enacting legislation barring the issuance of a certificate of navigation to any boat operating at high steam pressures. Although a bill was reported out of committee, it was not passed due to lack of time for mature consideration.[25]

In the same year, the Franklin Institute was founded in Philadelphia for the study and promotion of the mechanical arts and applied science.[26] The institute soon issued its *Journal*, and, from the start, much space was devoted to the subject of boiler explosions. The necessity of regulatory legislation dealing with the construction and operation of boilers was discussed, but there was a diversity of opinion as to what should be done. Within a few years, it became apparent that only a complete and careful investigation of the causes of explosions would give sufficient knowledge for suggesting satisfactory regulatory legislation. In June 1830, therefore, the Institute empowered a committee of its members to conduct such an investigation and later authorized it to perform any necessary experiments.

The statement of the purpose of the committee reflects clearly the nature of the problem created by the frequent explosions. The public, it said, would continue to use steamboats, but if there were no regulations, the needless waste of property and life would continue. The committee believed that these were avoidable consequences; the accidents resulted from defective boilers, improper design, or carelessness. The causes, the committee thought, could be removed by salutary regulations, and it affirmed: "That there must be a power in the community lodged somewhere, to protect the people at large against any evil of serious and frequent recurrence, is self-evident. But that such power is to be used with extreme caution, and only when the evil is great, and the remedy certain of success, seems to be equally indisputable."[27]

Here is a statement by a responsible group of technically oriented citizens that public safety should not be endangered by private negligence. It demonstrates the recognition that private enterprise was considered sacrosanct, but it calls for a reassessment of societal values in the light of events. It proposes restrictions while still professing un-

[25] *Annals of Congress* (Washington: 18th Congress 1st session), pp. 2670, 2694, 2707, 2708, 2765.

[26] For the history of the Franklin Institute, see S. L. Wright, *The Story of the Franklin Institute* (Philadelphia, 1938).

[27] *JFI*, VI, N.S. (1830), 33.

willingness to fetter private industry. It illustrates a change in attitude that was taking place with respect to the role of government in the affairs of industry, a change that was necessitated by technological innovation. The committee noted that boiler regulation proposals had been before Congress twice without any final action. Congressional committees, it said, appeared unwilling to institute inquiries and elicit evidence from practical men, and therefore they could hardly determine facts based upon twenty years of experience with the use of steam in boats. Since Congress was apparently avoiding action, the committee asserted, it was of paramount importance that a competent body whose motives were above suspicion should shoulder the burden.[28] Thus, the Franklin Institute committee began a six-year investigation of boiler explosions.

From 1825 to 1830, there had been forty-two explosions killing about 273 persons, and in 1830 a particularly serious one aboard the "Helen McGregor" near Memphis which killed 50 or 60 persons, again disturbed Congress. The House requested the Secretary of the Treasury, Samuel D. Ingham of Pennsylvania, to investigate the boiler accidents and submit a report.[29] Ingham had served in Congress from 1813 to 1818, and again from 1822 to 1829. He was a successful manufacturer who owned several paper mills; he was acquainted with the activities of the Franklin Institute and had written to the *Journal* about steam boiler problems.[30] Ingham was thus in a unique position to aid the Franklin Institute committee which had begun its inquiries. Before his resignation from Jackson's cabinet over the Peggy O'Neill Eaton affair, Ingham committed government funds to the Institute to defray the cost of apparatus necessary for the experiments.[31] This was the first research grant of a technological nature made by the federal government.[32]

Ingham attempted to make his own investigation while still secretary of the treasury. His interim report to the House in 1831 revealed that two investigators, one on the Atlantic seaboard and the other in the Mississippi basin, had been employed to gather information on the boiler explosions. They complained that owners and masters of boats

28 *Ibid.*, 34.

29 *Congressional Debates* (Washington: 21st Congress, 1st session), VI, Part 2, 739.

30 *Dictionary of American Biography* (New York, 1932), IX, 473; *JFI,* IX, N.S. (1832), 12 (communicated Oct. 21, 1830).

31 *JFI,* VII, N.S. (1831), 42.

32 Arthur V. Greene, "The A.S.M.E. Boiler Code," *Mechanical Engineer,* LXXIV (1952), 555; A. Hunter Dupree, *Science in the Federal Government* (Cambridge, Mass., 1957), p. 50.

seemed unwilling to aid the inquiry. They were told repeatedly that the problem was purely individual, a matter beyond the government's right to interfere.[33] In the following year, the new secretary, Louis B. McLane, circulated a questionnaire among the collectors of customs, who furnished information and solicited opinions about the explosions. Their answers formed the basis of McLane's report to Congress. They mentioned the many causes of boiler explosions. One letter noted that steamboat trips from New Orleans to Louisville had been shortened from twenty-five to twelve days since 1818 without increasing the strength of the boilers. A frequent remark was that the engineers in charge of the boilers were ignorant, careless, and usually drunk.[34]

This report prompted a bill proposed in the House in May 1832. It provided for the appointment of inspectors at convenient locations to test the strength of the boilers every three months at three times their working pressure, and the issuance of a license to navigate was made contingent upon this inspection. To avoid possible objections on the score of expense, inspection costs were to be borne by the government. To prevent explosions caused by low water supply, the bill provided that masters and engineers be required under threat of heavy penalties to supply water to the boilers while the boat was not in motion.

The half-hearted tone of the House committee's report on the bill hardly promised positive legislative action. The Constitution gave Congress the power to regulate commerce, the report noted, but the right of Congress to prescribe the mode, manner, or form of construction of the vehicles of conveyance could not be perceived. Whether boats should be propelled by wind, paddles, or steam, and if by steam, whether by low or high pressure, were questions that were not the business of Congress. No legislation was competent to remove the causes of boiler explosions, so that steam and its application must be left to the control of intellect and practical science. The intelligent conduct of those engaged in its use would be the best safeguard against the dangers incident to negligence. Besides, the report concluded, the destruction was much less than had been thought; the whole number of explosions in the United States was only fifty-two, with total casualties of 256 killed and 104 injured.[35] Supporters of the bill could not undo the damage of the

[33] Secretary of the Treasury, *Report,* March 3, 1831 (House of Representatives document 131, ser. 209 [Washington: 21st Congress, 2d session]), p. 1.

[34] *Doc. 478,* p. 44.

[35] *Ibid.,* pp. 1–7. Actually, the committee depended for its statistics upon the estimate of William C. Redfield, agent for the Steam Navigation Company of New York, who could hardly have been expected to be impartial. Comparing Redfield's figures with those listed in *JFI,* IX, N.S. (1832), 24–30 and with the sources listed

watered-down committee report, however. The bill died, and the disasters continued.

In his State of the Union message in December 1833, President Jackson noted that the distressing accidents on steamboats were increasing. He suggested that the disasters often resulted from criminal negligence by masters of the boats and operators of the engines. He urged Congress to pass precautionary and penal legislation to reduce the accidents.[36] A few days later, Senator Daniel Webster proposed that the Committee on Naval Affairs study the problem. He suggested that all boilers be tested at three times their working pressure and that any steamboat found racing be forfeited to the government. Thomas Hart Benton followed Webster, stating that the matter properly was the concern of the Judiciary Committee. The private waters of states were involved, Benton said; interference with their sovereignty might result. In passing, Benton remarked that the masters and owners of steamboats were, with few exceptions, men of the highest integrity. Further, Benton said, *he* had never met with any accident on a steamboat despite the fact that he traveled widely; upon boarding he was always careful to inquire whether the machinery was in good order. Webster still carried the day, since the matter went to the Committee on Naval Affairs; however, Benton's attitude prevailed in the session, for the reported Senate bill failed to pass.[37]

V

A program of experiments carried out by the Franklin Institute from 1831 to 1836 was based largely upon the reports of circumstances surrounding previous boiler explosions, the contemporary design and construction of boilers and their accessories, and methods of ensuring an adequate water supply. The work was done by a committee of volunteers led by Alexander Dallas Bache, later superintendent of the U.S. Coast Survey, who, at the time, was a young professor of natural philosophy at the University of Pennsylvania. A small boiler, one foot in diameter and about three feet long, with heavy glass viewing ports at each end, was used in most of the experiments. In others, the zeal of the workers led them to cause larger boilers to burst at a quarry on the outskirts of Philadelphia.

---

in n. 14, it is clear that he omitted many minor accidents; where the number of casualties were unknown, they were not counted; where they were estimated, Redfield took the lowest estimate.

36 *Congressional Globe,* I (Washington: 23d Congress, 1st session), 7.

37 *Ibid.*, I, 49, 442.

The group's findings overturned a current myth, proving conclusively that water did not decompose into hydrogen and oxygen inside the boiler, with the former gas exploding at some high temperature. The experimenters demonstrated that an explosion could occur without a sudden increase of pressure. Another widely held theory they disproved was that when water was injected into a boiler filled with hot and unsaturated steam, it flashed into an extremely high-pressure vapor, which caused the boiler to rupture. The group proved that the reverse was true: the larger the quantity of water thus introduced, the greater the decrease in the steam pressure.

The Franklin Institute workers also produced some positive findings. They determined that the gauge cocks, commonly used to ascertain the level of water inside the boilers, did not in fact show the true level, and that a glass tube gauge was much more reliable, if kept free from sediment. They found the fusing points of alloys of lead, tin, and bismuth, and recommended that fusible plates be employed with caution, because the more fluid portion of an alloy might be forced out prior to the designated fusion temperature, thus leaving the remainder with a higher temperature of fusion.[38] They investigated the effect of the surface condition of the shell on the temperature and time of vaporization, and they determined that properly weighted safety valves opened at calculated pressures within a small margin of error. The results of their experiments on the relationship of the pressure and temperature of steam showed close correspondence with those of the French, although, at this time, values of the specific heat of steam were erroneous due to the inability to differentiate between constant volume and constant temperature conditions.[39]

Simultaneously, another committee, also headed by Bache, investigated the strength of boiler materials. In these experiments, a sophisticated tensile testing machine was constructed, and corrections were made for friction and stresses producing during the tests. The investigators tested numerous specimens of rolled copper and wrought iron, not only at ambient temperatures but up to 1,300° F. They showed conclusively that there were substantial differences in the quality of domestic wrought irons by the differences in yield and tensile strengths. Of major importance was their finding that there was a rapid decrease in the ultimate strength of copper and wrought iron with increasing

[38] In this series of experiments, the committee was actually investigating the solid solutions of these metals and determining points on what would later be called equilibrium diagrams.

[39] Franklin Institute, *Report,* March 1, 1836 (House of Representatives document 162, ser. 289 [Washington]).

temperature. Further, they determined that the strength of iron parallel to the direction of rolling was about 6 per cent greater than in the direction at right angles to it. They proved that the laminated structure in "piled" iron, forged from separate pieces, yielded much lower tensile values than plate produced from single blooms. Their tests also showed that special precautions should be taken in the design of riveted joints.[40]

Taken as a whole, the Franklin Institute reports demonstrate remarkable experimental technique as well as a thorough methodological approach. They exposed errors and myths in popular theories on the nature of steam and the causes of explosions. They laid down sound guidelines on the choice of materials, on the design and construction of boilers, and on the design and arrangement of appurtenances added for their operation and safety. Further, the reports included sufficient information to emphasize the necessity for good maintenance procedures and frequent proof tests, pointing out that the strength of boilers diminished as the length of service increased.

VI

The Franklin Institute report on steam boiler explosions was presented to the House through the secretary of the treasury in March 1836, and the report on boiler materials was available in 1837. The Franklin Institute committee also made detailed recommendations on provisions that any regulatory legislation should incorporate. It proposed that inspectors be appointed to test all boilers hydraulically every six months; it prohibited the licensing of ships using boilers whose design had proved to be unsafe; and it recommended penalties in cases of explosions resulting from improper maintenance, from the incompetence or negligence of the master or engineer, or from racing. It placed responsibility for injury to life or property on owners who neglected to have the required inspections made, and it recommended that engineers meet certain standards of experience, knowledge, and character. The committee had no doubt of the right of Congress to legislate on these matters.[41]

Congress did not act immediately. In December 1836 the House appointed a committee to investigate the explosions, but there was no action until after President Van Buren urged the passage of legislation

40 *JFI*, XVIII, N.S. (1836), 217, 289; XIX, N.S. (1837), 73, 157, 241, 325, 409; XX, N.S. (1837), 1, 73.

41 *JFI*, XVIII, N.S. (1836), 369–75.

in December 1837.[42] That year witnessed a succession of marine disasters. Not all were attributable to boiler explosions, although the loss of 140 persons in a new ship, the "Pulaski," out of Charleston, was widely publicized. The Senate responded quickly to Van Buren's appeal, passing a measure on January 24, 1838. The House moved less rapidly. An explosion aboard the "Moselle" at Cincinnati in April 1838, which killed 151 persons,[43] caused several Congressmen to request suspension of the rules so that the bill could be brought to the floor, but in the face of more pressing business the motion was defeated.[44] The legislation was almost caught in the logjam in the House at the end of the session, but on June 16 the bill was brought to the floor. Debate centered principally upon whether the interstate commerce clause in the Constitution empowered Congress to pass such legislation. Its proponents argued affirmatively, and the bill was finally approved and became law on July 7, 1838.[45]

The law incorporated several sections relating to the prevention of collisions, the control of fires, the inspection of hulls, and the carrying of lifeboats. It provided for the immediate appointment by each federal judge of a competent boiler inspector having no financial interest in their manufacture. The inspector was to examine every steamboat boiler in his area semiannually, ascertain its age and soundness, and certify it with a recommended working pressure. For this service the owner paid the inspector $5.00—his sole remuneration—and a license to navigate was contingent upon the receipt of this certificate. The law specified no inspection criteria. It enjoined the owners to employ a sufficient number of competent and experienced engineers, holding the owners responsible for loss of life or property damage in the event of

[42] *Congressional Globe,* IV (Washington: 24th Congress, 2d session), 29; VI (Washington: 25th Congress, 2d session), 7–9.

[43] The *Moselle* disaster was important because of its effect upon marine insurance policies. The estate of the captain and part owner, Isaac Perrin, sued for recovery under the policy (*The Administrators of Isaac Perrin* v. *The Protection Insurance Co.,* 11 Ohio [1842], 160). The defense gave evidence that Perrin was determined to outstrip another boat and that when passengers expostulated with him concerning the dangerous appearance of the boiler fires, he swore that he would be "that night in Louisville or hell." Despite proof of negligence on the part of the captain, the court ruled against the insurance company, stating that the explosion of boilers was a risk insured against. The insurance companies, thereafter, moved to exclude boiler explosions as a covered risk. See *Citizens Insurance Co.* v. *Glasgow, Shaw, and Larkin,* 9 Missouri (1852), 411, and *Roe and Kercheval* v. *Columbus Insurance Co.,* 17 Missouri (1852), 301.

[44] *Congressional Globe,* VI, 342.

[45] *Ibid.,* VI, 455.

a boiler explosion for their failure to do so. Further, any steamboat employee whose negligence resulted in the loss of life was to be considered guilty of manslaughter, and upon conviction could be sentenced to not more than ten years imprisonment. Finally, it provided that in suits against owners for damage to persons or property, the fact of the bursting of the boilers should be considered prima facie evidence of negligence until the defendant proved otherwise.[46]

This law raises several questions, because the elimination of inspection criteria and the qualification of engineers rendered the measure ineffectual. Why was this done? Did Congress show restraint because it had insufficient information? Did it yield to the pressure of steamboat interests who feared government interference? Such questions cannot be definitely answered, but there are clues for some tentative conclusions.

The bill, as originally introduced, was similar to the Franklin Institute proposals, so that the Senate committee to which it was referred possessed the most recent informed conclusions as to the causes of boiler explosions and the means of their prevention. The President's plea to frame legislation in the face of the mounting fatalities undoubtedly persuaded the Democratic majority to act. They were unmoved by a memorial from steamboat interests urging the defeat of the bill.[47] But the majority was not as yet prepared to pass such detailed regulations as had originally been proposed. In response to a question as to why the provision for the qualification of engineers had been eliminated, the Senate committee chairman stated that the committee had considered this requirement desirable but foresaw too much difficulty in putting it into effect. Further, the Senate rejected an amendment to levy heavy penalties for racing, as proposed by the Whig, Oliver Smith of Indiana. The Whigs appear to have seen the situation as one in which the federal government should use its powers and interpose firmly. Henry Clay, R. H. Bayard of Delaware, and Samuel Prentiss of Vermont supported Smith's amendment, and John Davis of Massachusetts declared that he would support the strongest measures to make the bill effective. Those who had urged rapid action of the bill in the House were William B. Calhoun and Caleb Cushman of Massachusetts and Elisha Whittlesey of Ohio, all Whigs. But at this time the majority hewed to the doctrine that enlightened self-interest should motivate owners to provide safe operation. The final clause, specifying that the

[46] *U.S. Statutes at Large* (Washington: 25th Congress, 2d session, July 7, 1838), V, 304–6.

[47] *Congressional Globe*, VI, 265.

bursting of boilers should be taken as prima facie evidence of negligence until proved otherwise, stressed this idea.

The disappointment of the informed public concerning the law was voiced immediately in letters solicited by the secretary of the treasury, contained in a report that he submitted to Congress in December 1838.[48] There were predictions that the system of appointment and inspection would encourage corruption and graft. There were complaints about the omission of inspection criteria and a provision for the licensing of engineers. One correspondent pointed out that it was impossible legally to determine the experience and skill of an engineer, so that the section of the law that provided penalties for owners who failed to employ experienced and skilful engineers was worthless. One critic who believed that business interests had undue influence upon the government wrote: "We are mostly ruled by corporations and joint-stock companies. . . . If half the citizens of this country should get blown up, and it should be likely to affect injuriously the trade and commerce of the other half by bringing to justice the guilty, no elective officer would risk his popularity by executing the law."[49]

But there also was a pained reaction from the owners of steamboats. A memorial in January 1841 from steamboat interests on the Atlantic seaboard stressed that appropriate remedies for the disasters had not been afforded by the 1838 law as evidenced by the casualty figures for 1839 and 1840. They provided statistics to prove that in *their* geographical area the loss of life per number of lives exposed had decreased by a factor of sixteen from 1828 to 1838, indicating that the troubles centered chiefly in the western waters. But at the same time the memorial emphasized that the 1838 law acted as a deterrent for prudent men to continue in the steamboat business, objecting particularly to the clause that construed a fatal disaster as prime facie evidence of negligence. They argued that if Congress considered steam navigation too hazardous for the public safety, it would be more just and honorable to prohibit it entirely.[50]

However, it not only was the Congress that was reconsidering the concepts of negligence and responsibility in boiler explosions. The common law also searched for precedents to meet the new conditions, to establish guidelines by which to judge legal actions resulting from technological innovation. A key decision, made in Pennsylvania in 1845, involved a boiler explosion at the defendant's flour mill that killed the

[48] *Doc. 21.* [49] *Ibid.*, p. 396.

[50] *Memorial*, Jan. 23, 1841 (House of Representatives document 113, ser. 377 [Washington: 26th Congress, 2d session]).

plaintiff's horse. The defense pleaded that any negligence was on the part of the boiler manufacturer. The court, however, ruled otherwise, stating that the owner of a public trade or business which required the use of a steam engine was responsible for any injury resulting from its deficiency.[51] This case was used as a precedent in future lawsuits involving boiler explosions.

VII

Experience proved that the 1838 law was not preventing explosions or loss of life. In the period 1841–48, there were some seventy marine explosions that killed about 625 persons. In December 1848 the commissioner of patents, to whom Congress now turned for data, estimated that in the period 1816–48 a total of 233 steamboat explosions had occurred in which 2,563 persons had been killed and 2,097 injured, with property losses in excess of $3 million.[52]

In addition to the former complaints about the lack of proof tests and licenses for engineers, the commissioner's report included testimony that the inspection methods were a mockery. Unqualified inspectors were being appointed by district judges through the agency of highly placed friends. The inspectors regarded the position as a lifetime office. Few even looked at the boilers but merely collected their fees. The inspector at New York City complained that his strict inspection caused many boats to go elsewhere for inspections. He cited the case of the "Niagara," plying between New York City and Albany, whose master declined to take out a certificate from his office because it recommended a working pressure of only 25 p.s.i. on the boiler. A few months later the boiler of the "Niagara," which had been certified in northern New York, exploded while carrying a pressure of 44 p.s.i. and killed two persons.[53]

Only eighteen prosecutions had been made in ten years under the manslaughter section of the 1838 law. In these cases there had been nine convictions, but the penalties had, for the most part, been fines which were remitted. It was difficult to assemble witnesses for a trial, and juries could not be persuaded to convict a man for manslaughter for an act of negligence, to which it seemed impossible to attach this degree of guilt. Also, the commissioner's report pointed out that damages were given in cases of bodily injury but that none were awarded for loss of life in negligence suits. It appeared that exemplary damages might be effective in curbing rashness and negligence.[54]

[51] *Spencer* v. *Campbell,* 9 Watts & Sergeants (1845), 32.

[52] *Doc. 18,* p. 2. [53] *Ibid.,* pp. 18, 78, 80. [54] *Ibid.,* pp. 29, 52–53.

The toll of life in 1850 was 277 dead from explosions, and in 1851 it rose to 407.[55] By this time Great Britain had joined France in regulatory action, which the Congress noted.[56] As a consequence of legislation passed in 1846 and 1851, a rejuvenated Board of Trade was authorized to inspect steamboats semiannually, to issue or deny certificates of adequacy, and to investigate and report on accidents.[57] The time had come for the Congress to take forceful action, and in 1852 it did.

John Davis, Whig senator from Massachusetts, who had favored stricter legislation in 1838, was the driving force behind the 1852 law. In prefacing his remarks on the general provisions of the bill, he said: "A very extensive correspondence has been carried on with all parts of the country . . . there have been laid before the committee a great multitude of memorials, doings of chambers of commerce, of boards of trade, of conventions, of bodies of engineers; and to a considerable extent of all persons interested, in one form or another, in steamers . . . in one thing . . . they are all . . . agreed—that is, that the present system is erroneous and needs correction."[58]

Thus again, the informed public submitted recommendations on the detailed content of the measure. An outstanding proponent who helped shape the bill was Alfred Guthrie, a practical engineer from Illinois. With personal funds, Guthrie had inspected some two hundred steamboats in the Mississippi valley to ascertain the causes of boiler explosions. Early in the session, Senator Shields of Illinois succeeded in having Guthrie's report printed, distributed, and included in the Senate documents.[59] Guthrie's recommendations were substantially those made by the Franklin Institute in 1836. His reward was the post as first supervisor of the regulatory agency which the law created.

[55] *Congressional Globe* (Washington: 32d Congress, 1st session), Appendix, 287.

[56] *Ibid*., p. 2426.

[57] Public and General Acts, 9, 10 Victoria (1846), chap. l; 14, 15 Victoria (1851), chap. lxxix.

[58] *Congressional Globe* (32d Congress, 1st session), p. 1669. Organizations of experienced steamboat engineers were formed in many cities during the 1840's to promote safe operation and had attempted on previous occasions to influence Congress to improve the 1838 law, particularly with respect to providing for proof tests, better inspection methods, and the establishment of boards to qualify engineers. See *Relative to Steamboat Explosions* (House of Representatives document 68, ser. 441 [Washington: 28th Congress, 1st session]), which is a petition from a body in the city of Cincinnati.

[59] *Memorial of Alfred Guthrie, a Practical Engineer*, Feb. 6, 1852 (Senate miscellaneous document 32, ser. 629 [Washington: 32d Congress, 1st session]).

After the bill reached the Senate floor, dozens of amendments were proposed, meticulously scrutinized, and disposed of. The measure had been, remarked one senator, "examined and elaborated . . . more patiently, thoroughly, and faithfully than any other bill before in the Senate of the United States."[60] As a result, in place of the 1838 law which embodied thirteen sections and covered barely three pages, there was passed such stringent and restrictive legislation that forty-three sections and fourteen pages were necessary.[61]

The maximum allowable working pressure for any boiler was set at 110 p.s.i., and every boiler had to be tested yearly at one and one-half times its working pressure. Boilers had to be fabricated from suitable quality iron plates, on which the manufacturer's name was stamped. At least two ample safety valves—one in a locked grating—were required, as well as fusible plates. There were provisions relating to adequate supply of boiler feedwater and outlawing designs that might prove dangerous. Inspectors were authorized to order repairs at any time. All engineers had to be licensed by inspectors, and the inspectors themselves issued certificates only under oath. There were stiff monetary penalties for any infractions. The penalty for loading a safety valve excessively was a two hundred dollar fine and eighteen months imprisonment. The fine for manufacturing or using a boiler of unstamped material was five hundred dollars. Fraudulent stamping carried a penalty of five hundred dollars and two years imprisonment. Inspectors falsifying certificates were subject to a five hundred dollar fine and six months imprisonment, and the law expressly prohibited their accepting bribes.

A new feature of the law, which was most indicative of the future, was the establishment of boards of inspectors empowered to investigate infractions or accidents, with the right to summon witnesses, to compel their attendance, and to examine them under oath. Above the local inspectors were nine supervisors appointed by the President. Their duties included the compilation of evidence for the prosecution of those failing to comply with the regulations and the preparation of reports to the secretary of the treasury on the effectiveness of the regulations. Nor did these detailed regulations serve to lift the burden of presumptive negligence from the shoulders of owners in cases of explosion. The explosion of boilers was not made prime facie evidence as in the 1838 law, but owners still bore a legal responsibilty. This was

[60] *Congressional Globe,* (32d Congress 1st session), p. 1742.

[61] *U.S. Statutes at Large* (Washington: 32d Congress, 1st session, Aug. 30, 1852), X, 61–75.

made clear in several court decisions which held that proof of strict compliance with the 1852 law was not a sufficient defense to the allegations of loss by an explosion caused by negligence.[62]

The final Senate debate and the vote on this bill shows how, in thirty years, the public attitude and, in turn, the attitudes of its elected representatives had changed toward the problem of unrestricted private enterprise, mainly as a result of the boiler explosions. The opponents of the bill still argued that the self-interest of the steamboat companies was the best insurance of the safety of the traveling public.[63] But their major argument against passage was the threat to private property rights which they considered the measure entailed. Senator Robert F. Stockton of New Jersey was most emphatic:

> It is this—how far the Federal Government . . . shall be permitted to interfere with the rights of personal property—or the private business of any citizen . . . under the influence of recent calamities, too much sensibility is displayed on this subject . . . I hold it to be my imperative duty not to permit my feelings of humanity and kindness to interfere with the protection which I am bound, as a Senator of the United States, to throw around the liberty of the citizen, and the investment of his property, or the management of his own business . . . what will be left of human liberty if we progress on this course much further? What will be, by and by, the difference between citizens of this far-famed Republic and the serfs of Russia? Can a man's property be said to be his own, when you take it out of his own control and put it into the hands of another, though he may be a Federal officer?[64]

This expression of a belief that Congress should in no circumstances interfere with private enterprise was now supported by only a small minority. One proponent of the bill replied: "I consider that the only question involved in the bill is this: Whether we shall permit a legalized, unquestioned, and peculiar class in the community to go on committing murder at will, or whether we shall make such enactments as will compel them to pay some attention to the value of life."[65] It was, then, a question of the sanctity of private property rights as against the duty of government to act in the public weal. On this question the Senate voted overwhelmingly that the letter course should prevail.[66]

[62] *Curran* v. *Cheeseman,* 1 Cincinnati Rep. (1870), 52.

[63] *Congressional Globe* (32d Congress, 1st session), pp. 1741, 2425.

[64] *Ibid.*, pp. 2426, 2427.

[65] *Ibid.*, p. 2427.

[66] The strength of the vote can be gauged by the defeat, forty-three to eight, of a motion to table the bill by Senator Stockton just prior to its passage. The eight

Though not completely successful, the act of 1852 had the desired corrective effects. During the next eight years prior to the outbreak of the Civil War, the loss of life on steamboats from all types of accidents dropped to 65 per cent of the total in the corresponding period preceding its passage.[67] A decade after the law became effective, John C. Merriam, editor and proprietor of the *American Engineer*, wrote: "Since the passage of this law steamboat explosions on the Atlantic have become almost unknown, and have greatly decreased in the west. With competent inspectors, this law is invaluable, and we hope to hail the day when a similar act is passed in every legislature, touching locomotive and stationary boilers.[68]

There was, of course, hostility and opposition to the law immediately after its passage, particularly among the owners and masters of steamboats.[69] It checked the steady rise in the construction of new boats, which had been characteristic of the earlier years.[70] The effect, however, was chastening rather than emasculating. Associations for the prevention of steam boiler explosions were formed; later, insurance companies were organized to insure steam equipment that was manufactured and operated with the utmost regard for safety. In time, through the agency of the American Society of Mechanical Engineers,

---

were: Bayard (D., Del.), Butler (States Rights D., S.C.), Clemens (D., Ala.), Hale (Antislavery D., N.H.), Hunter (D., Va.), James (Protective Tariff D., R.I.), Pratt (Whig, Md.), and Stockton (D., N.J.). Although these senators represented only states along the eastern seaboard and in the South, it would be difficult to interpret their vote on a geographical basis, since eighteen senators from the same group of states voted against the motion. One might be tempted to ascribe some partisan basis to the vote, since only one Whig joined seven Democrats in supporting the motion. On the other hand, twenty-six Democrats and seventeen Whigs constituted the majority. Of those not voting–seven Democrats and four Whigs –by their comments during prior debates on the measure, Brodhead (D., Pa.) and De Saussure (D., S.C.) appear to have favored the bill, while Gwin (D., Calif.) was against it. The conclusion seems justified that the movement and final step toward positive regulation found support from congressmen of all political postures and from all geographical areas, that it was prompted by the recognition of the inadequacy of the 1838 law as evidenced by the continued severe loss of life, and that congressmen were urged to pass the legislation by constituents who were able to recognize how the problem could be solved.

[67] *10th Census*, IV, 5.

[68] L. Stebbins, pub., *Eighty Years' Progress of the United States* (New York, 1864), p. 243.

[69] Lloyd M. Short, *Steamboat Inspection Service* (New York, 1922), p. 5.

[70] Department of the Interior, *op. cit.*, IV, 5.

uniform boiler codes were promulgated and adopted by states and municipalities.[71]

Thus, the reaction of the informed public, expressed by Congress, to boiler explosions caused the initiation of positive regulation of a sector of private enterprise through a governmental agency. The legislation reflected a definite change of attitude concerning the responsibility of the government to interfere in those affairs of private enterprise where the welfare and safety of the general public was concerned. The implications of this change for the future can be seen by reference to the Windom Committee report of 1874, which was the first exhaustive study of the conditions in the railroad industry that led ultimately to the passage of legislation creating the Interstate Commerce Commission. One section of this report was entitled: "The Constitutional Power of Congress to Regulate Commerce among the Several States." The committee cited the judicial interpretation of the Constitution in *Gibbons* v. *Ogden,* that it was the prerogative of Congress solely to regulate interstate commerce, and also referred to the decision of Chief Justice Taney in *Genesee Chief* v. *Fitzhugh,* wherein it was held that this power was as extensive upon land as upon water. The report pointed out that no decision of the Supreme Court had ever countenanced the view that the power of Congress was purely negative, that it could be constitutionally exercised only by disburdening commerce, by preventing duties and imposts on the trade between the states. It fact, the report argued, Congress had already asserted its power positively. Referring to the acts of 1838 and 1852, it stated that "Congress has passed statutes defining how steamboats shall be constructed and equipped."[72] Thus, the legislation that was provoked by bursting boilers was used as a precedent to justify regulatory legislation in another area where the public interest was threatened.

Bursting steamboat boilers, then, should be viewed not merely as unfortunate and perhaps inevitable consequences of the early age of steam, as occurrences which plagued nineteenth-century engineers and which finally, to a large degree, they were successful in preventing. They should be seen also as creating a dilemma as to how far the lives and property of the general public might be endangered by unrestricted private enterprise. The solution was an important step toward the inauguration of the regulatory and investigative agencies in the federal government.

[71] Greene, *op. cit.*

[72] *Report of the Select Committee on Transportation to the Seaboard* (Senate Report No. 307, Ser. 1588 [Washington: 43d Congress, 1st session]), pp. 79–92.

Who Turned the Mechanical Ideal into Mechanical Reality?

ROBERT B. GORDON

In 1884 Charles Fitch described interchangeable manufacture as a *mechanical ideal* accomplished by American inventors, entrepreneurs, and mechanicians who in fifty years had transformed the United States from an agricultural nation dependent on imported manufactured goods into a country that was exporting sophisticated production machinery to European customers.[1] The work of Thomas Warner and Cyrus Buckland at the Springfield Armory in the 1840s and 1850s is prominent in Fitch's account because he believed they were most instrumental in converting the abstract concept of interchangeability—a late-18th-century idea officially adopted as an ideal by the Ordnance Department in 1813—into a working system of manufacture at the national armories.[2] The evolution of interchangeable manufacture continues to interest historians because it is one of the roots of American success in large-scale manufacturing. Moreover, because interchangeability implies reliance on machines, they regard it as underlying the removal of traditional artisan skills from the production

Dr. Gordon is professor of geophysics and applied mechanics and a member of the Council on Archaeological Studies at Yale University. He has benefited from extended discussions of the history of the Springfield Armory with Lennox Beach, Carolyn Cooper, Patrick Malone, and Michael Raber. He would like to thank Stuart Vogt for assistance with the study of lock mechanisms at the Springfield Armory Museum National Historic Site, Harry Hunter for helping with the examination of the gages for the M1841 rifle at the National Museum of American History, Smithsonian Institution, and the late Merrill Lindsay for identifying weapons made at the Whitney Armory. He has received valuable perspective on the skills required in making small arms from master toolmaker Arthur Goodhue, who began his career at the Marlin Arms Company and who prepared the sections of lock parts used in this study.

[1]Charles H. Fitch, "The Rise of a Mechanical Ideal," *Magazine of American History* 11 (1884): 516–27.

[2]Simeon North's contract of 1813 for manufacture of pistols was the first to require interchangeable parts. S. N. D. North and R. H. North, *Simeon North: First Official Pistol Maker of the United States* (Concord, N.H., 1913). Most scholars today would add John Hall too.

process. Authors who have studied American production machinery include Fitch, Robert Woodbury, and Edwin Battison.[3] Others, interested in American entrepreneurs and inventors, have cast their favorites in heroic molds and exposed the blemishes of the favorites chosen by others.[4] Within the last decade Merritt Roe Smith has added a new dimension to these studies, the role of the communities in which the entrepreneurs worked, and David Hounshell has explored the evolution of mass production from the 19th to the 20th century.[5]

Noticeably absent from this scholarly research is consideration of the role of the individual artificers who had to use the methods of Fitch's mechanical ideal to make material products. I will examine their contribution here.

The Mechanical Ideal

Accounts of the mechanical ideal have been based largely on statements by contemporary observers. These observers emphasized how making parts to gage with the self-acting, power-driven machinery introduced in American factories from the beginning of the 19th century onward eliminated much of the need for skilled artificers.[6]

[3]Charles H. Fitch, *Report on the Manufactures of Interchangeable Mechanism,* Tenth Census of the United States, 1880, deals with forging, woodworking, and metal-cutting machinery; Robert S. Woodbury, *Studies in the History of Machine Tools* (Cambridge, Mass., 1972), covers lathes, milling machines, and grinders; Edwin A. Battison, "Eli Whitney and the Milling Machine," *Smithsonian Journal of History* 1 (Summer 1966): 9–34, is a study of the origin of milling technology. Battison's article is of particular interest because of his use of evidence from artifacts to supplement the rather sparse and unreliable documentary record.

[4]Eli Whitney's reputation has undergone a particularly large share of revision and counterrevision, a reaction to the persisting, uncritical assessment of him as the originator of interchangeable manufacture.

[5]M. R. Smith, *Harpers Ferry Armory and the New Technology* (Ithaca, N.Y., 1977); David A. Hounshell, *From the American System to Mass Production 1800–1932* (Baltimore, 1984), pp. 38, 44.

[6]According to Robert Woodbury the system of manufacture of interchangeable parts that finally matured in the United States includes use of (1) precision machine tools, (2) precision gaging, (3) uniformly accepted standards of measurement, and (4) techniques of mechanical drawing ("The Legend of Eli Whitney and Interchangeable Parts," *Technology and Culture* 1 [1960]: 235–53). Paul Uselding added uniform materials to this list ("Henry Burden and the Question of Anglo-American Technological Transfer in the Nineteenth Century," *Journal of Economic History* 30 [1970]: 312–37); on the problem of the uniformity of iron see also R. B. Gordon, "Materials for Manufacturing: The Response of the Connecticut Iron Industry to Technological Change and Limited Resources," *Technology and Culture* 24 (1983): 602–34. Throughout most of the 19th century interchangeable manufacture did not entail Woodbury's items 3 and 4; products were made interchangeable by the duplication of a master pattern or a model rather than by manufacture to absolute dimensions specified on drawings. I use the term

George Talcott's report on the Springfield Armory written in 1841, for example, asserts:

> Soon after the war [1812] the system of *piece-work*, instead of day work, was extensively introduced. Previous to this, an *armorer* was a very different kind of mechanic: the skill of the eye and the hand being highly valued and indispensable. A "lock filer" filed up and fitted all the parts of a lock. The change of the system caused each one to devote his skill and energies to the completion of some single part and in time it was difficult to find many men who were able to file up all the parts equally well; and whenever it became necessary to change men from one limb of the lock to another . . . much difficulty occurred. Machines for performing the work (that was formerly done by the skill of the eye and the hand) have been gradually introduced from time to time, until at length the machines perform nearly all the work leaving the workman nothing to do but fix the article in a proper position, apply the necessary oil, and set the machine in motion. A great portion of the severe hand labor is thus dispensed with. The machines are usually so constructed as to stop when the work is done without the aid of the overseer. In this way a man can attend two or more machines. The excellence of this mode of working is fully exemplified at Hall's rifle works at Harpers Ferry, where machines are generally attended by *boys* and *young men*, who place and replace the pieces to be wrought, and only a *few men* are required to keep the machine in good order. In this way the skill of the armorer is but little needed: his "occupation's gone." A boy does just as well as a man. Indeed, from possessing greater activity of body, he does better.
>
> The difficulty of finding good armorers no longer exists; they abound in every machine shop and manufactory throughout the country. The skill of the eye and the hand, acquired by practice alone, is no longer indispensable; and if every operative were at once discharged from the Springfield armory, their places could be supplied with competent hands within a week.[7]

Again, in a letter to Julius Rockwell of Pittsburgh, August 7, 1845, Talcott states: ". . . The fact that hand labor is dispensed with and everything being now effected with machinery puts at rest all fears of

---

"artificer" to mean a skilled mechanic who makes metal or wooden products with hand or power tools. Inventories of the equipment at the Springfield Armory between 1834 and 1838 list the tools used by "artificers" separately from those used by "armorers and smiths," and it appears from these lists that the "artificers" were actually millwrights.

[7]Stephen V. Benet, ed., *A Collection of Annual Reports and Other Important Papers, Relating to the Ordnance Department* . . . (Washington, D.C., 1878), 1:395.

the kind [of a shortage of skilled workers]. Indeed, the skill of the eye and the hand of the old practical armorer is entirely dispensed with, and any good mechanic from a machine shop can perform the work at an armory, the machines having effected a total revolution, improving the quality of the work and reducing the cost there of . . . ."[8]

Felicia Deyrup has concluded from Talcott's statements that "The skill of the average arms worker continued to decline as a result of increased mechanization . . ." and that ". . . the development of machine tools was clearly recognized at the time as a factor of major importance in the reduction of skill. . . ."[9] Hounshell has carried this notion further with his assertion that at the Springfield Armory Thomas Blanchard's new machines "had eliminated the use of skilled labor in stockmaking by 1826" and that by 1850 "virtually all of the fabrication of the musket (except barrel welding) was carried out by machines . . . ."[10] A popular way of summarizing these claims is to say that "skill was built into the machines" and that the introduction of these machines began the process of deskilling.

The motives of those who wrote about the new manufacturing methods should be considered before their statements are accepted as objective descriptions of the way mechanical work was carried on. Talcott's letter to Rockwell is his response to an attempt to keep Lemuel Pomeroy of Pittsfield at work on contract arms, which Talcott opposed. He prepared his "Notes on the Springfield Armory" in August 1841, on the basis of a ten-day visit to the armory to investigate the high cost of labor there. He lists an expenditure of $21,829.67 on labor-saving machinery from 1820 to 1840, which, according to the master armorer, reduced the actual labor required to make a musket by one-third. Talcott wanted to realize a savings in labor costs from this investment by reducing wage rates.[11] Though he was an experienced inspector of arms and armories, Talcott was not an artificer; his descriptions of manufacturing are really attempts to rationalize administrative actions he sought to bring about.

I wish to challenge the notion that the precision of machines replaced the mechanical skills of artificers in the 19th century. To do this, I will use material evidence to show how representative products were made. Then I will define what I mean by skill and ask how much of the realization of the mechanical ideal was dependent on the performance of the artificers who used the new methods and to what

[8]Ibid., 1:52.

[9]Felicia J. Deyrup, *Arms Makers of the Connecticut Valley* (Northampton, Mass., 1948), p. 160.

[10]Hounshell (n. 5 above), pp. 38, 40.

[11]Deyrup (n. 9 above), p. 196.

degree the mechanical ideal was built into the new machinery instead of the abilities of the artificers. I will argue that the new methods of manufacturing metal products introduced in the 19th century not only fully engaged the traditional mechanical skills of artificers but also made new demands on their skills. The development and learning of these skills took many years and was, in fact, the factor that limited the progress of the new technology.

The role of the artificer's skills has been overlooked in previous research not perhaps for want of scholarly interest but for want of evidence. Documentary sources reveal little about what was required of the artificers engaged in manufacturing in the 19th century. The problem must be approached by archaeological methods. The surviving examples of 19th-century products are a source of information about the work of those who made them; they give us a kind of direct contact with the individual artificers that cannot be attained in any other way. Heretofore there has been little progress in interpreting many of these artifacts, for want of appropriate methods of analysis. But now appropriate methods are available.

Archaeologists have developed methods of identifying the way that tools have been used or objects made from analysis of lithic materials.[12] More recently, similar methods have been developed for analyzing metallic artifacts.[13] Reconstruction experiments, in which equipment of ancient design is built and operated in the laboratory or in the field, have proved to be a useful way of discovering the skills required to carry on old industrial processes.[14] But little use has yet been made in industrial archaeology of the many opportunities offered us by this technique; hence, the interpretation of artifacts is currently our best source of information on manufacturing technology.

Artifacts Examined

Many of the surviving examples of 19th-century manufactured goods could serve our purpose, but most suitable would be a manufactured product of some complexity, difficult to make, meeting demanding service requirements, and produced in large quantities at a place where new manufacturing technology was devised and for which documentary evidence has survived. Examples should be datable and

[12]Brian Hayden, ed., *Lithic Use-Wear Analysis* (New York, 1979).

[13]R. B. Gordon, "Laboratory Evidence of the Use of Metal Tools at Machu Picchu (Peru) and Environs," *Journal of Archaeological Science* 12 (1985): 311–27.

[14]For an excellent example of the investigation of ancient technology through the use of reconstructed equipment, see R. F. Tylecote, J. N. Austin, and A. E. Wraith, "The Mechanism of the Bloomery Process in Shaft Furnaces," *Journal of the Iron and Steel Institute* 209 (1971): 342–63.

there should be no major design changes in the time interval under consideration. Possibilities include clocks, cylinder locks, sewing machines, edge tools, railway equipment, textile machinery, and artillery, but military small arms best fit the above criteria. Although over-represented in discussions of the history of manufacturing technology, they nevertheless have some important advantages for our purposes. Arms collectors have identified date and place of manufacture for many museum specimens. The lock mechanism, stock, and barrel of the military small arms made in the United States, principally at the Springfield Armory, changed little in basic design until the bolt-action rifle was adopted in 1892.[15] Small arms also were made in the first part of the 19th century by a number of contractors, whose work can be compared to that done at Springfield.

There are about twelve essential parts in a percussion lock (illustrated in fig. 1) and a few more in a flintlock, and it is helpful to pick one part that can be studied in detail. I have chosen the tumbler (marked *T* in fig. 1). It has a complex shape and is subject to large forces as the lock operates. The hard service to which tumblers were subjected in use is evidenced by the breakage rate in the middle of the 19th century, about 2 percent per year.[16]

The form of the tumbler and its function are illustrated in figure 2. When force *T* is applied to the sear by pulling the trigger, the tumbler accelerates through about a quarter turn under the action of the force *F* applied by the mainspring, thereby driving the flint against the frizzen or the hammer against the percussion cap. To operate successfully, the tumbler must fit properly into the lock mechanism. Some aspects of this fit are particularly important. The axes of the spindle and the pivot of the tumbler must be concentric (see fig. 2) if it is to turn in its rigid metal bearings without developing large internal stresses. The notch engaged by the sear must be parallel to the axis of rotation and well hardened (if it is not, the notch and sear will be subject to damage as the tumbler is released), and the safety notch must be undercut sufficiently to hold the sear in place when the trigger is pulled. The flats on the spindle must be a close fit on the square hole in the cock or hammer; if they are not, working of this joint will damage both parts. In addition, the shoulder at the end of the spindle must be perpendicular to the axis of rotation, sufficient relief must be provided to keep the side from rubbing on the lock plate, and the heat treatment must give a hard surface at the notch and spring bearing points but leave

[15]The adoption of percussion ignition in place of flint in 1842 and of breech loading after 1865 resulted in little change to the lock mechanism of Springfield arms.

[16]Anon., *Ordnance Manual* (Washington, D.C., 1850), p. 194.

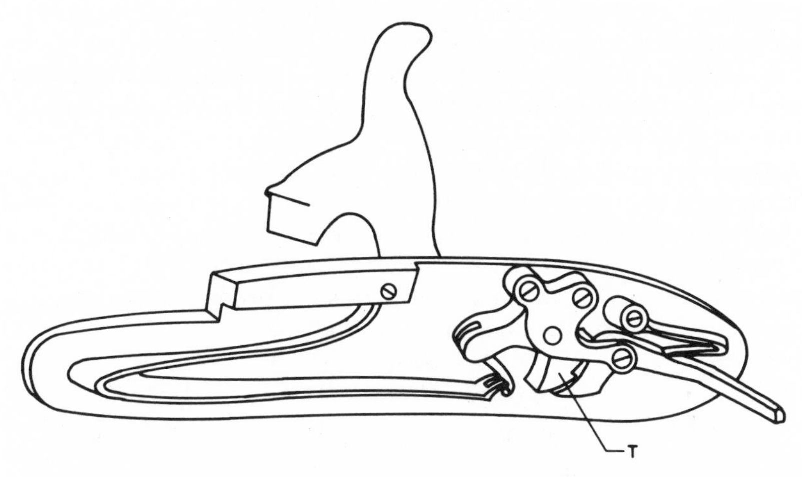

Fig. 1.—The mechanism of a percussion lock. The tumbler is marked *T*. There was little change in the design of this mechanism throughout the 19th century.

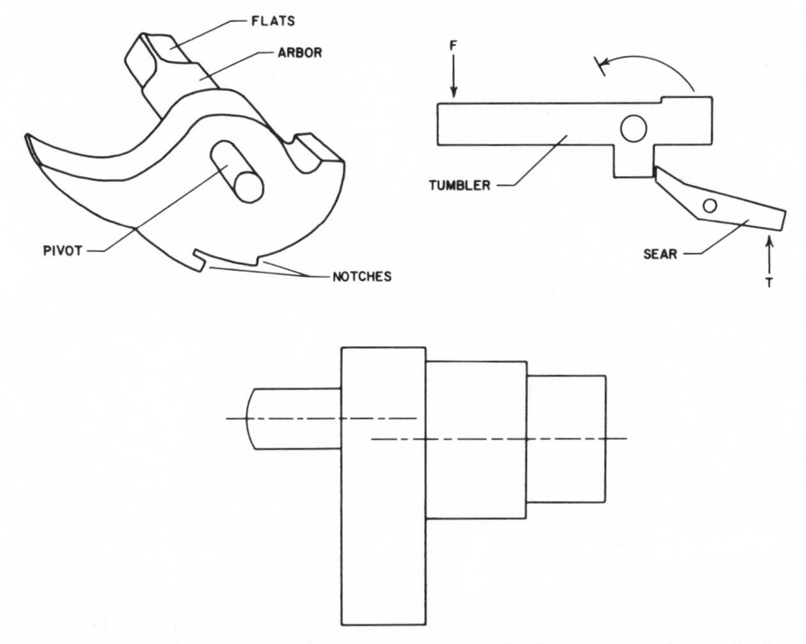

Fig. 2.—The principal parts of a tumbler (upper left). The hammer fits on the flats and the tumbler rotates on the arbor and pivot. Diagram of the mechanical action of a lock tumbler (upper right). Pulling the trigger (not shown) applies force *T* to the sear causing it to release the tumbler, which then rotates under the action of the force *F* applied by the mainspring. Offset of the centerlines of the arbor and pivot of an imperfectly made tumbler (bottom). The distance measured between these centerlines on artifacts is called the "eccentricity."

the metal sufficiently tough to resist the shock generated at the end of its rotation. Less important characteristics of the tumbler are the exact profile of its edge (most of it does not bear on any other part), the exact radial distance to the notches (the sear is spring loaded), and the central placement of the threaded hole for the retaining screw for the hammer or cock.

An artificer will encounter several difficulties in making a tumbler. The profile is an irregular curve, but flat, square, and round surfaces also have to be formed; a deep hole has to be drilled and tapped; and the part has to be heat treated to the requisite degree of hardness and toughness. Examination of the finished product will show how these manufacturing problems were handled.

Documentary Evidence

Surviving documents on the manufacture of lock mechanisms in the 19th century contain little technical detail. One of the most complete accounts is a list of the operations at the Springfield Armory in 1878.[17] According to this account, the tumbler was first forged to rough shape ("blocked") from bar stock with a trip-hammer, then drop-forged in closed dies to bring it closer to its final form, then milled on both sides and across the edges. The arbor and pivot were clamp milled and the end split with a circular saw (probably also on a milling machine). A hole was drilled and tapped in the arbor to accommodate the retaining screw for the hammer. The tumbler was then jig filed and hardened. The amount of filing done and its purpose are not specified.

Earlier descriptions are less complete but indicate that the basic process of making tumblers remained unchanged through the 19th century. In 1810 the steps were described as forging, milling (otherwise unspecified), and filing.[18] In 1819–25 the process was the same but the filer was said to have been provided with a gage.[19] There were forty-two lock filers (who used one file per lock) and only nine machinists (who did all the drilling, milling, and turning on locks) at the Springfield Armory in 1819. Wrought iron was used for tumblers at Springfield until about 1850; the finished parts were hardened by

[17]The report on the manufacture of the M1873 Springfield names, but does not describe, each operation and illustrates some of the machines used. Stephen V. Benet et al., *The Fabrication of Small Arms for the United States Service* (Washington, D.C., 1878).

[18]John Whiting's report to the Secretary of War on the Springfield Armory is reproduced in James E. Hicks, *United States Ordnance* (Mount Vernon, N.Y., 1940), 2:129.

[19]James Dalliba, "Armory at Springfield," *American State Papers, Class V, Military Affairs II* (Washington, D.C., 1860), pp. 543–54, Document 246; C. Meade Patterson, "Musket-Making Operations at Springfield Armory in 1825," *The Gun Report*, April 1980, pp. 44–48.

carburization followed by quenching and tempering.[20] By 1852 the proportion of machine work to handwork had changed; there were twenty-four lock filers (using one-third of a file per lock) and twenty-six artificers engaged in milling. The filing jigs and gages had become quite sophisticated.[21] The decrease in both the number of files used per lock and in the ratio of filers to millers shows that by 1852 relatively more of the metal was cut away with machines than had been the case in 1819. Yet the amount to be removed from the forged workpiece was still quite large. For example, a forging for an M1855 lock plate examined by E. A. Dixie in 1908 was 1/8-inch oversize on its perimeter; this excess would need to have been cut away by machining and filing.[22]

Inventories of armorer's and smith's tools and of machinery at the Springfield Armory, which exist for most years between 1834 and 1844, provide further clues on the manufacturing process.[23] Files are by far the most abundant type of tool. Gages and filing jigs are the only other tools present in large enough numbers for each artificer to have had at least one. None of the inventories shows enough calipers, dividers, squares, or straightedges to supply more than about half of the artificers engaged in bringing parts to final dimensions. No graduated scales or rules of any kind are listed. Until 1839 the armory had eighteen milling machines. (Indirect evidence suggests that these were all for clamp and hollow milling.) Between 1839 and 1841, when the armory was being retooled for production of a new model musket, the number of milling machines increased to thirty. Several of the machines on the 1841 list were designated for milling the sides and edges of parts such as lock plates and tumblers, and we can take this as the date of introduction of machines using formed cutters for milling profiles. The number of "mills" (which I take to mean milling cutters) per milling machine increased from nine in 1834 to twenty-three in 1843. This increase implies that the machines

[20]An attempt to use steel for tumblers was made in 1832 but most of them cracked during hardening or when assembled into locks. In 1850 the armory ordered 2,500 pounds of 3/4-inch octagon cast steel for tumblers. Steel tumblers are quenched and tempered rather than case-hardened. Joseph Weatherhead to Roswell Lee, March 30, 1832; James Ripley to Fullerton & Raymond, July 3, 1850; documents at the Springfield Armory National Historic site.

[21]Anon., *Bessey's Springfield Directory for 1851–2* (Springfield, Mass., 1851), pp. 157–67; Jacob Abbott, "The Armory at Springfield," *Harpers New Monthly Magazine* 5 (July 1852): 145–61; E. A. Dixie, "Some Old Gages and Filing Jigs," *American Machinist* 31–32 (1908): 381–83.

[22]Dixie (n. 21 above).

[23]There are manuscript inventories of all the tools and equipment for several years between 1834 and 1844 in the Springfield Armory National Historic Site Library.

TABLE 1
DISTRIBUTION OF MAN-HOURS USED TO MAKE A TUMBLER IN 1864

	Percentage
By type of work:	
Forging	6.3
Machining	39.2
Hand	54.5
By labor grade:	
Ordinary blacksmith	6.3
Ordinary mechanic	39.2
Good mechanic	2.1
First-class mechanic	52.4

NOTE.—Percentages are calculated from the data in Dyer's table of the hours of labor required to manufacture a Springfield rifle musket (see n. 24).

were being worked harder in 1843 so that more spare cutters were required. The number of filing jigs per artificer remained about the same throughout this period, which suggests that, while there was increased use of power tools for roughing cuts on forged parts, finishing to gage was still done by filers.

Some quantitative information on the manufacture of tumblers in 1864 can be extracted from a table of the mechanical power and labor required to manufacture 500 rifle-muskets in ten hours. This table was prepared by A. B. Dyer from information supplied by the foremen at the Springfield Armory.[24] Eighteen operations and seven types of power-driven machinery were used in making a tumbler, but one operation, hand filing by "first class mechanics," accounts for more than half of the man-hours required (see table 1). Similar proportions hold for the other components of the lock (except for some of the screws) and the stock. The high proportion of work done by hand and by the most skilled class of artificers is clearly at variance with the claims made by Talcott. To find out why this was so, we must turn to the material evidence.

The tumblers studied are listed in table 2. Most of these are from weapons in the collections of the Springfield Armory Museum, and the distribution of dates reflects, in part, the distribution of arms in this collection. First, surficial markings left from manufacturing operations were examined to identify the tools and techniques used for the final shaping of each part studied. Second, the principal dimensions of the tumblers were measured. Third, hardness measurements

[24]Benet (n. 7 above), 4:859–77.

were made when possible to show the success of heat treatment. Fractography and metallographic examination were applied to broken examples that were available for study.

Material Evidence: Surficial Markings

In the 19th century the exterior parts of small arms were usually polished after they were shaped, but the interior surfaces usually retain the markings left by the last metal-cutting operation performed on them, the one that brought the part to its final dimensions. Parts that have been altered, damaged, or replaced must be avoided, and I have relied on the curators of the collections studied to help detect such parts.

Different types of tools used to shape metal or wood leave distinctive surficial markings on the workpiece. The identification of tools and methods by such markings is a large subject only now being explored by industrial archaeologists, and we lack a general demonstration of the extent to which a specific tool and method of work can be uniquely associated with a given surficial mark. For our immediate purpose, however, two factors enable us to use surficial markings. First, research in the last few decades on the mechanism of metal cutting provides a theoretical basis for the interpretation of tool markings. Second, we need ask only two specific questions of the interpretation of the tool marks found on lock parts. First, can handwork be positively distinguished from machine cutting? Second, can certain specific milling processes, such as hollow milling or clamp milling, be recognized from the marks that they leave? These questions were answered by making reference surfaces with files and mills for comparison with the lock parts studied. It was found that filed surfaces can be distinguished with confidence from those finished by machine tools, that hollow and end milling can be easily identified, and that clamp milling can also be recognized, but with somewhat less confidence.[25] I will describe the surficial markings on two of the tumblers studied and then note important differences observed on the other examples.

M1812 musket made circa 1820.—This musket was made at the Whitney Armory after Eli Whitney had gained twenty years' experience in the manufacture of firearms. The tumbler, shown in figure 3, has been damaged by corrosion, but the tool marks on its surface are still visible. The edges of the tumbler were formed by transverse cuts with a coarse file, probably without a filing jig since the curves are not well blended and the tilts of the edge are not all in the same direction.

[25]R. B. Gordon, "Material Evidence of the Manufacturing Methods Used in 'Armory Practice,'" *IA: Journal of the Society for Industrial Archeology* 14, no. 1 (1988): 22–35.

FIG. 3.—Two views of a tumbler from an M1812 musket made at the Whitney Armory about 1820. Note the flat facets on the pivot caused by hand filing. A crack running through the arbor and flats is visible in the upper photograph. (Weapon from a private collection; photographs by William Sacco.)

The final finish on the top edge was made by curved strokes with a fine file; the areas that are more difficult to reach retain coarse file marks. The face of the tumbler on the pivot side shows no machine marks; file cuts in various directions were used to form this face. The face on the arbor side shows the characteristic marks of hollow milling. The cutter used chattered badly, producing deep gouges. At least two hollow mills were used; one formed the shoulder and the arbor, the other the relief on the face. The latter is a more demanding task, and the performance of the milling machine was poorer. The cylindrical surface of the arbor was formed by the hollow mill, but the flats that engage the square hole in the cock were filed. The pivot was made by transverse strokes with a hand file rather than by milling. Each stroke has left a facet on the surface of the pivot, which can be seen in the lower photograph of figure 3. (The arbor and pivot are off center by .030 inches.)

The evidence of the use of a hollow mill to form the arbor face is very similar to that described by Battison for a tumbler from the same model musket.[26] Comparison of his figure 4 with the tumbler studied here shows that the pattern of file marks on the edges of the two tumblers is different, which implies that there was no standard way of working among the makers of tumblers at the Whitney Armory at this time.[27]

M1841 rifle made 1851.—The M1841 rifle was made by a number of contractors, including the Whitney Armory, which was then under the management of Eli Whitney, Jr. Comparison of figures 3 and 4 shows that the quality of the mechanical work on the 1851 tumbler is greatly improved over that of 1820. The superior regularity of form evident in the picture is confirmed by the dimensional data presented below. Four types of surficial markings resulting from the use of coarse and fine files with longitudinal and transverse strokes can be recognized around the edge of this tumbler. Both faces were formed by hollow milling, probably with a double milling machine.[28] An additional hollow mill with a larger hole was used to make the shoulder on the arbor side. The accuracy of the work was sufficient to form a shoulder only 0.008 inch high, but the surface finish attained was poor and it was cleaned up with a file. Circular grooves with a slight inclination due to the feed rate of the mill are observed on the arbor and pivot and were probably made by the same milling operation that was used to form the faces. The flats were filed with transverse strokes.

[26]Battison (n. 3 above).

[27]Although Battison implies that the pivot on the example he studied was formed with a hollow mill, it may have been filed, as in the example described above. I believe that the Whitney Armory did not have a double milling machine in the lifetime of E. Whitney the elder and that Whitney's deathbed sketch of a tumbler mill (reproduced as fig. 23 in Battison's article) was his unexecuted plan to rectify this deficiency in his equipment.

[28]Battison (n. 3 above), p. 30, points out that "double milling machines" were included in the equipment made by Robbins & Lawrence for the Enfield Armoury in the 1850s and suggests that these machines had two spindles on the same axis for milling the arbor and pivot of the tumbler.

Fig. 4.—Tumbler of an M1841 rifle made at the Whitney Armory in 1851. Both the pivot and the arbor have been milled, but hand filing was used on the other surfaces as in the 1820 example. Comparison of figs. 3 and 4 shows the improvement in the quality of the filing achieved between 1820 and 1850. (Weapon from a private collection; photograph by William Sacco.)

Except for the milling of the pivot, no more machine work to final gage is present on the 1851 tumbler than on the 1820 example, but the quality of the handwork is enormously improved. The use of the hand file for the final shaping of the profile of the tumbler shows that the machines in use in 1851 at the Whitney Armory were not capable of finishing a product that would be to gage; handwork was required.

Springfield Armory.—Both faces of a tumbler made in 1803 were found to be hollow milled; it is filed on all except the arbor face, and the quality of the file work is somewhat better than on the Whitney 1820 example. An example for 1812 was made by the same methods as that for 1803, but the quality of file work is much lower. The number of artificers at the Springfield Armory doubled between 1808 and 1810. The decrease in quality suggests that the learning time for this work was at least two years.[29] There is a marked improvement in the

[29]There may have been a shortage of machinery as well. Whiting (n. 18 above) states that side screws were milled with waterpower, but the side screws in the 1812 Springfield musket examined in this study were hand filed.

appearance of Springfield tumblers by 1830; the example studied has been filed in the same places as the earlier examples, and the inner edge of the pivot has been damaged by careless filing (see fig. 5), but the filed faces are more nearly parallel, and the crown and slant of the edges are reduced. A tumbler for the same model musket made in 1839 is similar in appearance, but there has been a further improvement in workmanship, illustrated by the absence of damage to the edges of the pivot caused by the filing of the pivot face. Equivalent areas on the 1830 and the 1839 tumblers show similar patterns of file strokes, although within these areas there are differences in the angles and uniformity of the strokes. This suggests that a generally accepted way of doing the job had been established but that the details of carrying it out were left up to the individual filer.

The first indication of a difference in the machine work on the Springfield tumblers is found in the example for 1844, where it appears probable that clamp milling was substituted for hollow milling to form the cylindrical surfaces of the arbor and pivot. This change came at a time when the armory was making substantial additions to its machine-tool inventory. A tumbler for the same model musket

FIG. 5.—Tumbler of an M1816 musket made at Springfield in 1830. The inner end of the pivot has been damaged by hand filing of the tumbler face but the quality of the workmanship is much improved over that found in arms made in the first two decades of the operation of the Springfield Armory. (Photographed at the Springfield Armory Museum.)

(1842) made in 1852 is very similar and has the same pattern of filing marks but is not numbered. This shows that the practice of assembling locks by groups of parts had been dropped and that a generally accepted procedure for filing was in place. The tool marks on the examples for 1873 and 1884 are nearly identical. The upper, front, and bottom edges of these tumblers are described as being milled after forging, but the surficial markings show that these surfaces were brought to their final dimensions by filing. This includes the notch, which we know to have been jig filed. The quality of the file work is very high; the crown of the edges is barely detectable and the edges are perpendicular to the faces. The need to file the edges of the tumbler shows that as late as 1884 the milling machinery was still not capable of bringing tumblers to gage dimensions. How good the filers were at this task will be shown in the discussion on dimensions.

Other armories.—Examination of tumblers of M1841 rifles made by Harpers Ferry, Tryon (Philadelphia), Robbins & Lawrence (Windsor, Vt.), Remington, and Whitney shows that the same set of manufacturing operations was used on all of them and that these operations were the same as those used at the Springfield Armory for the M1842 musket (except that there is no evidence of clamp milling of arbors and pivots outside of Springfield). Although the overall quality of workmanship at the different armories is about the same, there is evidence of minor differences in artificers' techniques. At Harpers Ferry, for example, both faces of the tumbler were left as milled (this was noted on an M1842 Harpers Ferry musket tumbler also), and the milled faces have been deeply gouged by chatter marks. The Tryon tumbler is the only example stamped with letters, probably the initials of the artificer who made it. The milled surfaces of the Whitney examples are rough. The differences between the two Whitney examples show that they were finished by handwork.

The pattern of filing in all the examples indicates only minor differences in technique; an accepted procedure for carrying out this task was in place at all the armories by 1850. Since it is unlikely that each maker independently reached exactly the same conclusions about the best way to go about making a tumbler or that such details of method were imposed by contract terms, the exchange of ideas through the network of mechanicians must have been very effective in diffusing an accepted "right way" to do this job.[30]

[30]Such information was transmitted by interchange of experienced mechanicians from one works to another, as e.g., Thomas Warner's move from Springfield to Whitneyville in 1842, as well as by the migration of artificers. A. F. C. Wallace, *Rockdale* (New York, 1978), pp. 211–26, discusses the effectiveness of communication among mechanicians in the first part of the 19th century.

Material Evidence: Tumbler Dimensions

Reproducibility of dimensions.—Figure 6 shows the dimensions that were measured on the tumblers. In addition, the squareness of the edges to the faces and the amount of crown on the edges of each tumbler were examined, and as many of the dimensions as could be determined on each tumbler listed in table 2 were measured. The

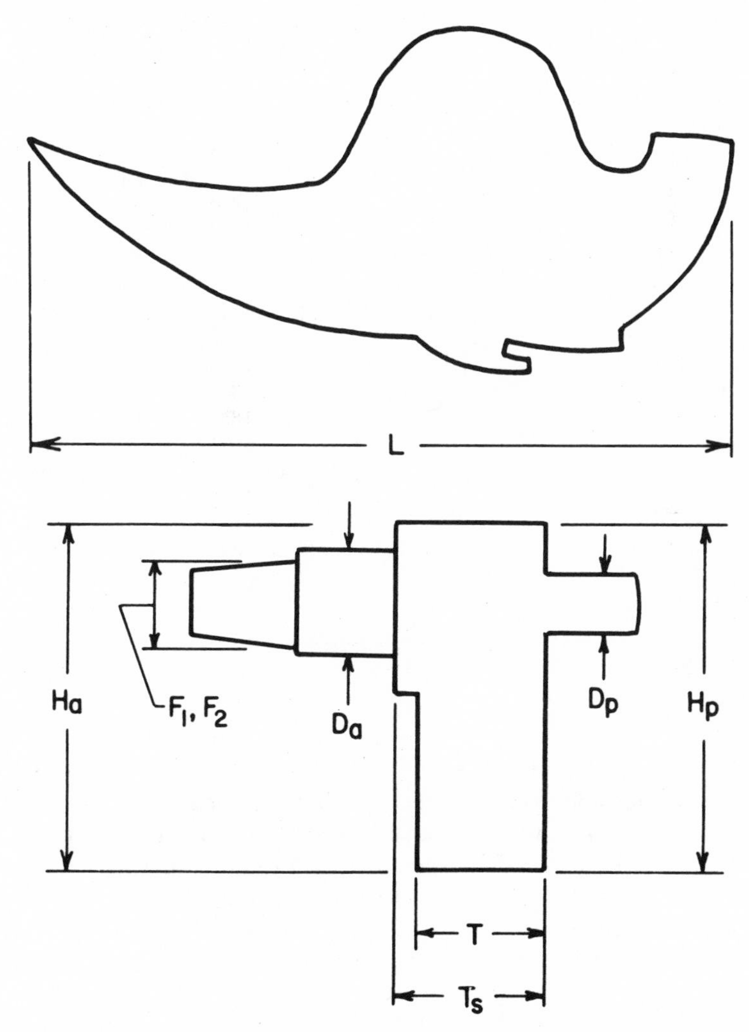

Fig. 6.—Diagram of the dimensions measured on the tumblers studied. Any difference between H_a and H_p shows that the edges are not square to the faces. F_1 and F_2 are the widths of the inner ends of the two pairs of flats.

TABLE 2
LIST OF TUMBLERS EXAMINED

Model and Maker	Date of Manufacture
M1795:	
Springfield	1803
Springfield	1812
M1812:	
Whitney	ca. 1820
M1816:	
Springfield	1830
Springfield	1839
M1842:	
Springfield	1844
Springfield	ca. 1850 (2 examples)
Springfield	1852
Harpers Ferry	1844
M1841:	
Harpers Ferry	1851
Tryon	1845
Robbins & Lawrence	1850
Remington	1853
Whitney	1851
Whitney	1854
M1855:	
Springfield	ca. 1855 (2 broken examples)
Tower (London)	1862
M1873:	
Springfield	ca. 1873
M1884:	
Springfield	1884

NOTE.—The locks for the M1873 and M1884 are identical.

mean of all the measurements of each dimension on all the examples of a given model lock was calculated. The average of the deviations of each dimension from the mean of all the measurements of that dimension on each model of lock is a measure of the consistency of size and shape attained in the manufacture of the tumblers. The calculated average deviations are shown in figure 7. The individual data points represent averages of from sixteen to forty-five measurements, depending on the number of examples available. The results show that there is a continuous improvement in the consistency of the dimensions along what may be described as a learning curve. By about 1880 the average deviation becomes less than 1/1000 inch.

Recently discovery was made of a manuscript report on the inspection of 100 Springfield muskets in 1828 by Benjamin Moor, mas-

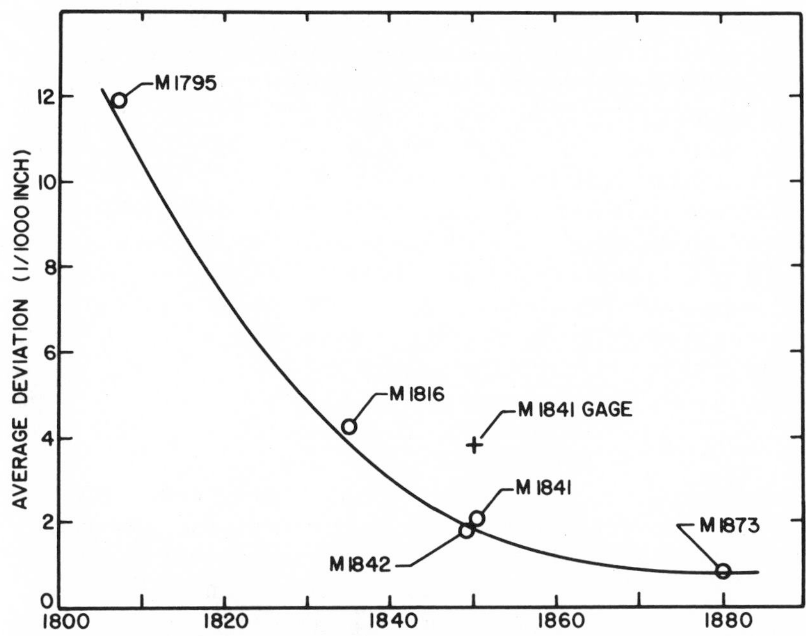

Fig. 7.—Average deviations (in units of 0.001 inch) of the dimensions of the tumblers for each model of lock examined from the mean of each dimension for that model. If all the tumblers of a given model were identical, the deviations would be zero. The average deviation is a measure of the degree to which the tumblers of a given model vary among themselves. The numbers of measurements averaged for each datum point are: M1795, 17; M1816, 14; M1841, 45; M1844, 30; and M1884, 16. The average departure of the measured dimensions from those set by the inspector's gages for the M1841 rifle is shown by the cross. It is slightly greater than the deviations of the tumblers among themselves.

ter armorer at the Pittsburgh Arsenal.[31] Moor tested the action of the locks and attempted to interchange parts among a sample of the muskets, finding that practical interchangeability had not been achieved. His report includes the greatest and least values of a large number of dimensions measured on the muskets examined. (Unfortunately, the individual measurements are not reported so a statistical description of them is not possible.) The dimensions that I measured

[31] Anon., *Report on the Inspection of 100 Springfield Muskets* (1828), manuscript at the Springfield Armory National Historic Site. A sample of 100 model 1816 muskets made in 1819 and 1820 at Springfield were inspected by Benjamin Moor. (See Smith, *Harpers Ferry* [n. 5 above], p. 278, for an account of Moor's technical qualifications.) The measurements on lock parts are given to either the nearest hundredth or half-hundredth of an inch. This suggests that the measurements were taken with a vernier caliper reading to hundredths of an inch and that the inspector estimated between divisions for the half-hundredths.

on the two M1816 tumblers at the Springfield Armory fall within the range of extremes given in Moor's report except for one measurement that was 0.01 inch smaller than the minimum reported.

Eccentricity.—The arbors and pivots of all the tumblers except one made at the Whitney Armory circa 1820 were milled, and any departure of their axes from coincidence (the "eccentricity," defined in fig. 2) shows a lack of accuracy in the machines used. The results of the measurements of eccentricity on the tumblers are shown in figure 8. There is a considerable scatter in the data for the early years, but the upper bound (lines in fig. 8) shows that there was a steady improvement in milling-machine accuracy, reaching 0.001 inch by the 1870s. The eccentricity of the tumbler of a British Tower musket of 1862 is shown for comparison; it is worse than that of contemporary American examples. The eccentricities (fig. 8) are substantially larger than the variability of the tumbler dimensions (fig. 7). This suggests that the accuracy built into the milling machines used to form the arbors and pivots was lower than the dimensional consistency achieved by the handwork of the artificers who filed the tumblers until about 1870, and even then did not exceed the accuracy attained by the filers.

Comparison with gage dimensions.—The dimensions of the M1841 rifle tumblers studied were compared with dimensions taken from gage no. 30, described as a receiver gage with holes and grooves,

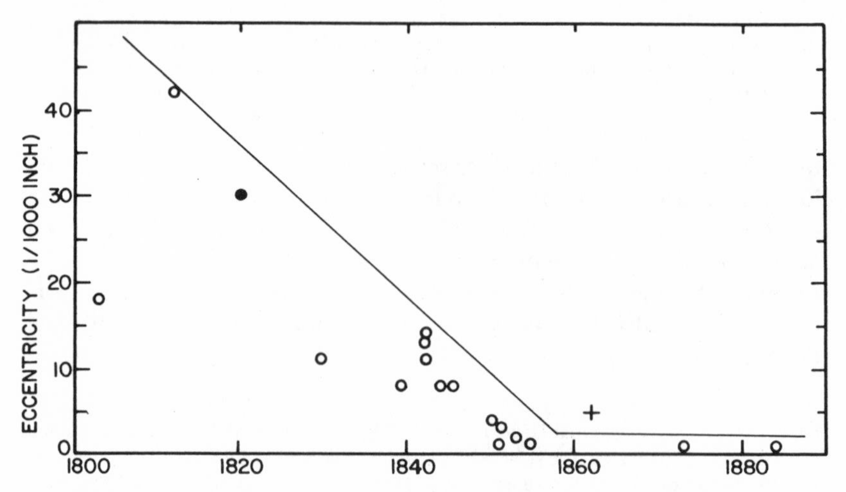

FIG. 8.—The eccentricities—the offsets of the centerlines of the arbors and pivots—of the tumblers examined. The datum for a British Tower musket is marked by a cross. The arbor and pivot are machined surfaces (except on one example among those studied, solid circle) and the eccentricity is a measure of the precision built into the machines used.

and no. 31, a tumbler pattern, in the set of inspection gages for the M1841 rifle now at the National Museum of American History. The dimensions shown in figure 6 as well as the lengths of the flats on the arbor were compared. The average departure of all dimensions of each tumbler from gage is shown in table 3.

In the fifty-four measurements made, the greatest departure from gage was 0.010 inch. On average, by 1850 the tumblers were being made to tolerances of better than 0.004 inch relative to gage dimensions. The accuracy of the work of the individual artificers who made the tumblers is actually better than this since the deviations in table 3 include errors due to deterioration the gages may have suffered, variations among the different gage sets for this model rifle, and any errors made in taking the dimensions of the gages. The average variation of the dimensions among themselves, a measure of the consistency to which the artificers worked, is only 0.002 inch.

Material Evidence: Metallurgy

Good mechanical work on lock parts goes for naught if the parts are not properly hardened after being brought to final size. In the 19th century the hardening had to be done without the aid of measuring instruments or scientific understanding of the processes used. Success depended on the judgment of temperature, control of the furnace atmosphere, and dextrous manipulation of the parts by the artificer entrusted with the work. Wrought-iron parts had to be carburized, and the carburized surface layer hardened; measurement of the depth of carburization and the surface hardness of artifacts can be used to determine the quality of the work, but a section cut from the tumbler is required to determine the depth of hardening. Hence, these data are available for only a few examples. Steel, used later for tumblers, was hardened by quenching and tempering after filing was

TABLE 3
COMPARISON WITH INSPECTOR'S GAGES

Maker and Year	Departure*
Harpers Ferry, 1851	4.2 (1/1000 inch)
Robbins & Lawrence, 1850	3.6
Remington, 1853	4.7
Tryon, 1845	2.6
Whitney, 1854	3.7
Average	3.8

*Average departure of tumbler dimensions from the dimensions of the inspection gages.

completed. To be hard enough to resist wear and simultaneously tough enough to sustain the shock loads to which they are subjected, steel or wrought-iron tumblers must be heat-treated to attain a hardness that falls in a narrow range. Difficulties may arise in attaining the correct temperatures for these operations or from decarburization of the surface of the part while it is at high temperature. The 1828 inspection report on Springfield muskets (see n. 31) states that 15 percent of the tumblers examined were too soft. Good control of the hardening process had not yet been attained.

Hardness data were taken on several tumblers that could be brought to the laboratory for study.[32] The surface hardness of the Whitney tumbler made in 1820 ranged from 1,033 to 560, very high hardness values that indicate a brittle surface. The combination of excessive hardness and poor surface finish would leave this tumbler susceptible to failure under the high stresses that would result because of the eccentricity of the arbor and pivot. This is the most likely cause of the crack that is present in the tumbler (see fig. 3). The Whitney tumbler made in 1851 has a surface hardness of 250–350. Since the hardness of iron is about 100, and fully hardened, high-carbon steel is about 900, this tumbler is hardly hardened at all. Metallurgical examination would be required to determine the cause; the most likely source of difficulty is surface decarburization due to exposure to an oxidizing atmosphere in the heating furnace.

Two M1842 musket tumblers were available for laboratory examination in section. One of them showed considerable wear on all its bearing surfaces and had a surface hardness of 137. When sectioned it was found to be only slightly carburized. The second, which was not worn from use, had a carburized case 0.002 inches thick with a hardness of 512. It had been correctly heat treated. Fragments of failed M1855 tumblers were found to be made of quenched steel only slightly tempered with a hardness of 635. They failed because the temper was not drawn enough; this left them susceptible to cracking caused by stress concentrations developed at the sharp corners at the shoulder with the arbor and by the poor finish of the milled surfaces.

[32]Microhardness measurements were used to avoid surface damage to the artifacts. Because the surfaces could not be polished for measurement, the data show more scatter than they would on prepared specimens, but comparison with data taken on sectioned specimens shows that the average hardness values obtained on the artifacts are reliable. The hardness numbers reported are Vickers microhardness (HV) measured with a 0.3-kg load on the indentor. Successful heat treatment of a tumbler requires that the hardness be between HV 520 and 580. (This corresponds to a Rockwell "C" hardness range of 50–54. The Rockwell C scale is frequently used today for hardness measurements on steel in manufacturing plants.)

An M1873 tumbler was found to have a hardness of 540, about correct for this type of service.

Before proceeding to an interpretation, I will summarize the material evidence. The tool marks and dimension measurements show that by 1850 artificers using hand files had learned to bring rough forged and machined parts of complex shape to final dimensions specified by gages to an accuracy of a few thousandths of an inch in routine production. This was achieved by artificers at the national armories and at the works of private contractors from Philadelphia to Windsor, Vermont. They achieved higher standards of accuracy than could be attained with machine tools throughout most of the 19th century. A generally accepted method of doing this handwork was followed at all of the armories. The metallurgical data suggest that hardening remained a difficult problem for the makers of lock parts long after good dimensional control had been achieved in production. Difficulties with heat treatment of lock parts persisted at Springfield at least through 1855. The excessive hardness of the broken M1855 tumblers suggests that the armory artificers had trouble developing control of heat treatment of the newly introduced steel used to make tumblers.[33]

Machine Tools

The principal change in an artificer's task brought about by the increased use of machine tools for the manufacture of small arms in the 19th century appears to have been reduction in the amount of physical labor required in removing metal. Calculations based on documentary and material evidence confirm this inference.

While no examples of forged but unmachined lock tumblers have yet been found, the amount of metal that had to be removed by machining and filing can be estimated from finished artifacts. A polished and etched cross section of an M1842 Springfield musket tumbler is shown in figure 9. The curved bands of slag and pearlite revealed by etching would have been parallel in the wrought-iron bar stock before forging. The metal flowed laterally when upset in the forging dies; the resultant displacement of the slag and pearlite bands shows the amount of flow and can be used to estimate the shape of the forging before it was machined. The estimated shape is sketched

[33]Heat treatment was still causing serious problems at Springfield as late as 1917, when there were difficulties with the bolts and receivers of the M1903 rifle. Julian S. Hatcher, "Metallurgical Improvements in Springfield Rifle," *Army Ordnance* 2 (1922): 351–53. This is another demonstration of the much greater difficulty that manufacturers in the United States had with metallurgical than with mechanical problems. See also Gordon (n. 6 above).

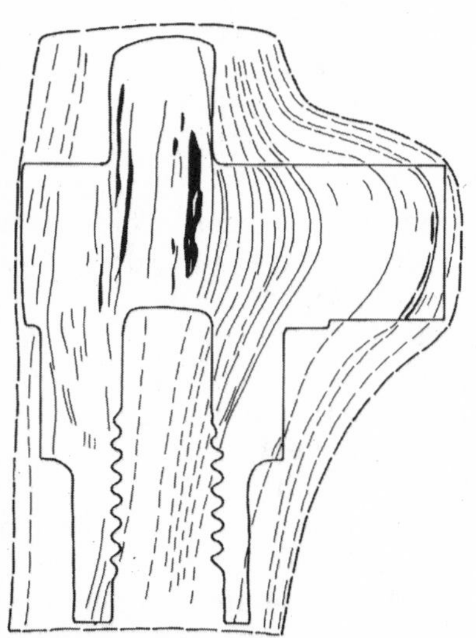

FIG. 9.—Polished and etched section cut through the center of the arbor and pivot of a tumbler for an M1842 Springfield musket lock made about 1850 (left). Black streaks are slag inclusions; grey bands are pearlite (regions of high carbon content) in the wrought iron. Flow of the iron during forging has displaced these streaks and bands. The shape of the forging before machining and filing inferred from these displacements is shown by dashed lines on the sketch at the right. (The scale of the picture is shown by the diameter of the arbor, 0.444 inch.)

in figure 9 and shows that between 1/8th and 1/4th of an inch of metal had to be removed from the forged piece to bring it to final dimensions. This is consistent with Dixie's report that the lock plate for the M1855 Springfield was forged 1/8th of an inch oversize (see n. 22).

Records of iron purchased and arms made show that a large part of the iron brought into the Springfield Armory never emerged in finished products and must, therefore, have been reduced to chips and scrap. From data on the number of arms made and the weight of iron in each, I estimate that only 31 percent of the iron purchased in 1852 emerged from the armory as components of weapons. This is an improvement on the 24 percent attained in 1823.[34] While the aggregated data do not show how the losses were apportioned among the different parts of a musket or the different processes used—and

[34]Data on the number of arms made, the weight of iron in each part, and the amount of iron purchased for 1852 are given in *Bessey's Directory* (n. 21 above). I assumed that there were no changes in the stock of iron held at the armory this year. The amount of iron required to make muskets in 1823 was estimated by Roswell Lee; see Deyrup (n. 9 above) p. 86.

some of the loss would have been in forging—they confirm that a substantial amount of metal removal was required of the armory artificers.

The table of machine and labor requirements for making rifle muskets prepared at the Springfield Armory in 1864 (see n. 24) lists the power and time required to machine each part; these data can be used to calculate that the physical work entailed in milling a tumbler was about 124,000 foot-pounds. From the number of hours of filing required as well as Rankine's estimates of the physical work done in various manual tasks,[35] I estimate that not more than 14,000 foot-pounds of work was required to file a tumbler after it was milled. Thus, the introduction of milling machinery before 1864 had resulted in at least a tenfold decrease in the physical labor required to bring a forged blank to the final dimensions of a tumbler.

The observations above suggest that the machine tools used at the Springfield Armory in the first two-thirds of the 19th century were capable of useful amounts of metal removal but were not capable of the same precision that could be attained by filing. Documentary evidence is scant, few examples of machines survive, and none of these have been subjected to careful analysis that would reveal likely sources of poor machine-tool precision. One possibility is inadequate support of the tool and workpiece because of loose bearings and light frames. Precise machining also requires accurate lead screws with graduated dials to control the positioning of the cutter and workpiece. Careful measurements of the lead screws in surviving machine tools dating from the first two-thirds of the 19th century have not yet been made, but it is likely that errors in pitch would have made precise graduation impossible. Finally, and probably most important, the carbon steel cutting tools then in use, while capable of precise work in cuts of short duration done at low speed, required frequent resharpening. The Springfield Armory had 548 cutters on hand for use on thirty milling machines in 1844, when only three models of arms were in production.[36] This implies a large reserve of cutters for each machine and the need for frequent resharpening, which was a difficult task until Brown invented the formed cutter, which retained its contour when reground.[37]

Artificers' Skills

The material evidence presented above shows that bringing lock parts to gage required handwork with a file at least through 1884.

[35]William Kent, *The Mechanical Engineer's Pocket-Book* (New York, 1903), p. 433.
[36]Springfield Armory inventory (n. 23 above) for 1844.
[37]Joseph W. Roe, *English and American Tool Builders* (New Haven, Conn., 1916), p. 206.

The dimensional tolerances and the quality of workmanship achieved by the artificers improved continuously along a learning curve. The improvement in product quality attained was primarily due to the superior mechanical skills developed among the artificers who made the lock parts, although better organization of the work and manufacturing procedures helped facilitate development of these skills. Clearly, the skills required to achieve interchangeability in the new system of manufactures was not "built into the machines" but remained in the hands of the artificers.

During the 19th century, science and engineering principles began to influence the development of mechanical technology in areas such as heat engines, water motors, bridge building, shipbuilding, structures, and, toward the end of the century, metallurgy. But the principles of metal cutting were not studied until the 20th century, and science and engineering theory had little effect on early manufacturing technology. This remained the domain of the mechanicians and artificers who solved problems through an understanding of the interaction between tools and materials. We call this "skill" because it cannot be adequately described in verbal or quantitative terms.

The skills in question here are those that have been described as "genuine" rather than "socially constructed."[38] I like to describe the skills required of mechanical artificers in terms of four components: dexterity, judgment, planning, and resourcefulness. *Dexterity* is the ability to manipulate tools with facility, and *judgment* is the capacity to gage size and shape by eye and goodness of fit between mechanical parts by feel. *Planning* and *resourcefulness* relate to decisions an artificer must make in organizing the way in which a task is to be undertaken and in responding to the unanticipated complications that inevitably crop up owing to inhomogeneity of the materials used or wear of the tools or machinery employed. All these elements of skill were required of artificers making small arms, but the relative importance of the four components changed throughout the course of the 19th century.

Planning.—Planning out a sequence of work is important when all the parts of a mechanism have to be made and fitted together without guidance from gages. Both documentary and material evidence show that the scope for the exercise of planning skills by artificers in the armories was reduced by division of labor, improved communication between mechanicians, and use of filing jigs. These factors required the artificer to focus on the narrower problem of making one part to gage according to a generally accepted method of working instead of

[38]Genuine and socially constructed skill are defined and discussed by Charles More, *Skill and the English Working Class* (London, 1980).

planning the way that the parts of an entire mechanism were to be fitted together.[39]

Resourcefulness.—The task of 19th-century artificers who specialized in making one particular lock part to gage never became entirely repetitious. Inhomogeneities in the material used, variations in the amount of metal left from the forging and rough machining, and wear of the files and jigs created contingencies to be dealt with differently in each successive part made. There is abundant evidence of deficient homogeneity in both the wrought iron and the American-made steel available to the armories in the first two-thirds of the century.[40] Nonuniformly distributed slag particles in wrought iron can cause small pieces of metal to break away during machining or filing, thereby thwarting attempts to attain the required dimensions. Although the uniformity of the iron may have improved during the century, the dimensional accuracy required of the filers also increased; it is likely that the vagaries of iron remained a problem for artificers until it was replaced by steel in the 1870s.

Dexterity.—The dexterity of an artificer is more than an ability to manipulate small objects, such as is tested in standard psychological tests; it includes the capacity, learned by practice and experience, to manipulate hand tools so as to produce work of superior quality. The dexterity achieved by the artificers who made lock mechanisms is directly visible in what they made: features such as smooth, continuous curves, flat surfaces, right angles at corners, and good finish. Many

[39]Use of dimensions in the manufacture of mechanisms was impractical until the micrometer caliper became widely available late in the 19th century. This will be discussed later. In a recent paper, John Harris has described the difficulties encountered in attempts to transfer technology in the 18th century with drawings or descriptions or even with purchase of machines. Transfer was difficult for the mechanical arts, but almost impossible for metallurgy. The same situation applied through much of the 19th century and placed a distinct limit on the effectiveness of communication between designers and artificers. The importance of the presence of a skilled artificer in the transfer of new technology is illustrated by the experience at the Springfield Armory with barrel welding in rolls. There were several unsuccessful attempts to introduce this process, which had been developed in England, before 1855. Success was achieved only when the machinery, the bar iron to be welded, and an experienced artificer were all imported from England. The production of welded barrels per welder went up tenfold the year after roll-welding was introduced. J. R. Harris, "Industrial Espionage in the Eighteenth Century," *Industrial Archaeology Review* 7 (1985): 127–37; Deyrup (n. 9 above), table p. 247; R. B. Gordon, "English Iron for American Arms: Laboratory Evidence on the Iron Used at the Springfield Armory in 1860," *Journal of the Historical Metallurgy Society* 17 (1983): 91–98.

[40]Difficulties with wrought iron were particularly severe in barrel making; much of the iron used for locks was left over from the barrel shops. *Bessey's Directory* (n. 21 above).

people find such features pleasing, but there is a scientific as well as an aesthetic basis for the standards of workmanship used to judge mechanical dexterity; smooth curves and the absence of deep tool marks reduce the stress concentrations that are the most common cause of cracks in the metal parts of mechanisms.

The introduction of filing jigs, which reduced the need for planning skills, placed new demands on the filer's dexterity.[41] The jigs necessarily contained hardened steel templates, and contact with these would quickly ruin a file. To achieve gage, the artificer had to manipulate the file so as to bring the work to the same profile as the template without touching the latter.

The poor workmanship found in the locks made before about 1830 shows that there was an absolute scarcity of artificers with the requisite mechanical dexterity. This observation is confirmed by the inspection report of 1828 (see n. 31); Benjamin Moor complains of poorly formed notches, variability of the curved edge, and deviations in the angular setting of the flats on the arbor in relation to the bearing of the mainspring. The quality of the workmanship observed in the locks improves as the dimensional consistency (shown in fig. 6) improves. By 1880, the quality of the filing done at the Springfield Armory was so good that it requires observations with modern micrometers and microscopes to distinguish this work from 20th-century machining.

Judgment.—A measure of the level of judgment skill required of an artificer is the number of variables that must be dealt with in completing a task relative to the amount of explicit information about these variables that is available. Little judgment is required when an artificer is supplied with exact information about the requirements to be met and the condition of his work at any stage of its progress—as, for example, when a dimension is gaged with a digital readout against established, numerical limits. A high level of skill is required when many decisions have to be taken, but only limited information is supplied by the designers of the product being made or the instrumentation available to monitor the progress of the task. In the first two-thirds of the 20th century, designers conveyed their intentions by means of drawings and dimensions (Woodbury's last two conditions; see n. 6). In much of the 19th century, however, models and, later, gages were used rather than drawings and dimensions; the artificer was required to duplicate the form and size of the model or gage in

[41] The term "jig" is supposed to have been derived from the complex, dancelike motions required of an artificer using this appliance. In the hands of the unskilled it was said to be "death to files." F. G. Parkhurst, "Manufacture by the System of Interchangeable Parts," *American Machinist* 24 (1901): 39–43.

the part being made, perhaps with the aid of fixtures and jigs.[42] I will argue that the methods of measurement and gaging introduced in the 19th century, instead of eliminating personal judgment, actually demanded increasingly sophisticated levels of judgment skills from artificers.

Reliable scales or rules were not available to American artificers before 1850, when J. R. Brown made a linear dividing engine suitable for graduating steel rules for shop use. In 1851 Brown began manufacture of a vernier caliper that made it possible for mechanical artificers to measure to 0.001 inch. But manufacture of the micrometer caliper, the instrument most useful in precision shop work, began in America only in 1868.[43] The difficulty of specifying dimensions in the absence of adequate measuring instruments is illustrated by the descriptions of bore gages issued by the Ordnance Office in 1822; the gages are required to match the diameter of a sphere of lead of specified weight rather than a specified diameter.[44] The limit of the sensitivity of measurement with the unaided eye and a steel rule is about 0.005 inch, yet we have seen that by 1850 the dimensions of lock parts were being held to within 0.002 inch and, by 1873, to better than 0.001 inch in routine production at works hundreds of miles from each other.[45] Such precision was possible because an artificer can determine deviation of a fraction of a thousandth of an inch by the feel of how a part fits into a gage. This was the skill that was used throughout most of the 19th century to transmit dimensional requirements and to inspect the work of artificers by gages.

The steady increase in the use of gages is shown in the records of the Springfield Armory. No gages are mentioned in Whiting's report of 1810. The first experiments with gages are said to have begun in about 1817, and Dalliba's report, widely quoted by modern writers on manufacturing technology, asserts that in 1819 the principle of making parts to gage was being perfected and that each artificer was

[42]Making gages, jigs, and fixtures required increasingly sophisticated application of planning skills on the part of the artificers who made this new equipment. This is discussed below.

[43]Roe (n. 37 above), pp. 202–13.

[44]Ordnance Office letter of July 25, 1822 quoted in C. M. Green, undated MS at the Springfield Armory Museum National Historic Site, Springfield, Mass.

[45]One commonsense reason for the continued preference of American artificers for the inch rather than the millimeter is that the inch is of a length that can be read to hundredths by eye and thousandths by vernier; the vernier on a metric caliper reads to 1/50 rather than 1/100 of a millimeter and is less convenient to use. French ideas of rationalism that inspired the interest in interchangeable manufacture also saddled practical men with impractical units of measure.

supplied with a gage for his own part.[46] This may be a statement of intention rather than a report of an accomplishment. It would have been nearly impossible to supply gages in a span of just two years even for the 118 artificers employed in bringing musket parts to final dimension at Springfield. The government requirements published in 1823 for the inspection of contract arms, which were supposed to be held to the same standards as those made at the national armories, lists only eleven gages.[47] The 1828 inspection report (see n. 31) shows that the gages used failed to detect several tumbler faults considered important by the inspector: The cock notch had not been filed perpendicular to the tumbler face in 70 percent of the examples examined, and in 25 percent the squares were described as "wrenched."

By 1850, fifty-six gages were used, fifteen for the barrel, fifteen for the lock, and twenty-six for the mountings and stock; by 1878 the number had increased to 154.[48] Although there had been a large increase in the number of gages, there was little change in the way they were made and used.[49] The National Museum of American History's complete set of gages for the M1841 rifle makes use of the following basic gage elements: (1) *patterns,* which show some aspect of the form of a part; (2) *receiving gages,* into which a part must fit (a receiving gage shows the outer boundary of a part but not an inner bound; how well the part fits once it is in the gage is a matter of judgment); (3) *groove, hole, and plug gages,* which are used to check one dimension by the goodness of fit determined by the sense of touch (most of these gages do not provide for limits, and it is left to the artificer's judgment to determine if a part is made to gage dimensions); and (4) *location gages,* which show the relative positions of different parts of the weapon, such as the position of the cone relative to the breech of the barrel. The tumbler gage from this set incorporates groove, hole, receiving, and pattern elements (see fig. 10).

Making these gage sets was a formidable task. First, a master weapon had to be made. Gages for inspection and for the use of the artificers were then designed, made, and fitted to the master parts. Consistency had to be checked by judgment of feel of fit and by eye. The level of success achieved depended on both the quality of the gages, which was determined by the artificers who made them, and the design of

[46]Smith (n. 5 above), p. 109; Whiting (n. 18 above); Dalliba (n. 19 above).

[47]Anon., *Regulations for the Inspection of Small Arms* (Washington, D.C., 1823).

[48]Anon., *Ordnance Manual* (Washington, D.C., 1850); Benet et al. (n. 17 above).

[49]No examples of the gages used before about 1840 have survived but they were probably similar in design to those used later. The method of making and using gages had changed very little even by 1917. The gages used in 1917 are described in Fred H. Colvin and E. Viall, *United States Rifles and Machine Guns* (New York, 1917).

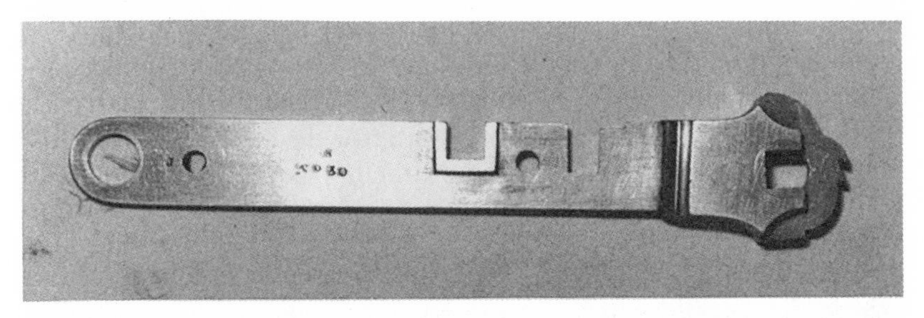

FIG. 10.—A tumbler gage for the M1841 rifle. The form of the edge of the tumbler is shown on one end of the gage; a tumbler can be inserted in the gage for comparison. The large notch in the center shows the width and length of the flats on the arbor. The smaller notches show the thickness of the tumbler and the length of the cylindrical part of the arbor. The fit of a tumbler to these slots must be judged by feel and by eye; no limits are specified. (Photographed at the National Museum of American History.)

the gages—their capacity to measure all critical dimensions and avoid requiring unnecessary precision where not needed.[50] Examination of the gages for the M1841 rifle shows that, except for holes and cylindrical surfaces, they were made entirely by hand filing. The scribed lines used for layout are still visible on some of them. The gages used at Springfield are described by Dalliba as made of hardened steel, and Roswell Lee proposed that gages be case-hardened. But these gages appear not to have been hardened (except for the barrel-plug gages) and so would be subject to loss of accuracy through wear in use.[51] The increased reliance on gages placed new demands on the judgment skills of the artificers who used them; making gages called for the highest standards of workmanship in all of the components of mechanical skill. Gage making and maintenance was an important task at the national armories, and the practitioners of this special art were the progenitors of the toolmakers of the 20th century.

Only one gage in the set for the M1841 rifle is described in the Ordnance Manual as a limit gage. It is a plug gage for the bore of the barrel and is 0.009 inches larger than the standard plug gage. The gages that were used to control the absolute accuracy of lock parts to better than 0.004 inch by 1850 were not limit gages; they depended entirely on the artificer's judgment of the fit of the part he was making to the dimension set by the gage.

[50]A critical study of the design of the gages for the M1841 rifle has not been made yet but it appears that precision was called for in many places where it was not needed. This would be a critical concern in determining the utility of armory practice in commercial manufacturing.

[51]Dalliba (n. 19 above); G. S. Cesari, "American Arms-Making Machine Tool Development" (Ph.D. diss., University of Pennsylvania, 1970).

So, as manufacturing methods developed through the 19th century there was less call for planning skills but heavier demands on the artificer's dexterity and judgment. Superior mechanical skills were developed and disseminated among a growing number of artificers. The mechanical requirements and the methods of attaining them were established early in the century, and after that there was little change in the basic methods of making parts for lock mechanisms other than increased use of machine tools for the heavy labor of roughing cuts, which allowed filers to concentrate on the final, precise shaping of parts to gage. There was a continuing improvement in the quality of work done with hand tools. The rate of progress in the attainment of interchangeable manufacture in the first two-thirds of the 19th century may have been limited as much by the rate of learning and transmission of the requisite mechanical skills among artificers as by the rate of invention of mechanical appliances to aid their work.

Economic History of the Springfield Armory

The Springfield Armory was not required to be commercially profitable, but did have to be able to demonstrate that its production of small arms was economical in order to serve its function as a "yardstick" against which the prices charged by the private armories could be evaluated and to deflect attempts by private contractors to capture its business.[52] One method of investigating whether decisions about manufacturing technology at the armory were made in response to economic considerations is econometric analysis of the aggregated data on expenditure and output of arms. Such analysis is framed in terms of *production functions,* which relate quantity of arms produced to *factors of production,* such as expenditure on labor and equipment.

In a 1968 article Edward Ames and Nathan Rosenberg addressed the problem of devising production functions that reflect technological change and that could be used to explain why small-arms producers made more use of specialized machinery in the United States than in England.[53] In order to describe the consequences of using

[52] The division of small- arms procurement between government armories and private contractors was in dispute throughout the 19th century. The best organized and most determined attempt to capture all of the government small arms business was made by the Association of Manufacturers of Arms, Ammunition, and Equipments of the United States in the late 1870s. The association's efforts were deflected with the aid of data on the cost of arms production at the Springfield Armory developed by Superintendent J. G. Benton. C. Meade Patterson, "Springfield on Trial," *Gun Report,* July 1963, pp. 15–23, 44; August 1963, pp. 14–17; September 1963, pp. 10–13.

[53] Edward Ames and Nathan Rosenberg, "The Enfield Arsenal in Theory and History," *Economic Journal* 78 (1968): 827–42.

specialized production machinery, they asserted that machine-made arms are interchangeable while craft-made arms are not. (More accurately, the assertion is that interchangeability could be attained at reasonable cost by using production machinery but not by craft methods.) This led them to describe, for example, gunstocks made after 1820 as interchangeable (because they were made with specialized machinery) and to state that a higher degree of interchangeability was attained through the use of milling machines than through hand filing with jigs. This formulation of technological change in small-arms manufacture is contrary to the material evidence discussed above. A better identification of the factors of production with the physical processes actually used is needed if econometric analysis is to define how decisions on technology may have been based on costs of production.

A step in this direction was made by Paul Uselding in an analysis intended to test the importance of including materials as well as labor and capital in the production function for arms making.[54] He showed that the growth of labor efficiency at the Springfield Armory was particularly great between 1840 and 1850; during this time there was no marked increase in the ratio of wage rate to rental of capital, which argues against the labor-scarcity hypothesis that has been suggested to explain the preference of American armories for specialized machine tools. Instead, Uselding finds evidence of improvement in labor quality at the armory. Since the analysis is based on aggregated data, this generalization cannot be identified with actual physical processes, but Uselding's results appear to be in accord with the material evidence that shows a marked improvement in the quality of the mechanical work done by armory artificers in the first two-thirds of the 19th century.

Traditional methods of armory work established early in the 19th century persisted well into the 20th. This was a subject of criticism when the methods used at the Springfield Armory were viewed by observers experienced with the production methods used in other industries in 1917. Colvin and Viall found that a great many of the gage requirements for the M1903 Springfield rifle were unnecessarily tight and that it made little sense to make a part like the trigger guard by machining and filing away a forging that initially weighed 3½ pounds to less than ½ pound; a stamping would have served as well.[55] The design of the gages used in 1917 had changed little from those used

[54]Paul J. Uselding, "Technical Progress at the Springfield Armory, 1820–1850," *Explorations in Economic History* 9 (1972): 291–316. This paper is discussed by V. Kerry Smith, "The Ames-Rosenberg Hypothesis and the Role of Natural Resources in the Production Technology," *Explorations in Economic History* 15 (1978): 257–68.

[55]Colvin and Viall (n. 49 above), p. 155.

in 1850, and the last step in many of the sequences of operations for the rifle parts was still filing to gage. (One reason for the continued use of the file may have been the antiquity of much of the machinery at the armory, however.) Clearly, leadership in manufacturing technology had passed from the armories to other industries sometime in the last third of the 19th century.

Conclusions

A large reduction in the physical labor of shaping wood and metal parts was achieved by the armories in the 19th century by use of power-driven machine tools. Yet what made possible the attainment of interchangeable manufacture was the remarkable growth in the skills of the artificers who brought the machined parts to final dimensions with hand tools. This has been demonstrated by study of the tumblers of the lock mechanisms of military small arms, but examination of the other lock parts and of the stocks of the weapons studied shows that the same conclusions apply to these.[56]

Manufacture of arms for the government is not subject to the economic constraints that apply to products that must be sold on commercial markets; hence, results obtained from a study of developments at the armories making military arms are not necessarily applicable to other branches of manufacturing. Once the technical capacity to manufacture to certain standards exists, the decision on the degree to which this capacity is used in making any given product is an economic one.[57] The very wide variation in the quality of the mechanisms in commercial arms made in the 19th century shows that not all manufacturers found the federal armory standards of quality economically useful. However, the skills developed in small-arms making by the armory artificers were available when required in other industries. Study of artifacts other than small arms will be required to show how much these skills were drawn on in other branches of American manufacturing. I anticipate that they will be found to have been fully exploited by midcentury in making both machine tools and measuring equipment. Thus, the superior quality of the machinery

[56]In view of the claims made on behalf of John Hall's machinery, it would be of particular interest to apply the archaeological methods of examination to surviving samples of Hall rifles. I think it likely that such examination will show results not appreciably different from those found for the products of the other armories.

[57]E. Buckingham, *Principles of Interchangeable Manufacture*, 2d ed. (New York, 1941). On the lack of incentive to use interchangeable manufacture in the nonmilitary small arms industry, see R. A. Howard, "Interchangeable Parts Reexamined: The Private Sector of the American Arms Industry on the Eve of the Civil War," *Technology and Culture* 19 (1978): 633–49.

sent to the Enfield Armoury by Robbins & Lawrence and the Ames Manufacturing Company in the 1850s, some of which remained serviceable for over 100 years, is as much due to the skills of American mechanical artificers as to the design capabilities of American mechanicians. And, once manufacture of vernier and micrometer calipers began in the United States, dependence on imported instruments for factory use was quickly eliminated, and the cost of accurate tools was reduced enough so that each artificer could have his own.

The new levels of skill attained in the armories were not economical in cases where a more crudely made mechanism was serviceable to most customers, as with agricultural and sewing machinery throughout much of the 19th century. Nevertheless, the availability of skilled artificers and accepted methods of working to close tolerances, together with new machinery such as the cylindrical and centerless grinders, made possible the production of new products, such as bicycles, steam turbines, and, early in the 20th century, aircraft engines, in which precision manufacturing was required for success. It was only in the 1920s that precision filing began to become unnecessary in training programs for apprentice toolmakers; the introduction of more rigid machine tools with better bearings and of high-speed cutting tools finally made machine work more accurate than the handwork of even the most skilled mechanical artificers.

A mechanical ideal was indeed achieved in 19th-century American manufacturing, but it was not the one proposed by Fitch. It was the superior standards of workmanship and mechanical skill of American artificers.[58]

[58]Since this paper was written in 1985, examination of the Hall pattern rifles and carbines in the collection of the Springfield Armory Museum has shown that, while some surfaces in the lock mechanisms that are neither mating nor bearing carry milling marks, all surfaces where a fit to another part is required have been filed to their final dimensions.

From Iron to Steel: The Recasting of the Jones and Laughlins Workforce between 1885 and 1896

DAVID JARDINI

> Technological progress has been one of the most potent forces in history in that it has provided society with what economists call a "free lunch," that is, an increase in output that is not commensurate with an increase in effort and cost necessary to bring it about. [JOEL MOKYR, *The Lever of Riches* (New York, 1990)]

A "free lunch," or Pandora's box of evils? Scholars have long debated which idea better characterizes technological change and its social consequences. Over the past two centuries, the extent and rapidity of technological change have been perhaps the most dramatic features of the world economy. This was especially true during the "Second Industrial Revolution" of the late 19th and early 20th centuries when mechanization, the adoption of mass-production techniques, and evolving organizational structures multiplied both worker productivity and the gross productive achievement of industrialized nations. Despite the remarkable gains in output and productivity realized during this period, however, historians have been sharply divided in their appraisals of technology's net social value.

Prominent in this debate has been the issue of technology's effects on the industrial workforce. Here, scholarship concerning the consequences of growing capital intensity on workforce skill content has been characterized by a distinct polarization. On one hand, many

DR. JARDINI completed a Ph.D. at Carnegie Mellon University in 1996. His dissertation studies the history of the RAND Corporation, focusing on RAND's development of analytical methodologies and its diversification from national security into social welfare research during the 1960s. He is currently a vice president in the metals section of Mellon Bank's Corporate Banking Department and a postdoctoral fellow in Carnegie Mellon's Center for History and Policy. The author wishes gratefully to acknowledge the comments and criticisms of David Hounshell, John Komlos, Richard Oestreicher, the members of the Pittsburgh Center for Social History, and the *Technology and Culture* referees, as well as the technical assistance of Stephen Verba.

writers argue that rising capital intensity stimulated disproportionately large increases in demand for skilled workers, thereby enhancing workers' overall living standards and work experiences. These authors find that entrepreneurs, motivated to pursue technological advance by the attraction of greater productive efficiency, replaced the preponderance of brutish labor with highly productive semiskilled operatives. On the other hand, a second body of scholarship concludes that technological change had a generally deskilling effect and converted the workforce into a more or less homogeneous mass of laborers. These historians argue that capitalist-sponsored technology was a thinly veiled weapon used to break labor's bargaining power by destroying its knowledge and control of production processes. It was thus strategically employed to crush labor organization and multiply profits.

The first school of thought is composed primarily of economic historians and historians of technology. They find that the mechanization of production technology has created a general upward shift in skill content across the occupational structure and that capital and skill are complementary factors of production.[1] More recently, work in this vein has focused on the idea that complex, mechanized production systems increase the degrees of responsibility and flexibility required of workers.[2] At the same time, technical change is also argued to have raised the general level of workers' technical understanding to a more advanced state despite reducing the demand for specialized craft skills.

The scholarship that takes a more pessimistic view of the consequences of technological change is founded largely on the Marxist

[1] For the development of ideas of general skill uplifting, see Simon Kuznets, "Quantitative Aspects of the Economic Growth of Nations: Industrial Distribution of the National Product and Labor Force," *Economic Development and Cultural Change* 5, no. 4 (1957): 3–111; Robert Blauner, *Alienation and Freedom: The Factory Worker and Industry* (Chicago, 1964); and Daniel Bell, *The Coming of Post-industrial Society* (New York, 1973). The argument that capital and skill are complementary factors of production is largely founded on H. J. Habakkuk's *American and British Technology in the 19th Century: The Search for Labour-Saving Inventions* (Cambridge, 1967). Jeffrey G. Williamson and Peter H. Lindert, in *American Inequality: A Macroeconomic History* (New York, 1980), argue that "capital can substitute for unskilled labor more readily than for skilled labor, which is needed as a complement to machinery. This meant that a greater proportion of unskilled labor than skilled labor was replaced by mechanization, at least at the macro level" (p. 286).

[2] Larry Hirschhorn, *Beyond Mechanization* (Cambridge, Mass., 1984); and Michael Nuwer, "From Batch to Flow: Production Technology and Work-Force Skills in the Steel Industry, 1880–1920," *Technology and Culture* 29 (October 1988): 808–38.

critique that technology degrades labor and leaves in its wake a virtual sea of unskilled, underemployed workers. This thesis was developed most extensively by Harry Braverman in his seminal *Labor and Monopoly Capital,* which argues that capitalist-controlled technology has historically been used to transfer knowledge and control of production out of workers' hands through the minute division of labor and mechanization.[3] Scholarship here has been conducted largely by labor historians such as David Montgomery, David Brody, and Katherine Stone, who find that technology has been repeatedly used as a weapon by capitalists to degrade workers and subvert their shop-floor power.[4] Among historians of technology, David F. Noble has most stridently supported this position, arguing that "technology serv[es] at once as the vehicle and mask of domination."[5]

While a bipolar division of such a vast and sophisticated literature ignores both important subtleties in the respective authors' theories and points where the opposing schools may converge, it illustrates a fundamental disagreement among historians of industrialization. This article seeks to shed light on this disagreement by analyzing the changing skill, wage, and productivity characteristics of an American iron and steel plant's workforce as that plant underwent revolutionary technological change during the late 19th century. It finds that although the adoption of mass-production technology in a traditional industry dramatically reduced the manual skill content of the workforce, workers were by no means reduced to a homogenized pool of interchangeable operatives. Instead, integrated, large-scale production techniques placed great numbers of workers in strategically vital positions along an operational continuum, thus allowing them to maintain at least the potential for effective resistance to managerial power. This implies that despite their commanding social position

[3] Harry Braverman, *Labor and Monopoly Capital: The Degradation of Work in the 20th Century* (New York and London, 1974).

[4] See David Montgomery, *Workers' Control in America: Studies in the History of Work, Technology, and Labor Struggles* (New York, 1979), and *The Fall of the House of Labor: The Workplace, the State, and American Labor Activism, 1865–1925* (New York, 1987); David Brody, *Workers in Industrial America: Essays on the 20th Century Struggle* (New York, 1980), and *Steelworkers in America: The Nonunion Era* (New York, 1960); and Katherine Stone, "The Origins of Job Structures in the Steel Industry," *Radical America* 7 (November–December 1973): 19–64. Also, among economic historians, see Richard Edwards, *Contested Terrain: The Transformation of the Workplace in the 20th Century* (New York, 1979); and David M. Gordon et al., *Segmented Work, Divided Workers: The Historical Transformation of Labor in the United States* (New York, 1982).

[5] David F. Noble, *Forces of Production: A Social History of Industrial Automation* (New York, 1986), p. 326.

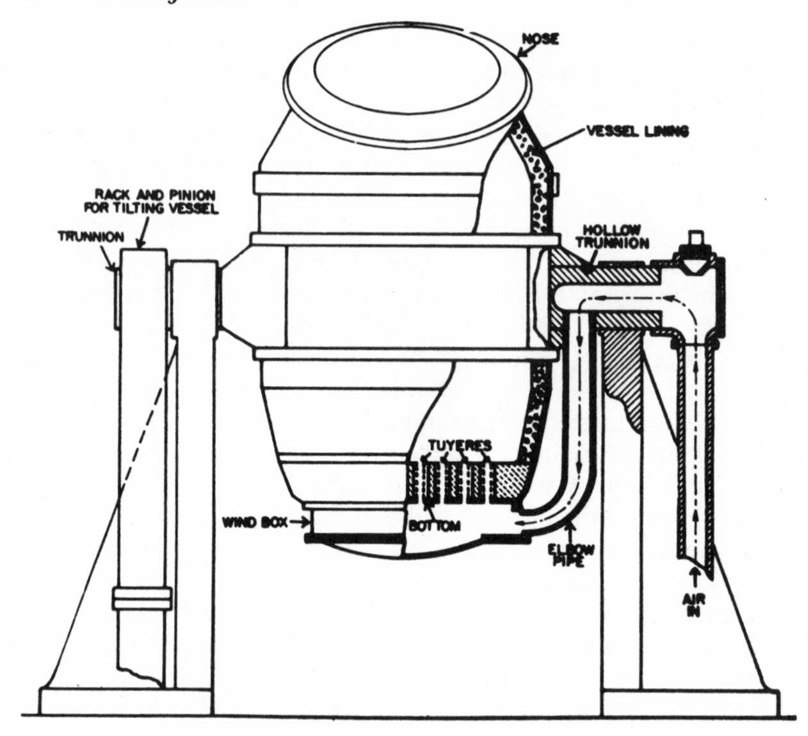

FIG. 1.—Line drawing of a Bessemer converter. (*The Making, Shaping, and Treating of Steel,* 6th ed. [Pittsburgh, 1951], p. 381.)

and their apparent control over capital investment strategy, industrial employers in the iron and steel industry were unable to use technological development as an unambiguous means of seizing control of the shop floor.

For several reasons, the American iron and steel industry of the late 19th century offers an excellent opportunity to study the consequences of technological change for that period's industrial workforce. First, the replacement of the puddling process for raw iron refinement by Bessemer steel production was a quintessential technological development of the period.[6] (See fig. 1.) This innovation revolutionized basic metal production in the United States, replacing the traditional, small-batch techniques of the early industrial period with

[6] Although the "puddling" and "boiling" processes for refining pig iron into malleable wrought iron differed slightly in their technical characteristics, these terms may be considered synonymous for present purposes.

coordinated, mass-production technology. While Bessemer steelmaking in the late 19th century never achieved the engineering ideal of continuous processing, prodigious strides were made in this direction.[7] Second, many of the scholars debating the ramifications of technological change for the American industrial workforce have used analyses of the conversion from puddled wrought iron to steel production as the foundation of their theories. For example, Michael Nuwer, Stone, Montgomery, and Brody all founded their arguments on studies of the late-19th- and early-20th-century iron and steel industry.

The conversion of American industry from wrought-iron puddling to Bessemer steel production in the closing decades of the 19th century allowed the mass production of refined iron for the first time.[8] Pig iron, which is the metallic product smelted from iron ore in blast furnaces, contains substantial amounts of nonferrous elements such as carbon, sulfur, silicon, and phosphorus. These substances give cast iron a very brittle character and make it unsuitable for many industrial purposes. Both puddling and Bessemer converting were processes used to remove these components and to produce a malleable metal that resisted shattering. Wrought iron, the product of puddling refinement, is iron which retains almost no carbon and is therefore relatively soft and workable. Steel, on the other hand, is refined iron which generally contains carbon in the range of 0.1 percent and 1.0 percent, is malleable yet stronger than wrought iron, and is more suitable to bearing heavy loads. Although not perfect substitutes, wrought iron and Bessemer steel were competitors in many of the nation's most lucrative metal products markets during the late 19th century.[9]

[7] For an excellent discussion of the transformation from batch-driven wrought-iron production to continuous steel processing, see Nuwer (n. 2 above). While focusing on open-hearth rather than Bessemer steel processing, Nuwer's work vividly describes the vast dissimilarities of the traditional and mass-production operations.

[8] For discussions of the history and processing details of iron puddling and Bessemer steelmaking, see J. M. Camp and C. B. Francis, *The Making, Shaping, and Treating of Steel* (Pittsburgh, 1919); William T. Hogan, *Economic History of the Iron and Steel Industry in the United States* (Lexington, Mass., 1971); Brody, *Steelworkers in America* (n. 4 above); and Nuwer (n. 2 above).

[9] Many scholars writing on the iron and steel industry of this period incorrectly characterize wrought iron and Bessemer steel as perfect substitutes for one another. In his 1985 dissertation, Steven Usselman argues convincingly that steel represented a significant technological advance over wrought iron, particularly in railroad applications. He points out that as early as 1866, tests conducted by the Pennsylvania Railroad indicated that steel rails lasted eight times as long as iron rails but cost only twice as much. This was in addition to steel's considerably superior performance as a construction material. See Steven Walter Usselman, "Running the Machine: The Management

FIG. 2.—Interior of a wrought-iron puddling mill showing (extreme right) a puddler at his furnace; (center) a muck-ball squeezing mechanism; (extreme left) a rolling mill. In a typical wrought-iron works such as Jones and Laughlins, there would be a multiplicity of puddling furnaces. (*Every Saturday*, March 18, 1871, p. 160.)

Wrought-iron puddling, the method of raw iron refinement that characterized the initial phase of industrialization, was a skill-intensive handcraft industry whose technological roots reached into the 18th century. It was a batch-oriented production process carried out in a series of discrete steps that included the refining of the pig iron in puddling furnaces, the initial rolling of the refined metal into long bars, the shearing of the rolled bars into short pieces, the stacking and bundling of the sheared pieces, and the subsequent delivery of these bundles to the finishing mills for reheating and rolling. (See fig. 2.) Between each of these steps, stockpiles of material were created which fed subsequent operational needs and cushioned the impact of isolated breakdowns in the chain of processing. This discontinuous nature of puddling operations, combined with the multiplicity of refining furnaces, encouraged the adoption of atomized management structures within wrought-iron factories. Indeed, most of

of Technological Innovation on American Railroads, 1860–1910" (Ph.D. diss., University of Delaware, 1985).

the skilled workers in puddling mills were "inside contractors" who used company-owned equipment and materials but hired, trained, and paid their own crews.[10]

Wrought-iron production was an exceedingly hot and dangerous operation that both depended on and was limited by the physical strength of the craftsmen who manned the puddling furnaces. Each furnace was attended by two men, the puddler and his assistant, who loaded the furnace with pig iron, manually worked the molten metal with long iron rods as it was "brought to nature," and extracted balls of refined iron from the furnaces. This reliance on human manipulation restricted the output of the furnaces to amounts that could be efficiently handled by one or two men. A seemingly endless stream of inventions throughout the late 19th century sought to remove this constraint by mechanically reproducing the motions of the puddler. All efforts proved unsuccessful, however, due to their inability to mimic the craftsman's judgment and dexterity.[11] Thus, the scale of wrought-iron production could not expand beyond the physical limits of the individual workers. (See figs. 3 and 4.)

Unlike puddling, the more capital-intensive Bessemer steelmaking process was a mass-production technology in which the metal was refined in huge converters, cast into ingots, and delivered to the primary "blooming" mill without being allowed to cool. Whereas the segmentation of puddled-iron production insulated the overall operation from stoppages in any component procedures, a Bessemer steel department could be entirely paralyzed by an isolated breakdown in its more continuous production sequence. The concentration of the central refining function in most Bessemer departments on only two converting vessels compounded this vulnerability. Additionally, the integrated nature of Bessemer steel operations required the concentration of operational direction in the hands of relatively few workers. As a result, the crews of the steel department's three operational areas, the converting shop, the soaking pits, and the blooming mill, were under the immediate direction of a handful of experts who coordinated the entire operation. (See figs. 5 and 6.)

The survival of extensive records for the American Iron and Steel Works of Jones and Laughlins, Ltd. (later the Jones and Laughlin Steel Corporation) offers a unique opportunity to microanalyze the

[10] For an excellent discussion of the history and nature of the inside contract system, see Ernest J. Englander, "The Inside Contract System of Production and Organization: A Neglected Aspect of the History of the Firm," *Labor History* 28 (1987): 429–46.

[11] Brody, *Steelworkers in America* (n. 4 above), p. 8; and Camp and Francis (n. 8 above), pp. 242–43.

FIG. 3.—Puddled-iron ball being removed from the furnace at Ludlum Steel Company, Dunkirk, New York, in the early 1920s. This photograph illustrates the extreme physical demands placed on the puddler and his assistant, working in close proximity to the heat and glare of the furnace. A puddler's output of refined metal was limited by his ability to manipulate the molten bath and its product. (Courtesy of the Smithsonian Institution.)

consequences for the workforce of the technological shift from wrought-iron to Bessemer steel production.[12] One of the few large-scale, fully integrated steelworks that began operations as a traditional wrought-iron plant, this facility was opened originally as the American Iron Works in 1853 by Benjamin Franklin Jones and several partners.[13] Its initial operations included just four puddling furnaces

[12]Jones and Laughlin Steel Corporation Archives, Archives of Industrial Society (AIS), University of Pittsburgh. Records include payroll ledgers, 1861–1901; diary of founder and president Benjamin Franklin Jones, 1875–1901; and memoranda book on properties, 1871–94. The archives also include numerous company histories and publications describing the various departments of the mill.

[13]Of the eleven Bessemer steelworks in the United States in 1896 with an annual production capacity of 300,000 tons or greater, only two were originally opened as puddled-iron plants. These were the Cambria Iron Works of Johnstown, Pennsylvania, and the Jones and Laughlins American Iron and Steel Works. See the *Directory of Iron and Steel Works of the United States* (Philadelphia, 1896).

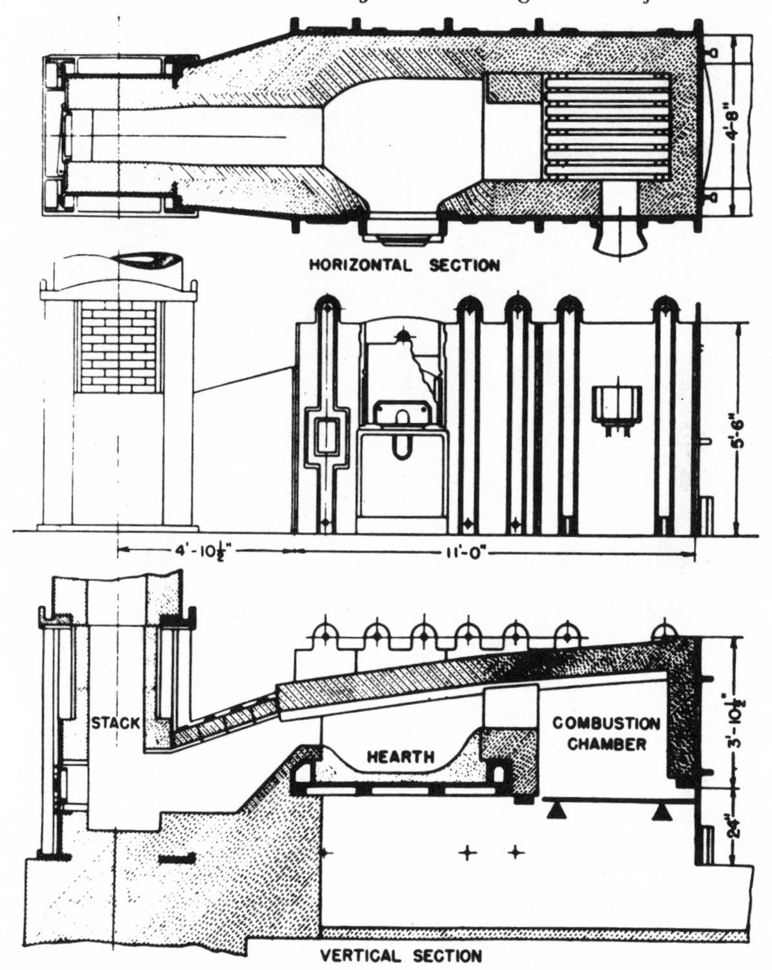

FIG. 4.—Diagrammatic sections and center elevation of a puddling furnace. (*The Making, Shaping, and Treating of Steel,* 6th ed. [Pittsburgh, 1951], p. 357.)

capable of producing 7 tons of wrought iron per day. By 1885, the plant had become the largest ironworks in Pittsburgh and the second largest facility of its kind in the United States, producing annually over 50,000 tons of wrought iron. In that year the works straddled the Monongahela River, with the refining and rolling operations of the South Side plant connected by a railroad bridge to their pig iron source, the smoldering giants of the Eliza blast furnace complex which lined the river's northern bank. The South Side shops alone employed more than 2,200 workers and included seventy-nine pud-

Fig. 5.—Interior of the converting department of Bessemer Mill No. 2, Pennsylvania Steel Company, 1884. The scale and degree of mechanization evident here greatly exceed that of the wrought-iron operation in figure 2. With the Bessemer operation, output was no longer constrained by the physical strength of the workers. (Courtesy of the Hagley Museum and Library.)

dling furnaces; thirty heating furnaces; nineteen rolling mills; nail, chain, and bolt factories; a cold rolling mill; a foundry; sheet and plate iron mills; more than forty steam engines; and a coal mine.[14]

In August 1886, the American Iron Works entered the age of steel with the erection of an integrated Bessemer steelmaking department. This facility was located in the South Side plant and included two 7-ton Bessemer converters, a bank of soaking pits, and a blooming mill designed to roll steel ingots. Steel production at the site grew rapidly, first as a complement to puddled-iron production, but soon as its replacement. In its first full year of operation, 1887, the Besse-

[14] During this period, Jones and Laughlins operated a coal mine whose portal was on the face of a steep hillside less than a mile from its American Iron and Steel Works. This mine was connected to the works by a railway that passed through the residential neighborhood adjacent to the plant. See diary of B. F. Jones, May, 13, 1884 (AIS); Charles Longenecker, "Jones and Laughlin Steel Corporation: In the Service of the Country for Almost a Century," *Blast Furnace and Steel Plant* (August 1941), pp. 858–93; A. L. Holley and Lenox Smith, "American Iron and Steel Works," *Engineering* 26 (November 1, 1878): 347.

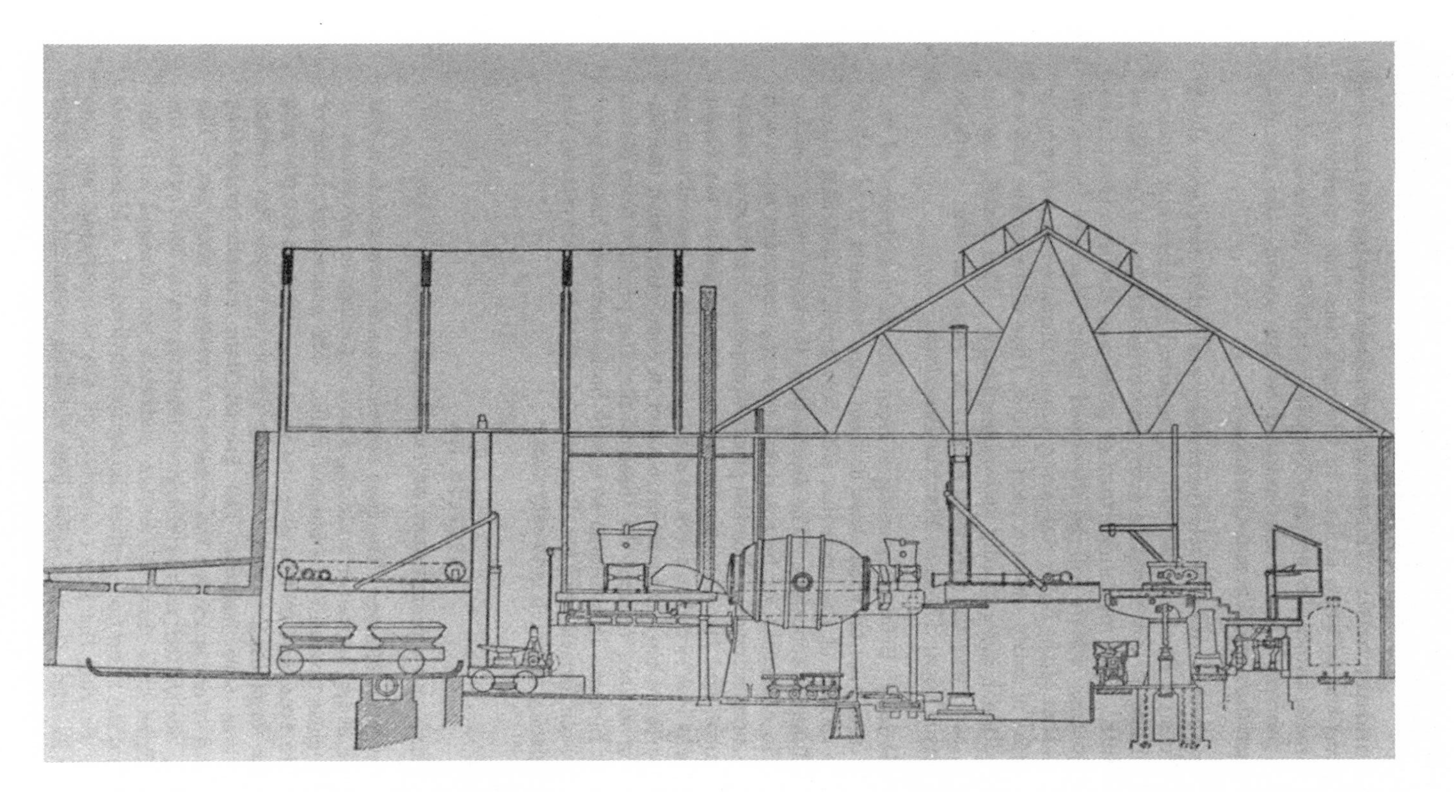

Fig. 6.—Cross-section of a Bessemer converting plant. Bessemer steel departments were designed to facilitate the flow of materials through the process steps. A central objective in this design effort was to overcome the discontinuous nature of the operation and achieve continuous mass production. (J. M. Camp and C. B. Francis, *The Making, Shaping, and Treating of Steel,* 2d ed. [Pittsburgh, 1920], p. 178.)

mer department produced 63,397 tons of steel; by 1895 annual production had risen to 279,271 tons.[15] Wrought-iron production, meanwhile, peaked at 56,861 tons in 1886, remained stable until 1892, and then fell precipitously.[16] In late 1895 the company announced that, as of January 1, 1896, it would produce only steel products.[17] As the last puddling furnaces were dismantled on February 8, 1896, the transformation of the plant from the old iron puddling technology to the new Bessemer steelmaking process was complete.

The American Iron and Steel Works records used most extensively for this study have permitted a close analysis of the changing skill, wage, and productivity characteristics of the mill's workforce as operations were transformed during the years between 1885 and 1896.[18]

[15] Diary of B. F. Jones, January 3, 1888, and January 7, 1896.

[16] Ibid., various dates.

[17] Ibid.

[18] The payroll ledgers included in the Jones and Laughlins archives were the primary data source for the present study and are organized into two-week pay periods. For each of these periods, the workers were listed according to the department or functional areas in which they worked. Every worker had a separate entry which recorded his or her occupation, the number of days or hours worked (or tons produced for tonnage workers), wage rate, gross amount earned, amount the worker owed the company (for rent, supplies, etc.), and net amount of pay. (All monetary amounts for the study's various periods were converted into constant 1880 dollar amounts in order to allow accurate comparison.) Although rich in information, the payroll records suffer from several deficiencies. First, the collection of ledgers recovered by AIS is not complete but, instead, includes a random sampling of departments and pay periods over the 1861–1901 period. As such, it was possible to construct full workforce profiles of the departments being analyzed only during a discrete number of pay periods. Fortunately, the records available were more than adequate for the purposes of the present study. They allowed the analysis of the department during the years immediately prior to introduction of steelmaking and the Bessemer converting department during nearly all of its first ten years of operation. A second weakness of the records is their failure to include any payroll records for the iron puddlers or their assistants. The puddlers were inside contractors who were paid according to uniform tonnage production rates firmly established in annual contract negotiations held between the workers' union, the Amalgamated Association of Iron and Steel Workers, and a manufacturers' association. In order to overcome this imperfection in the payroll records, the following system was used to estimate the wages earned by the puddlers and their assistants. The October 24, 1887–November 19, 1887 pay period is used as an example: (1) The amount of wrought iron puddled during a given pay period was taken from B. F. Jones's diary and divided by the standard tonnage produced by one puddler in one shift: 4,258 tons / 1.1 tons = 3,870.9 shifts worked in the pay period. (2) The number of shifts worked was divided by the number of days in the pay period to determine the number of crew shifts worked per day: 3,870.9 / 23 days = 168 shifts per day. Thus, during this period there were, on average, 168 puddlers employed at the works. (3) The average pay for each puddler was estimated using information supplied by the Jones diary. This document defines the pay of the assistant puddle boss, for whom pay amounts are recorded in the available ledgers, to be equal to the

In order to focus on the effects of technological change, the scope of this study was limited to the production segment that was most directly and dramatically altered by the advance of iron-refining technology. This segment included the refining of pig iron, either by the puddling or Bessemer converting process; the primary rolling operations that produced semifinished metal pieces; and the subprocesses attendant to these two operations. The semifinished pieces thus produced would subsequently have been delivered to one of the diverse finishing departments of the plant for further processing into final products.

On review of the Jones and Laughlins records, the most apparent effect of the adoption of Bessemer technology was a vast improvement in the productive efficiency of the plant. For example, figure 7 demonstrates the wide disparity in the amount of labor required to produce a ton of metal between the puddling and Bessemer converting operations. Even in the earliest months of steel production, the man-hours required to produce a ton of semifinished metal were less than half those required in the wrought-iron operation. By early 1892, only 3.0 man-hours of labor were required for each ton of steel produced, compared to 29.7 man-hours required for each ton of wrought iron in 1887.

Simple efficiency in labor use does not tell the entire story because of the massive capital expenditures required for the Bessemer operation. This raises the issue of whether heightened capital costs may have exceeded labor savings and thus resulted in a net loss of total factor productivity. A comparison of total factor productivity indexes for the two processes helps to answer this question and further demonstrates the superior efficiency of the Bessemer technology. These indexes are computed by finding the ratio of metal output to the

---

average puddler's pay for that period. The assistant puddle boss's gross pay amount was therefore used as the average two-week income for puddlers. (4) The income of puddlers' assistants was calculated using the industry standard for the percentage of a puddler's wages he passed on to his assistant: puddler's income × (1/3 + 5%) = assistant's share. Finally, while incomplete in their coverage of the entire period, the payroll records generally allowed the study of several months of pay periods within a given year. It was therefore necessary to select the specific two-week pay periods to be used in the analysis from among this range of alternatives. The periods actually studied were selected according to the following criteria: (1) The periods studied must be representative of the years from which they were drawn. Primarily, the amount of metal produced during the period should be near the average week's production for that year. (2) No strikes or unusual events should have altered production patterns or characteristics during the periods chosen. (3) The periods should be chosen such that they are complemented to the greatest extent possible by supporting information from the Jones diary and other archival sources.

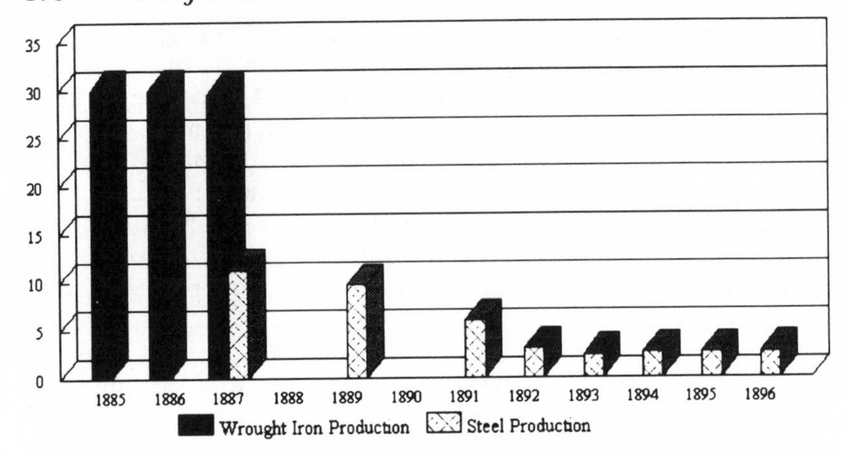

FIG. 7.—Iron and steel production at Jones and Laughlins, 1885–96: man-hours per ton of metal.

average amounts of labor and capital used in the two operations. In this way, they allow a comparison of the total efficiency of production in the two technologies given their usages of both capital and labor, rather than of labor alone.[19] The index values of total factor productivity calculated for iron and steel production at Jones and Laughlins

[19] "Total" factor productivity analysis is used by economists to measure and compare efficiency in production. When used to compare the productive efficiency of two different products, such analyses generally focus on the relative levels or rates of change for the two products in order to avoid comparing apples to oranges. In the present study, however, total factor productivity indexes are used to illustrate the efficiency advantages of Bessemer steel production over wrought iron. Although the two products studied were not perfect substitutes, the similarities in performance and usage between wrought iron and Bessemer steel are of sufficient degree that absolute levels are useful for comparative purposes. While the index numbers presented must be regarded as rough estimates, they are sufficiently accurate for the purpose at hand. The geometric indexes of total factor productivity used in this article were calculated by finding the ratio of metal produced to the average amount of inputs employed as follows:

$$\text{Index Value} = \frac{Q}{L^{al} \times K^{ak}}$$

where Q = output of metal, L = input of labor in man-hours, K = value of capital utilized, and al, ak = shares in value of output corresponding to labor and capital. The values for Q in the four calculations performed correspond to the tons of semifinished metal produced by the respective operations during the four pay periods analyzed. The values for K correspond to the estimated values of plant and equipment used in the respective operations. For wrought-iron production this equaled the number of furnaces available multiplied by $675 (the cost of new furnace construction in are: iron, October 1887, .022; steel, March 1887, .012; steel, August 1891, .030; steel, November 1892, .066.

These values indicate that the steel department, in its earliest years of operation, was not as efficient as the iron department in its combined use of labor and capital. This is because the enormous capital cost of the Bessemer operation swamped its relatively low output during the earliest years of production. As output accelerated toward full capacity, however, the efficiency of the steel plant equaled and then approximately tripled that of the older operation. It should also be noted that even this disparity underestimates the superiority of Bessemer production since archival limitations prevent the inclusion of fuel costs in the factor productivity analysis. Unlike Bessemer converters, puddling furnaces consumed large quantities of coal, which made fuel expenditures a considerable portion of their total costs. The absence of these costs from the present analysis causes an overstatement of the efficiency of puddling and, therefore, an understatement of the disparity between wrought-iron and steel production efficiency.

How then did the shift from iron puddling to Bessemer steelmaking at Jones and Laughlins affect the skill content of the workforce? In order to answer this question, skill index numbers, on a scale from 0 (unskilled) to 10 (highest skill), were assigned to each of the occupations engaged in the wrought-iron and Bessemer steel production segments. These index numbers were assigned according to the number of months of training required to gain a proficiency at each occupation, with the index numbers for both wrought-iron and steel production corresponding to six-month intervals. A skill value of 0, for example, represents occupations which required less than six months of training, while a skill value of 1 would represent jobs requiring from six to less than twelve months of training. Each successive skill index number corresponds to subsequent six-month intervals with the highest assigned skill value, 10, indicating an occupation which required sixty or more months of training.[20]

1887, per Jones's diary) plus $75,000 (per the properties memoranda book). *K* for the Bessemer department was taken from *Iron Age* (August 19, 1886), which estimated the cost of Jones and Laughlins's new Bessemer steel plant to have been $300,000. The share of output value attributable to labor, *al,* was calculated using total production cost estimates for wrought-iron and steel production taken from B. F. Jones's diary. The company estimated that labor accounted for approximately 30.5 percent of total wrought-iron production costs, while the same statistic for steel was 21.6 percent. The shares of output value attributable to capital, *ak,* in the two operations are assumed to be the complements (1 - *x*) of the respective *al* percentages.

[20]The association of six-month training intervals with consecutively numbered skill levels introduces some degree of distortion into the workforce skill measurements.

In order to determine the appropriate training periods for each occupation, and thereby to make accurate skill index assignments, the content of all wrought-iron and Bessemer steel occupations was defined using descriptive sources such as technical manuals, trade journals, and labor publications. Training period assignments were then made based on these definitions and their supporting data. In the case of the steel department occupations, these sources were supplemented by steel industry job descriptions created in 1943 by the American Iron and Steel Institute (AISI) for the United States government.[21] These AISI records both defined job responsibilities and estimated the training periods necessary to attain proficiency in all steel industry occupations in order to establish occupational draft deferment priorities during World War II. The information contained in the records has proven suitable for the present study's purposes due to the absence of fundamental change in the Bessemer steelmaking process during the first decades of the 20th century.

The assignment of skill values for puddlers and their assistants provides an example of the procedure followed for this analysis. The descriptive information concerning the content of these occupations was gleaned both from contemporary sources and from more recently written secondary works.[22] Following this survey of sources, a typical puddler's career path was determined to be such that in his youth, he would often serve from one to two years as a second helper to an older puddler who was nearing the end of his career. The individual would then work from three to five years as a first assistant,

---

The present method of analysis implies a linear relationship between months of training and the amount of skill acquired—that the amount of skill acquired in the last month of an individual's training was equal to the amount gained in the first month. Learning curves, however, are not linear; the rate of learning declines as training proceeds. As such, the use of a linear learning function in this analysis tends to overstate the skill levels of both wrought-iron and Bessemer steelworkers at the upper end of the skill range. Since this distortion is consistent across the two technologies, it does not substantially impair the comparison of the respective workforces. It would, however, be an obstacle for scholars seeking to compare the skill levels of workers within each work group.

[21] American Iron and Steel Institute, *Iron and Steel Industry Position Descriptions and Classifications* (New York, 1943).

[22] Contemporary sources include James J. Davis, *The Iron Puddler: My Life in the Rolling Mills and What Came of It* (New York, 1922); Jesse S. Robinson, "The Amalgamated Association of Iron, Steel, and Tin Workers," in *Johns Hopkins University Studies in Historical and Political Science,* ser. 38, no. 2 (Baltimore, 1920); and Camp and Francis (n. 8 above). Secondary sources include Brody, *Steelworkers in America* (n. 4 above); Montgomery, *House of Labor* (n. 4 above); and John William Bennett, "Iron Workers in Woods Run and Johnstown: The Union Era, 1865–1895" (Ph.D. diss., University of Pittsburgh, 1977).

TABLE 1

WROUGHT-IRON PRODUCTION AT JONES AND LAUGHLINS, FOR THREE PAY PERIODS

Pay Period	Days in Operation	Number of Employees	Hours per Day	Total Output (in Tons)	Average Wages (in 1880 Dollars)	Mean Skill Level
July 6, 1885–July 20, 1885	12.0	405	11	1,776	11,536.23	3.85
October 25, 1886–November 8, 1886	12.0	413	11	1,812	12,241.64	3.82
October 24, 1887–November 7, 1887	12.0	499	11	2,220	15,867.22	3.81

accumulating the enormous skill and experience necessary to master the craft. At this point, the worker was usually ready to assume direction of his own furnace and would do so as the opportunity presented itself. Thus, the skill values assigned to these occupations were 2 (from twelve to eighteen months of training) for puddlers' assistants and 8 (from four to four-and-one-half years of training) for puddlers. This occupational life course was particularly evident in the years constituting the present study, which were the heyday of wrought-iron production in Pittsburgh. During the two decades following 1880, Bessemer steel production expanded rapidly and, in relative terms, eclipsed that of wrought iron. However, puddled-iron production continued to grow nationally in absolute terms and did not peak until 1902. As John Ingham notes, "Wrought iron was still king in Pittsburgh in the mid-1890s, even though the output of the Carnegie works alone exceeded that of all thirty-one iron mills combined."[23]

The payroll and miscellaneous records available for the Jones and Laughlins plant allow a comprehensive analysis of the wrought-iron workforce during three two-week pay periods: July 6, 1885–July 20, 1885, October 25, 1886–November 8, 1886, and October 24, 1887–November 7, 1887. During these periods the puddled-iron production segment embraced 28 occupations and employed between 405 and 499 men on two 10- to 11-hour shifts. (See table 1.) Although workforce skill profiles were constructed for each of the pay periods, the skill characteristics of the crews did not change appreciably, and the three profiles were virtually identical. These skill profiles are represented in figure 8 by that for the workforce of October 1887.

The wrought-iron skill profile presented in figure 8 corresponds closely to the descriptive accounts of puddling department operations

[23]John N. Ingham, *Making Iron and Steel: Independent Mills in Pittsburgh, 1820–1920* (Columbus, Ohio, 1991), p. 77.

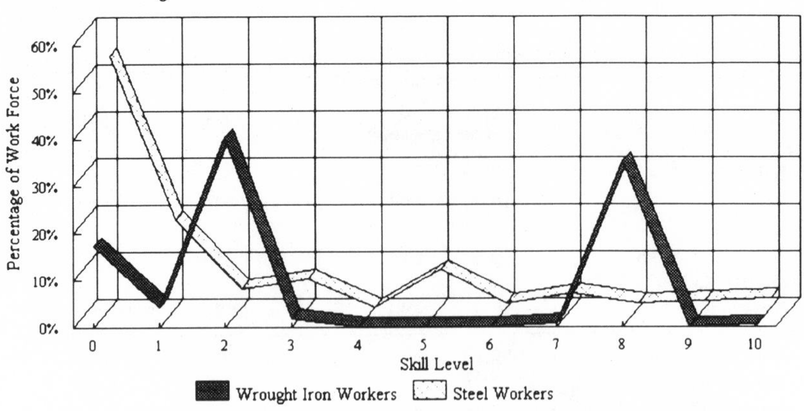

FIG. 8.—Jones and Laughlins 1887 iron and steel workforce skill profiles

during the 19th century. The profile is distinguished by large percentages of workers in both high-skill and lower semiskilled ranges. These peaks reflect the numerical prominence in the operation of the skilled puddlers and their semiskilled assistants. During the October 1887 pay period, for example, 336 of the 499 workers in the department fell into one of these occupations. Of the remaining workers, the preponderance was composed of virtually unskilled laborers whose occupations, such as ash wheelers, drag downs, and metal breakers, required intense manual labor under the most brutal of conditions.

In contrast to the puddling department, where more than one-third of the workers were highly skilled, the analysis of the steel department workforce indicates that a much smaller portion of that group commanded substantive manual skills. The archives available for the Jones and Laughlins steel department allowed the evaluation of this work group over two-week pay periods in nearly every year from 1887 to 1896.[24] (See table 2.) During these periods there were between 86 and 97 occupations involved in the steelmaking operations. In the first two periods (1887 and 1889) there were 195 and 203 men working on two 11-hour shifts. The workforces of the later periods averaged 335 men working on three eight-hour shifts. Despite minor differences, the general skill patterns of the steel workforces were quite similar across the decade and are represented in

[24] The pay periods analyzed are listed in table 2. On subsequent figures the data points for the November–December 1892 pay period are labeled "1893" in order to distinguish them from the March 1892 period.

TABLE 2

STEEL PRODUCTION AT JONES AND LAUGHLINS, FOR EIGHT PAY PERIODS

Pay Period	Days in Operation	Number of Employees	Hours per Day	Total Output (in Tons)	Average Wages (in 1880 Dollars)	Mean Skill Level
January 17, 1887–January 31, 1887	11.0	195	11	2,082	5,259.04	1.57
March 11, 1889–March 25, 1889	11.0	203	11	2,498	7,607.59	1.39
August 1, 1891–August 15, 1891	12.0	356	8	5,662	16,618.84	1.65
March 14, 1892–March 28, 1892	11.0	353	8	10,238	20,906.55	1.66
November 21, 1892–December 5, 1892	11.0	333	8	12,028	17,819.83	1.81
August 1, 1894–August 15, 1894	9.0	319	8	8,507	8,010.45	1.91
August 1, 1895–August 15, 1895	12.3	317	8	11,692	11,757.76	1.92
January 17, 1896–January 31, 1896	13.3	330	8	12,999	13,728.13	1.98

figure 8 by the skill profile of the steel production crews in January 1887.

As was the case with the wrought-iron analysis, the skill profile generated for the Bessemer department corresponds to both descriptive accounts and the findings of historians such as Brody. Most conspicuously, the general skill content of the new department's crews was substantially lower than than of the puddling crews. While the workforce did include skilled workers, such as Bessemer blowers, cupola foremen, chemists, and blooming mill rollers, these groups made up only a small fraction of the workers. Dominating the occupation structure were jobs such as cinder tappers, pitmen, and blooming mill hookers, occupations that required very little traditional skill. Indeed, almost three-quarters of the jobs in the steel department were in skill ranges 0 and 1, which represent occupations requiring less than twelve months of training. Of particular significance was the complete absence from the steel department of the two occupations that dominated the wrought-iron crews: the puddlers and their assistants. Figure 9, which shows the mean skill levels for the workforces during each of the study periods, further substantiates the impressive skill content differences between the puddling and Bessemer steel operations. While the mean skill level of workers in the puddling plant hovered around 3.8, the mean skill level of the steelworkers never reached 2.0.

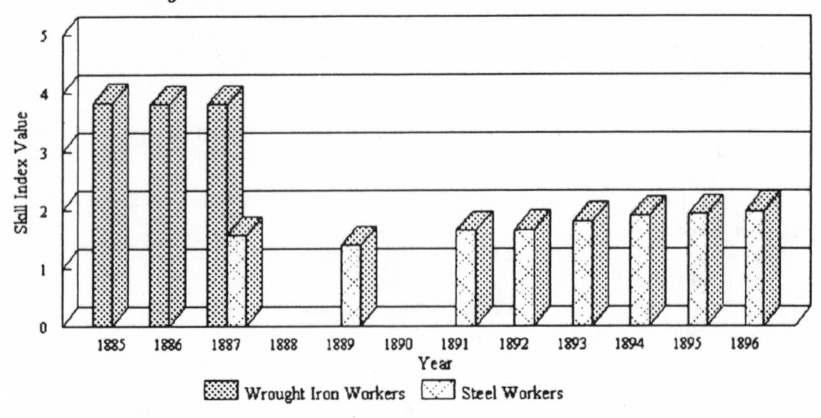

FIG. 9.—Iron and steel production at Jones and Laughlins, 1885–96: average worker's skill level.

At this point a note should be made of the nature of skill analysis performed in this study. While the content of each occupation was evaluated and assigned an index value individually, this aggregative approach presents potential problems. Primarily, it threatens to gloss over important differences in the work environments experienced by puddling mill and Bessemer department workers and to mask the effects of subtle dissimilarities in work structures. For example, the inside contracting form of labor organization was more prevalent in wrought-iron production than in steelmaking and may have had an impact on the shop-floor realities of occupational definitions. Nevertheless, the intention here is to compare the overall skill structures of the respective operations. In this pursuit it seems reasonable that weaknesses in the sensitivity of analysis created by the use of aggregated, quantitative data are overwhelmed by the scale of the skill structure differences identified.

This said, the dramatically reduced skill content of the Jones and Laughlins steel department seems to support the theoretical positions of those scholars who find that technology had a deskilling effect on the industrial workforce. Brody's study of the iron and steel industry before 1929, for example, found that "increasingly, mill workers became semiskilled operatives adept at routine mechanical duties which were not comparable to the virtuosity of hand-rolling steel. . . . The mass of rolling men slipped into the ranks of the semiskilled."[25] The skill profile of the steel workforce shown in figure 8 sustains his argument that the technological leap between wrought-iron and steel pro-

[25] Brody, *Steelworkers in America* (n. 4 above), p. 32.

duction shifted demand from skilled workers toward a combination of a few skilled hands and many slightly skilled.[26] As the figure indicates, the small percentage of skilled steelworkers was in fact vastly overshadowed by the legions of unskilled or slightly skilled workmen.

These results also seem to support the arguments of other scholars who find that technology replaced both day-labor gangs and skilled craftsmen with a vast, undifferentiated class of workers. For example, Stone's analysis of the iron and steel industries of the 1890s found that "the effects of the new technology were to eliminate the distinction between skilled and unskilled workers and create a largely homogeneous workforce," which was "a new class of machine operators known by the label 'semi-skilled.'"[27] She argues that this transformation of labor demand created a homogeneous workforce with inherent tendencies toward strong class-consciousness and that capitalists, responding to this threat of worker unity, fashioned artificial job hierarchies in order to create an illusion of occupational differentiation and encourage divisiveness within the workforce. Richard Edwards sustains this argument in a broader context by finding that employers were, in general, forced to invent artificial job hierarchies within the mass of undifferentiated and unskilled workers who lay in the wake of technological advance. These hierarchies created the illusion of technical differentiation among occupations and legitimized managerial authority.[28]

But is it true that the adoption of Bessemer technology reduced the workforce of the steelmaking department to a homogenized mass of interchangeable operatives who lacked any real shop-floor power? If this were the case, one would expect that the steelworkers' wage levels would have reflected their considerably inferior skill structures. Indeed, by treating the puddling and Bessemer departments separately, the data show that within the pay periods studied there were strong correlations between workers' skill levels and earnings.[29]

However, analysis of workforce wage structures at the American Iron and Steel Works during this period of technological transition indicates that despite dramatic reduction in traditional skill content, steelworkers' earnings were actually higher than earnings in the more highly skilled wrought-iron department. Figure 10 uses box-and-

[26] Ibid., p. 58.

[27] Stone (n. 4 above), p. 33.

[28] Edwards (n. 4 above).

[29] For example, regression analysis of the relationship between occupational skill levels and average two-week wages for wrought-iron and steel production in 1887 yielded R-squared values of .68 and .69, respectively.

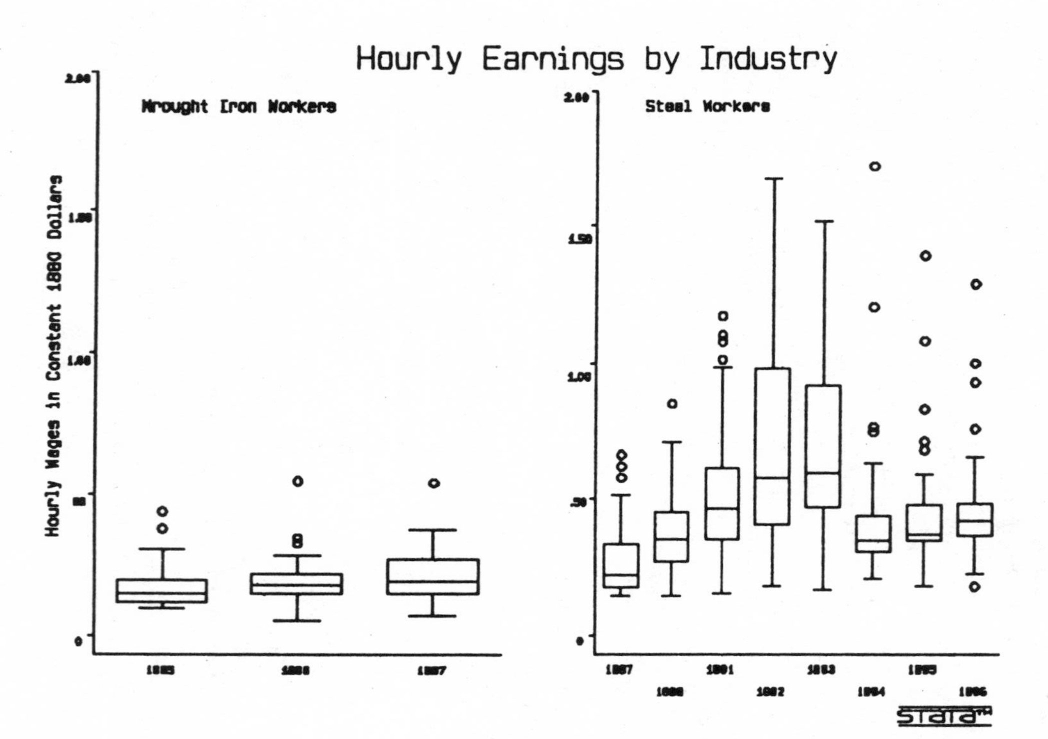

FIG. 10.—Jones and Laughlins's wrought-iron and steelworker hourly wages, 1885–96

whisker graphs to represent the hourly wage structure for each workforce studied during this decade.[30] Each workforce is represented by a box indicating the range of hourly wages that include the middle 50 percent of the workforce (those workers from the 25th to the 75th wage percentiles), with the horizontal line within each box indicating the median hourly wage earned for that pay period. The whiskers attached to each box embrace the range of values included in the normal distribution curves for the work groups. Data points shown as dots located beyond the whiskers are considered outliers. Thus, this figure presents not only the median wage levels for the respective work groups, which form the basis for subsequent wage level comparisons, but also expresses the shape and character of each working group's wage distribution.

As figure 10 indicates, initial median wages in the Bessemer steel crews were roughly equivalent to those earned by workers across the plant in the puddling departments. Between 1887 and 1893, however, a series of forces combined to destabilize the wage structure of the Jones and Laughlins steel department. During the few years immediately following 1887, production in the department accelerated rapidly as the bugs were worked out of the new operation and it was brought to full efficiency. Since many of the workers in the Bessemer department were paid on a per ton or "tonnage" basis, this expansion of output rapidly raised earnings for these workers. Adding to this phenomenon was the stickiness of the tonnage rates paid by the company, which remained virtually constant between 1887 and 1892. The result of these forces is shown in figure 10 by the widening of the workforce wage structures between 1889 and late 1892. After 1892, resetting of tonnage rates and economic depression pushed earnings down from their elevated levels. It is crucial to note, however, that earnings did not sink to a level consistent with the apparently minimal skill content of the workforce. In fact, at the end of the study period in 1896, constant dollar hourly wages in steel production were nearly double those enjoyed in iron production ten years earlier.

Hourly wage rates provide an imperfect medium of comparison between the wrought-iron and Bessemer workers since variations in the length of the work day, the number of days worked per week, and the pay status of various occupations as either wage or tonnage

[30] Constant 1880 wage levels were calculated using the cost-of-living index developed for the United States, 1860–1914, in E. H. Phelps Brown, *A Century of Pay* (New York, 1968), app. 3.

threaten to invalidate simple hourly wage comparisons. Figure 11 corrects for these variables by showing box-and-whisker representations for the two-week earnings structures of the wrought-iron and Bessemer steel workforces. This figure indicates that in terms of median two-week earnings, the steel workers still experienced considerably higher incomes. In fact, despite a shift from eleven- to eight-hour workdays, steelworkers' earnings in 1896 remained substantially higher than those of the more highly skilled ironworkers of the previous decade.[31] Furthermore, the normal shape of the distributions and the limited number of outliers shown by the figure demonstrate that the elevated median earnings of the Bessemer crews did not result from the presence of a handful of extremely well-paid individuals whose compensation levels artificially elevated the average levels.

This additional evidence that steelworkers' earnings did not suffer in proportion to their skill content raises the question of what factors may have caused the seeming incompatibility between changes in Jones and Laughlins' workforce skill and wage structures. One possible explanation for this divergence is that the Bessemer workforce was, for some reason, able to acquire skill at a substantially faster rate than the wrought-iron workforce. The establishment of a concerted training effort at Jones and Laughlins could have caused the steelworkers to gain skill at a much faster rate than their counterparts, thereby invalidating the assumption that the rate at which workers acquired skills was roughly equivalent in the two departments. If it were, in fact, the case that skill in the steel occupations was somehow acquired more rapidly, the workers' actual skill content across the two operations would vary despite similar or equal training periods. This would cause the present analysis to understate the skill content of the steel department workforce. Under such a scenario, the mill's management may have simply increased steelworkers' wages in response to a comparable increase of skill content. However, the extant Jones and Laughlins records make no mention of training programs established to facilitate worker skill acquisition. Despite the subjectivity of the skill analysis, it is unlikely that the rates at which workers acquired skill under the two technologies differed to such a degree as to create the observed disparity in skill content.

[31] It should be noted that throughout the decade-long study period, the workers in both the wrought-iron and Bessemer steelmaking departments worked under contracts negotiated between their union, the Amalgamated Association of Iron and Steel Workers, and their employer. As such, it is reasonable to assume that the wage levels in neither department benefited asymmetrically from union representation and that the wage levels at Jones and Laughlins did not represent an exception to industry norms.

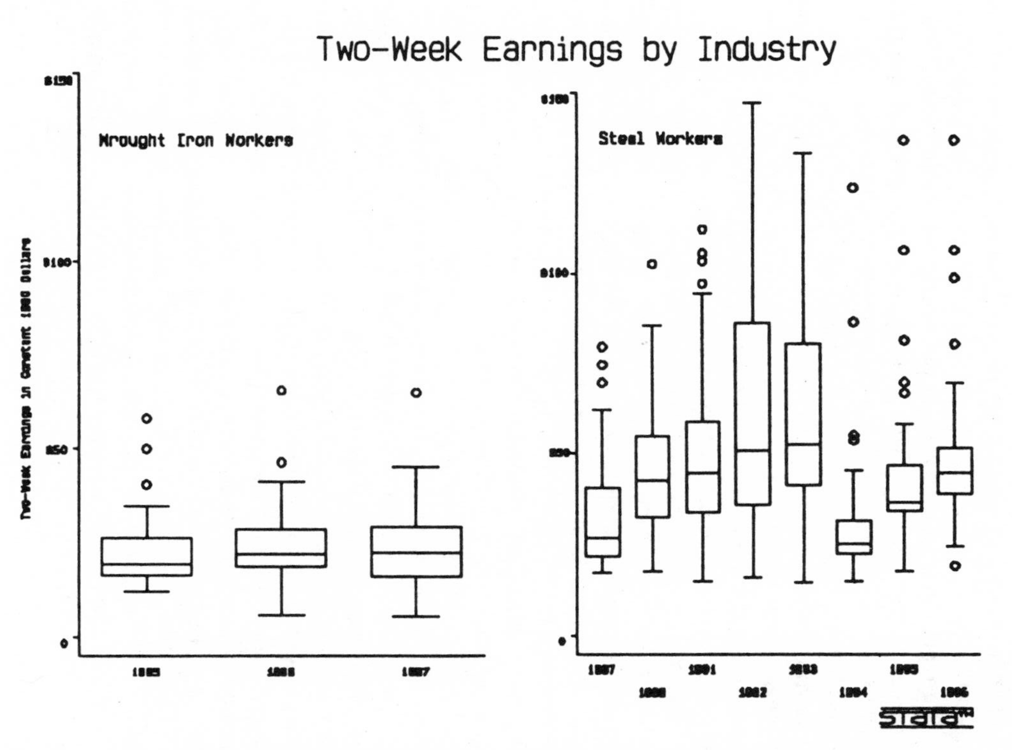

FIG. 11.—Jones and Laughlins's wrought-iron and steelworker two-week earnings, 1885–96

A second explanation of the apparent wage/skill disparity at the Jones and Laughlins mill lies in the potential for temporary labor market disequilibrium caused by the rapid expansion of the American Bessemer steel industry during the decade being studied. In these years, U.S. production of Bessemer steel increased from 1.7 million tons in 1885 to 4.6 million in 1892 and then to 8.4 million in 1899.[32] This expansion of output may well have created a demand for steelworkers that outran the supply of men who were familiar with the new technology. Firms such as Jones and Laughlins, which sought to staff large new operations with experienced workers, would have been forced to lure workers from other firms by offering wages that were in excess of normal returns to the workers' skill levels.

Research into the background of the skilled steelworkers at Jones and Laughlins lends credibility to the argument that many employees for the new plant were hired from distant sources.[33] Of the forty-one skilled workers on the payroll on February 12, 1887, only 17 percent could be located in the city directory for 1885, one year before the steel plant's construction.[34] Three years later, in 1888, 51 percent were found in the directory. This strongly suggests that Jones and Laughlins recruited some of the workers for the new steel operation from firms located outside the Pittsburgh area. Under such a scenario, it would be expected that the Bessemer workforce would enjoy a temporary period of heightened wage levels during the years of most rapid expansion. Following this initial period, steelworkers' wages would have receded to levels that were in closer accordance with skill levels. Research of steelworkers' wages over a longer period would prove useful in testing this possibility, but the records at hand do not offer such an opportunity.

The presence of labor market disequilibrium in the years studied does not, however, explain the observed differences between the skill and wage experiences of the Jones and Laughlins workforce. If the technological pessimists are correct in finding that the mass of workers was relegated to positions that were both unskilled and perfectly interchangeable, a temporary shortage of experienced steel hands would not have appreciably pushed up wage levels. Instead, employers would have been free to draw workers from any and all available sources. Disequilibrium and heightened wage levels would therefore

[32] Hogan (n. 8 above), p. 185.

[33] Skilled workers were defined for this purpose as workers in occupations with a skill index value of 5 or greater.

[34] The city directory used for this analysis was J. F. Diffenbacher, *Directory of Pittsburgh and Allegheny Cities* (Pittsburgh, 1885 and 1888).

indicate some degree of selectivity in hiring and indicate a differentiation of workers according to occupational experience. Yet, the present analysis indicates that the presence of traditional craft skills in the Bessemer department could not have provided the basis for such differentiation. Thus, a contradiction between the skill and wage findings remains and must be explained.

In his article concerning technology and skill content, Nuwer offers a plausible explanation for this apparent discrepancy. His research, unlike that of Stone and Edwards, finds that although the adoption of mass-production technology diminished the demands for craft-type manipulative skills, workers' occupations retained technical and functional distinctiveness.[35] Nuwer's thesis incorporates two component theories, the idea of "strategic workers" and that of "diagnostic skills," and applies these theories to the experience of the late-19th-century industrial workforce. First, Nuwer's research demonstrates the usefulness of the "strategic worker" concept in understanding the consequences of mass-production technology for the workforce. Here, he builds on the works of William Lazonick and Bernard Elbaum who argue that mechanization and the adoption of integrated mass-production techniques increased employers' dependence on the "reliability, dependability, attentiveness, and loyalty" of workmen.[36] Elbaum and Frank Wilkinson find that "although training and skill requirements may not have been great for machine-tending jobs, responsibility for machinery and materials could be considerable, as high throughputs in bigger plants increased the costliness of errors. Top production hands in both the U.S. and U.K. retained some supervisory authority even after subcontracting had been abolished, and thereby retained strategic importance. In these circumstances bargaining leverage of groupings of workers could well increase even as the skill and supervisory status of top hands declined."[37]

Of particular importance here are the heightened operational responsibilities assigned to steelworkers and the multiplication of error costliness. As noted above, wrought-iron production was a discontinu-

[35] Nuwer (n. 2 above).

[36] William H. Lazonick, *Competitive Advantage on the Shopfloor* (Cambridge, Mass., 1990), and "Technological Change and the Control of Work: The Development of Capital-Labour Relations in U.S. Mass-Production Industries," *Managerial Strategies and Industrial Relations: An Historical and Comparative Study,* ed. Howard F. Gospel and Craig R. Littler (London, 1983), p. 112, quoted in Nuwer (n. 2 above), p. 811.

[37] Bernard Elbaum and Frank Wilkinson, "Industrial Relations and Uneven Development: A Comparative Study of the American and British Steel Industries," *Cambridge Journal of Economics* 3, no. 3 (September 1979): 284.

ous operation which incorporated a large number of small-scale furnaces and roll stands. This compartmentalized system meant that isolated breakdowns were relatively inexpensive and did not threaten overall production. In the capital-intensive Bessemer steel department, however, a single worker's miscue could arrest the carefully synchronized production stream. For example, the error by even a slightly skilled "hooker" in the blooming mill could have sent a huge bloom crashing into a mill stand, wreaking tens of thousands of dollars worth of damage.

Scholars who find that growing capital-intensity inherently eroded workers' bargaining positions often fail to recognize the significant degree of leverage workers might have gained from their critical roles in industries such as Bessemer steelmaking. The argument is frequently made that technological change during the late 19th and 20th centuries substantially eroded workers' bargaining positions by eliminating the leverage of the skilled craftsmen and reducing the workforce to a virtually helpless mass of interchangeable parts. Such arguments overlook the vulnerability of continuous, synchronized mass production to disruptive efforts on the part of a unified workforce. Even though the preponderance of steelworkers did not perform highly skilled, autonomous functions, a worker's mistake, whether intentional or otherwise, could paralyze an extensive portion of the productive organism and prove extremely costly to management. As was the case with the blooming mill hooker, it was within the power of any single disgruntled employee or an organized group of workers to inflict severe damage on the employer. Therefore, the maintenance of high wage levels in the Bessemer department may have been an effort by management to compensate workers for their higher degree of responsibility and to forestall any efforts to take advantage of their strategic roles in the mill's operation.

Nuwer complements this concept of strategic workers by incorporating Larry Hirschhorn's idea of "diagnostic skills" in his argument for occupational differentiation.[38] Nuwer finds that technological change in the steel industry created a new demand for informal skills which allowed workers to orchestrate their attention and dexterity in the fast-paced continuous processing environment. He argues that technology "'deskilled' craft occupations in the sense that manipulative skill was greatly reduced" but that "the integrated system of machines relied on work-force skills that involved the diagnoses of product flows. Integrated production, such as that in the steel industry, 'reskilled' industrial occupations in the sense that diagnostic skills

[38] Hirschhorn (n. 2 above).

were increased."[39] These informal diagnostic skills, which were not reflected in the months of training required for each job, "strengthened the bargaining position of the workforce, even when unorganized, and thereby restricted managerial prerogative over control and coordination."[40]

Scholars who believe technology's effects to have been generally degrading, however, would find Nuwer's hypothesis unconvincing and argue that evidence from the changing conduct of labor relations in the iron and steel industry during the late 19th century militates against the significance of diagnostic skills in strengthening the workers' bargaining position. Brody, for example, argues that the importance of skill in wrought-iron production made strikes a waiting game for both labor and management. Faced with a stoppage by puddled ironworkers, management's strategy was to try to hold out until the internal divisions of the workforce eroded worker cohesiveness and caused the strike's failure. In the new environment of the steel plants, however, Brody finds that employers quickly realized that they could restart stricken facilities by using inexperienced men led by a handful of loyal skilled workers.[41] The Carnegie Steel Company, for example, followed just such a strategy in 1890 at its Duquesne Works, which incorporated revolutionary continuous rolling technology and synchronized processing.[42] Here management was able to defeat a strike by operating the plant with a staff of purportedly inexperienced workers. Two years later, Carnegie Steel was likewise able to use replacement workers in winning the decisive Homestead strike.

However, such arguments do not demonstrate the ineffectiveness of strategic occupations and diagnostic skills, as opposed to manipulative craft skills, to serve as a foundation for worker organization and bargaining leverage. The use of inexperienced workers as strikebreakers had long been a tradition among 19th-century iron masters who repeatedly employed inexperienced puddlers' helpers to replace striking puddlers. In this strategy, employers were assisted by the fact that the helpers were often excluded from union membership by their more highly skilled brethren. Thus, the tactics of the Carnegie

[39] Nuwer (n. 2 above), pp. 837–38.

[40] Ibid., p. 811.

[41] Brody, *Steelworkers in America* (n. 4 above), p. 58.

[42] Joel Sabadasz, "Duquesne Works: Overview History," unpublished research document prepared for the Historic American Engineering Record, summer 1991. Sabadasz found that the mill in 1890 "not only employed the latest in existing steel mill technology, but also introduced new, more powerful mill engines which made it possible to roll rails directly from the soaking pits without any intermediate reheating steps" (p. 3).

interests at Duquesne and Homestead can be seen as an extension of patterns of struggle from one technological environment into a new one.

What was different at Duquesne and Homestead was the nature of organization needed for the workers to translate their strategic positions and diagnostic skills into effective bargaining leverage. As the Jones and Laughlins analysis indicates, a traditional, craft-based union of a single trade, puddlers and their assistants, could organize two-thirds of the primary operations' workforce and thus exert considerable pressure. In the far more occupationally diverse Bessemer department, however, an organization structured around a discrete skill or group of craft skills could not hope to harness more than a small fraction of the workforce. The workers at Homestead realized this by 1889 and confronted the steel company with a labor organization that cut across skill and occupational lines.[43] This threat to production at the enormously expensive Homestead works elicited a strategy by the company that was indeed novel: the application of overwhelming state power. Only through the use of state militia and court measures was Carnegie Steel able to bring into play the old formula of strikebreakers in 1892. In fact, the employment of state military and judicial power, not the use of replacement workers, marks a shift in employers' labor strategy during the late 19th and early 20th centuries. Furthermore, it seems reasonable to conclude that employers resorted to this strategic change not because their technological strategy had crushed the skill-founded power of the workers, but, at least in part, because the installation of capital-intensive mass-production technologies had heightened their perceived vulnerability to concerted labor action.

In conclusion, it may be fairly asked of this study, "so what?" How useful for historians of technology is the conclusion that technological change in American ferrous metal refining during the late 19th century did not, through skill destruction, eliminate the potential for worker bargaining leverage? Perhaps quite useful. Many historians of technology would agree that the course of technological development is not randomly determined but is a social process in which outcomes are the result of interaction among a variety of forces.[44] It

[43] Paul Lewin Krause, "The Road to Homestead" (Ph.D. diss., Duke University, 1987).

[44] This argument has been most fully elaborated in studies of the social construction of technology. See, e.g., Wiebe E. Bijker, Thomas P. Hughes, and Trevor Pinch, eds., *The Social Construction of Technological Systems: New Directions in the Sociology and History of Technology* (Cambridge, Mass., 1987).

seems apparent that a wide range of factors, including political and cultural circumstances, the relative influence of pertinent social groups, and the momentum and trajectory of technological development, must all be analyzed as one seeks to understand the emergence of a particular technological system.[45]

This article has sought to address a question that lies at the intersection of the history of technology and labor history: to what extent can a particularly powerful or well-organized interest group manipulate the course of technological change so as to serve best its own narrowly defined interests? In the case of the conversion from puddling to Bessemer steel production, a prominent group of historians has argued that industrial capitalists were able to control the course of innovation and investment so as to exorcise workers' control from the shop floor. The findings of the present article contradict this interpretation. While the productivity advantages of Bessemer steelmaking were clear to industrialists, the implications of the new technology for labor relations were ambiguous at best and presented a balance of new opportunities checked by increased vulnerabilities. It remains for historians to understand better the ways in which social groups reacted to the uncertainties created by innovation and the ways in which these reactions led to unanticipated yet profound social changes.

[45] For a discussion of "momentum" in technological development, see Thomas P. Hughes, *Networks of Power: Electrification in Western Society, 1880–1930* (Baltimore, 1983), pp. 15–16. The idea of technological "trajectory" is developed in Giovanni Dosi, "Technological Paradigms and Technological Trajectories," *Research Policy* 11 (1982): 147–62.

Technology and the Market: George Eastman and the Origins of Mass Amateur Photography

REESE V. JENKINS

> The manifest destiny of the Eastman Kodak Company is to be the largest manufacturer of photographic materials in the world or else to go to pot. [GEORGE EASTMAN, 1894][1]

From the time of the introduction of commercial photography in 1839 until the late 1870s, the technical complexities of the photographic process were so great that only professional photographers and a very few avid amateurs chose to pursue the practice. In the 1870s the photographer had, for example, to prepare the photosensitive materials; adjust the camera settings; expose, develop, and fix the glass-plate negative; and print and fix the positive paper copy. Twenty years later nearly anyone interested in obtaining photographs, regardless of his practical knowledge of optics or of photographic chemistry, could at least press the button on a simple hand camera, remove the exposed film from the camera, and in a few days obtain finished prints from a local photographer or distant factory. The change in practice of photography from the dominance of the professional to that of the amateur revolutionized both the photographic industry and the social role of photography. The market and organizational revolution which occurred in the photographic indus-

DR. JENKINS is currently a professor in the Department of History at Rutgers University. He provides the following acknowledgments: The Eastman Kodak Company of Rochester, New York, has generously made available to me its extensive collection of George Eastman correspondence and other historical materials, and a number of people associated with the company have provided valuable assistance, in particular, William S. Vaughn, Thomas Robertson, Dolores Stover, Donald Ryon, and Gail Freckleton. The research for this paper and for my general study of technology and the photographic industry was supported by the National Science Foundation (GS-1726), the Newcomen Society of North America, the Graduate School of Business Administration of Harvard University, and the Division of Humanities and Arts of Case Western Reserve University.

[1]George Eastman to Henry A. Strong, December 20, 1894, George Eastman Correspondence, Office of Corporate Information, Eastman Kodak Company, Rochester, N.Y. (hereafter cited as GEC).

try between 1880 and 1895 was, in contrast to much of American business, less a response to the growth of the national urban market[2] than it was an outgrowth of a series of interrelated changes in technology and marketing which culminated in a changed conception of who was to practice photography. Although several persons in Europe and America contributed to the initial phases of the revolution in the practice of photography, the Rochester entrepreneur George Eastman stood at the heart of the subsequent key conceptual and technical changes which created the mass amateur market and played the commanding role in the consequent transformation of the industry.

Because of the influence of the historical interpretation of the romantic-heroic school, the linking of the name of George Eastman with the origins of mass amateur photography and of large-scale enterprise in the photographic industry will surprise no one. However, despite the many books and articles on the history of photography[3] and even on George Eastman,[4] the precise nature of his early work and the motivations behind his activities in photography have not been adequately pursued. In the literature treating Eastman's work, he is portrayed as seeking "simplification" of photography without regard to the meaning of "simplification" and without sensitivity to a broader range of motivational issues. Therefore, his role must be reexamined within the context of the state of the art and of the business climate of the time, focusing on the fundamental steps which led from the practice of wet-collodion photography by a small number of professionals to the practice of gelatin roll film photography by millions of novices.

[2]Alfred D. Chandler, Jr., has convincingly argued for the importance of the growth of the national urban market for American industry in general in his classic paper "The Beginnings of 'Big Business' in American Industry," *Business History Review* 33 (Spring 1959): 1–31.

[3]Among the most important general histories of photography are Helmut Gernsheim and Alison Gernsheim, *The History of Photography from the Camera Obscura to the Beginning of the Modern Era* (New York, 1969); Beaumont Newhall, *The History of Photography from 1839 to the Present Day* (New York, 1964); Robert Taft, *Photography and the American Scene: A Social History, 1839–1889* (New York, 1938); and Josef M. Eder, *History of Photography* (New York, 1945).

[4]The best discussion of Eastman's technological endeavors is Beaumont Newhall's "The Photographic Inventions of George Eastman," *Journal of Photographic Science* 3 (March–April 1955): 33–40. Other studies which refer to Eastman's technical activities include: O. N. Solbert, "George Eastman," *Image*, vol. 2 (November 1953); Donald C. Ryon, "Development of the No. 1 Kodak Camera," in *The Photographic Historical Society Symposium, September 19–20, 1970* (Rochester, N.Y., 1970), pp. 19–31; and Carl W. Ackerman, *George Eastman* (London, 1930).

I. *From Wet-Collodion to Dry-Gelatin Plates*

The wet-collodion process, which in the middle 1850s had replaced the popular daguerreotype process, employed a solution of nitrocellulose, called collodion, as a carrier for photosensitive halogen silver salts. The photographer prepared the negative photosensitive material by flowing the halogen-salted collodion onto the glass plate at the site of the picture taking. Because of the perishability of the photosensitivity as the collodion dried, the photographer prepared the negative material just prior to exposure. He also produced his own photosensitive print paper. Hence, every photographer served not only as the camera operator but also as a decentralized handicraft producer of photosensitive materials and of finished positive prints.

Photographers sought throughout the 1860s to produce dry collodion or some other relatively imperishable carrier for the photosensitive material. During the 1870s several British photographers[5] succeeded in employing a highly sensitive substitute for wet collodion: dry gelatin. By the late 1870s the British photographic journals were publishing gelatin formula information and, thereby, facilitated the change from collodion to gelatin. The dry-gelatin carrier not only heightened the photosensitivity of the silver salts but preserved the sensitivity for many months, allowing specialized factory production of the most critical material in photography. In the late 1870s several gelatin-dry-plate companies began production in Britain, and by 1880 at least four American firms were engaged in the commercial production of such negative plates: Cramer and Norden of Saint Louis; John Carbutt of Philadelphia; D. H. Cross of Indianola, Iowa; and George Eastman of Rochester. Soon Cramer, Carbutt, and Eastman attained commanding positions in the infant industry because of their superior emulsions and their exclusive sales agencies with the three national photographic jobbers: Anthony, Scovill, and Gennert.[6] One of these producers, Eastman, not only pioneered in the introduction of this new technology but ultimately prompted an even greater revolution in photography.

George Eastman (1854–1932) entered the photographic business in 1880 after less than three years of pursuing photography as a serious amateur. The son of the deceased owner and operator of a Rochester business school, Eastman had eight years of public and private schooling, had worked for a few years as a clerk in an insurance office, and

[5]Richard Leach Maddox, William B. Bolton, J. King, J. Johnston, and Charles Bennett (see Gernsheim and Gernsheim, pp. 327–32).

[6]Reese V. Jenkins, *Images and Enterprise: Technology and the American Photographic Industry, 1839–1925* (Baltimore: Johns Hopkins Press, in press), chap. 3.

then served as a bank clerk and assistant bookkeeper in the Rochester Savings Bank. Disappointed when a relative of one of the bank officers received preference in promotion over him, Eastman turned his avocational interests to the serious pursuit of photography, especially the new gelatin emulsion photography which he had read about in the British photographic journals in the late 1870s. He prepared his own dry plates by mixing silver halogen salts (usually silver bromide) with gelatin, cooking the gelatin for several days (ripening), and shredding (noodling) and washing the emulsion before finally pouring it onto glass plates. Once the dry plates were prepared, they retained their sensitivity for several months. They enabled Eastman to take a box of plates, his camera, and tripod on field trips where he could make exposures, return the exposed plates to the light-tight box, and return home to develop and fix the glass negatives and print the paper photographs in his own darkroom. His gelatin dry plates quickly attained a local reputation, and he began to consider commercial production. Soon the young man, who had been so impressed by the machinery at the Centennial International Exhibition in Philadelphia in 1876, designed in the late 1870s a machine for continuously coating the glass plates with gelatin emulsion.[7]

Following the advice of his friend George Selden,[8] another Rochester inventor and a patent attorney, Eastman not only obtained a U.S. patent on the coating machine but also developed a strategy to patent his machine in Britain and, later, on the Continent and then to sell his European patents, employing the returns as capital for his own plate business in the United States.[9] Despite a trip to Europe to supervise licensing of his British patent, Eastman barely secured enough funds to cover his basic patent expenses. Nevertheless, he spent the first half of 1880 preparing for the production of gelatin plates while maintaining his full-time position at the bank. He initiated production in the summer and soon thereafter obtained a sole agency contract for marketing his plates with E. and H. T. Anthony and Company of New York, the nation's leading photographic supply house and jobber.

During the next year he maintained his position at the bank and worked in the dry-plate business in his remaining waking hours. As the business grew, the capital requirements expanded beyond the limited savings of the parsimonious young man. Eastman's energy and business ability impressed Henry A. Strong, a very successful local buggy-whip manufacturer who, with his wife, roomed for a short time at the home of Eastman's mother. Strong joined Eastman

[7]U.S. Patent no. 226,503 (application filed September 9, 1879; issued April 13, 1880).

[8]Eastman to M. K. Eastman, October 8 and October 13, 1879, GEC.

[9]Eastman to R. Talbot, July 14, 1880, GEC.

as a partner in the business early in 1881. To the new Eastman Dry Plate Company, Eastman contributed his patents, the limited production facilities of his "old" firm, and his knowledge of product and production technology. Strong, who did not work in the business during the firm's first decade, provided $5,000 capital and the policy advice and reputation of an experienced businessman.

From 1881 to 1883 the Eastman Dry Plate Company established a national reputation and, as a consequence, expanded in sales and profits. However, the barriers to entry into the industry, despite the critical nature of the emulsion, were quite low, and the high profit margins attracted to the industry a large number of new companies from across the nation.[10] The pioneer firms retained their leadership because of their exclusive relationship with the national jobbers, but during 1883 and early 1884 this advantage waned as price competition became vigorous and profit margins fell precipitously.[11] This decline in profits stimulated Eastman to reexamine and reconceptualize his business.

II. *From Dry Plates to the Roll Film System*

The erosion of profits and the leveling off of demand threatened the young entrepreneur, who held a strong commitment to the goal of growth. Eastman reacted with both short-run and long-run strategies. For the short term, he joined with other leading manufacturers in forming the Dry Plate Manufacturers' Association, an organization which expressly sought to stabilize prices of dry plates. This could, however, provide only temporary relief from price competition; it could not allow the Eastman company to resume rapid growth in sales and profits. Eastman, therefore, also pursued a long-term strategy: "The prosecution of experiments, having in view the perfection of a system of film photography that would supplant the use of glass dry plates."[12] By extending his original patent strategy, developed in consultation with George Selden, Eastman sought to

[10]Jenkins, chap. 3.

[11]While dollar sales of the Eastman Dry Plate Company increased from 1881 to 1884 at the rate of 100 percent per year, from the middle third of 1883 the rate of increase began to fall. At the same time, total labor costs were increasing, which, assuming productivity and labor wages did not change markedly, indicates that the unit rate of production actually increased at the same time wholesale prices on plates fell by about 60 percent. Hence, the profit margins began to fall substantially from the middle third of 1883. This interpretation is based on data contained in the merchandise and labor accounts of the Eastman Dry Plate Company ledger for 1881–84, in the corporation records of the Eastman Kodak Company, Rochester, N.Y.

[12]Affidavit of George Eastman filed May 23, 1891, Eastman Co. v. Blair Camera Co., no. 2883, Records of U.S. Circuit Court D, District of Massachusetts, Boston.

develop a new product which would be well protected by patents. In his search for an alternative to glass plates, he collaborated with a local camera and dry-plate manufacturer, William Hall Walker, with whom he had been acquainted since 1880.[13] As the plans of Eastman and Walker advanced, Walker sold his small business and joined the Eastman firm in January of 1884 "to assist in experimenting on a new system of film photography."[14]

During 1884 Walker and Eastman devoted considerable energy to the developmental effort, with Walker working full time on it and Eastman devoting to it substantial time from his managerial and production responsibilities. To their joint project Walker brought his experience in the production of cameras and the design of new portable dry-plate cameras. In addition to the patenting of two plate cameras in the early 1880s, Walker had designed and introduced Walker's Pocket Camera, which, along with developing and printing kits, he directed to the growing amateur market.[15] Eastman brought to the project experience in the production of photographic materials and the design of production machinery, detailed knowledge of the American and foreign photographic literature, and the resources of an established company with a national reputation. But, most important, he brought a sensitivity to the interrelationship of product design, production methods, and market considerations.

From an early point in their relationship, Eastman and Walker committed themselves to developing a roll film system—an approach different from the traditional plate system of photography. Although the idea of a continuous, flexible base was an old one, it is unlikely that either man knew of the ideas of those who had worked in the 1850s with roll film systems.[16] These early systems were not commercial successes, largely because of the deficiencies inherent in the mechanism of the roll holder and in the quality of the photosensitive film. However, the two Rochesterians did know of Leon Warnerke's unpatented roll film system which had been introduced in Britain in the early 1870s. The Warnerke system consisted of a roll holder, which attached to the back of a plate camera, and of a continuous roll of collodion tissue which moved between the two rollers in the roll

[13]George Eastman testimony, transcript of record, Eastman Co. v. Blair Camera Co., no. 105, U.S. Circuit Court of Appeals, First Circuit, Boston, 1:210 (hereafter cited as Eastman v. Blair no. 105).

[14]Ibid., p. 211.

[15]U.S. Patent no. 259,064 (issued June 6, 1882) and no. 276,311 (issued April 24, 1883). Advertisement in *Brooklyn Advance* (June 1882), p. 9 (copy in the Society for Preservation of New England Antiquities, Boston).

[16]Captain Harry J. Barr, Joseph G. Spencer, and Arthur J. Melhuish (see Gernsheim and Gernsheim).

holder. The roll film employed during the 1870s was coated with relatively insensitive dry-collodion emulsion. Although the Warnerke system attracted the attention of the British photographic journals, it was not a commercial success. At least five criticisms were made of the system: (1) the film was costly to produce and was, therefore, too expensive for the popular market;[17] (2) the film required the awkward operation of installation and removal from the rollers while in the camera because the rollers were not removable;[18] (3) the film indicators for advancing the film and for cutting the film during printing inflicted damage on the film; (4) when Warnerke introduced gelatin film for the system in the early 1880s, changes in atmospheric humidity altered the tension on the film, resulting in buckled film and out-of-focus images;[19] and (5) the handicraft methods of production of the holders and film made the system highly expensive and unreliable.[20] Despite these problems and the system's lack of commercial success, it served as the basis from which Walker and Eastman began.

Eastman and Walker divided principal responsibilities for the design and development of the principal elements of the roll holder system: (1) the roll holder mechanism; (2) the roll film; and (3) the production machinery for the roll film. Walker, with his experience and background in camera design and production, assumed principal responsibility for the design of the roll holder; and Eastman, with his experience in the production of photosensitive materials and the design of production machinery, accepted major responsibility for the development of the film and for design of the machinery for its production. The two worked together and even shared credit on the patent applications on the three basic elements, recognizing the conceptual interrelationship of the elements and realizing that patenting each component of a total system helped to fortify against imitation of the whole.[21]

Early in 1884 Eastman, like so many photographers during the previous thirty years, sought a suitably flexible, tough, relatively inert, and transparent substitute for glass. At first he created thin films of gelatin or collodion, but they proved too fragile. Later efforts to pro-

[17]Ibid., p. 292.

[18]*British Journal of Photography* (1887), pp. 689, 802, 803.

[19]Eastman Kodak Co. v. Blair Camera Co., 62 *Fed. Rep.* 400–403.

[20]Leon Warnerke, "A New Departure in Photography," *British Journal of Photography* (1885), p. 603.

[21]George Eastman testimony, abridged transcript of record, U.S. v. Eastman Kodak Co., no. A-51, U.S. Supreme Court, Appellate Case File no. 25,293, Legislative, Judicial, and Diplomatic Records Division, National Archives, Washington, D.C., p. 267 (hereafter cited as U.S. v. Eastman Kodak Co.).

duce tougher films by multiple flows or thicknesses of collodion proved no more satisfactory. Consequently, he turned to the old idea of coating paper with collodion but then stripping the collodion film from the paper base after the development of the image. When the collodion did not prove strong enough, he tried gelatin and eventually produced a suitable process. By early March he had developed a stripping film which consisted of a photographic paper base on which he first flowed a water-soluble layer of gelatin and then over that the less water-soluble photosensitive gelatin emulsion. In the processing of this film, the paper base remained attached during exposure, development, and fixing; then the emulsion was detached from the paper and attached to a sheet of glass, which was varnished and from which positive prints were initially obtained.[22] While Eastman became convinced that this stripping film was the most promising form of roll film, he and Walker continued with efforts to improve it. During the next two months, Walker introduced the idea of coating the back of the paper film with gelatin so that humidity expansion and contraction of the gelatin would occur uniformly and, therefore, prevent curling.[23] In the processing of these "American Films," as they were called, Eastman replaced during the next two years the stripping varnish with a thick coating of gelatin which produced a gelatin skin on top of the developed and fixed emulsion. After drying, the gelatin skin and emulsion were detached from the glass and employed as a negative film for producing positive prints.

Meanwhile, they both also worked on modifications and improvements of the unpatented Warnerke roll holder. By the middle of 1884 Walker and Eastman had developed a holder, containing two film spools and two guide rollers, which superficially resembled the Warnerke holder but which differed from it in several important respects. The effect of atmospheric humidity changes on the tension of the film and, hence, on whether the film remained in the focal plane, prompted Walker and Eastman to design the roll holder so that tension could be maintained at all times. They equipped one spool with a brake and the other spool with a spring, which maintained constant tension. Moreover, they designed the holder with removable spools in order to facilitate loading and unloading the holder. They provided the holder with a measuring roll and click device which provided an audible

[22]George Eastman testimony, transcript of record (quoted from Celluloid Co. v. Eastman Dry Plate & Film Co.), Goodwin Film & Camera Co. v. Eastman Kodak Co., no. 194, U.S. Circuit Court of Appeals, 2d Cir., New York, pp. 322–33 (hereafter cited as Goodwin v. EKCo.).

[23]U.S. Patent no. 306,470 (issued October 14, 1884); Eastman to H. L. Aldrich, December 13, 1886, GEC.

indication for advancing the film and with a small perforating pin on the measuring roll which marked the film for later separation of the negatives. These constituted major improvements in the Warnerke design.[24]

Eastman and Walker, as their ideas for the design of the roll holder and film took shape, also gave consideration to the design of production machinery for the roll film. Walker had seen and described in detail to Eastman the carbon-paper coating machine of Allen and Rowell, a small photographic paper producer in Boston. Moreover, Eastman had seen a similarly designed coating machine in operation at the Anthony plant in New York.[25] Eastman may also have seen the Sarony and Johnson photographic paper coating machine in operation during the two-week period he worked in the photographic materials plant of Mawson and Swan in Scotland in 1882.[26] Although the Eastman-Walker machine resembled these machines (which consisted of a roller, trough, and drying device), the Rochester inventors made the significant addition of a hang-up drying mechanism that allowed the continuous coating and drying of paper. Eastman described the final machine as consisting of

> a trough in which there was a partially submerged roller. The roll of raw paper was hung in bearings behind that trough, and the paper led over an idler and under the roll into the emulsion, then over other rolls through a cooling device to set the gelatine. The gelatine will not stop running until it is set by the cold. Then it was hung on what is called a hang-up machine, a machine having movable slats, which took it as it came and hung it in loops in the drying chamber.[27]

This basic design for coating photographic paper was used for at least the next two generations.

By early fall of 1884 the three basic elements of the roll film system (holder mechanism, film, and production machinery) were developed, and patents on the elements were applied for in the United States and in several Western European countries. Clearly, the de-

[24]Eastman testimony, Eastman v. Blair no. 105, 1:209–11, 227.

[25]Eastman Co. v. Getz, 77 *Fed. Rep.* 412–20 and 84 *Fed. Rep.* 458–63; and Eastman testimony, transcript of record, Blackmore v. Eastman Kodak Co., 2:1001–2, 1005, 1034 (hereafter, cited as Blackmore v. EKCo.).

[26]D. Burton Payne, director, Mawson and Swan, Ltd., to Beaumont Newhall, director, George Eastman House, Rochester, December 23, 1957 (photocopy in Office of Corporate Information, Eastman Kodak Co., Rochester, N.Y.); and Newhall, "Photographic Inventions," pp. 36–37.

[27]George Eastman testimony, Blackmore v. EKCo., 2:1004. The hang-up technique of drying was called festooning.

velopment of a system of patents on the new form of photography was an integral part of the entire invention-innovation strategy of Eastman. Modifications and refinements in the roll holder mechanism made by Eastman employees, such as Willis A. and Louis H. Bannister, and by Walker and Eastman in the fall of 1884, were patented, and, in the case of the Bannisters, the patents were assigned to the Eastman company.[28] Early in 1885 George Selden, who now was serving as the company's patent attorney, learned that the film-marking feature of the Walker-Eastman roll holder likely infringed upon that feature in a roll film camera patent granted in October of 1881 to David H. Houston, then a farmer in Cambria, Wisconsin. Houston's film-marking design, no doubt prompted by that of Warnerke, differed from Warnerke's by employing a sharp point or stud to perforate the edge of the film.

Convinced of the importance of having patent control of every feature of the system in order to preserve the strength and integrity of the entire system, Eastman dispatched Walker to the Dakota Territory, where Houston then lived, to negotiate, if possible, the purchase of the Houston patent. The price demanded by Houston for the patent was much too high for the infant Rochester firm, and, therefore, it initially obtained for $700 a shop license for Monroe County, New York. Later, in 1888 and 1889, when the company's financial resources were somewhat improved, Eastman reopened negotiations with Houston, who then asked $30,000 for the patent. Eastman's hard bargaining resulted in the acquisition of the patent in the spring of 1889 for $5,000. This acquisition was part of his strategy of acquiring virtually all patents relating to roll film photography, a strategy which contributed substantially to the eventual dominant position of the Eastman Kodak Company in the mass amateur market.[29]

After a little less than a year of development work, Eastman and Walker had created a substantially modified version of the Warnerke roll film system. With this new system now ready, the Eastman company reorganized and began to pursue a new direction. In October 1884, Eastman and Strong dissolved their partnership and formed a new corporation, the Eastman Dry Plate and Film Company. It consisted of three principal stockholders—Strong, Eastman, and Walker—and eight smaller stockholders—Rochester businessmen who were acquaintances of Strong or Eastman. Indicative of the company's new direction, it severed the marketing relationship with Anthony in March of 1885 and established its own sales department.

[28]U.S. Patent no. 316,933 (application filed October 13, 1884; issued May 5, 1885).

[29]Eastman testimony, Eastman v. Blair no. 105, 1:229, 234, 252–54.

Moreover, Walker departed in May for London to establish and operate in Soho Square a wholesale outlet for the company for Europe. This initiated the international operations of the company.[30]

Late in 1884 attention was turned to production facilities for the new film system. The company subcontracted the production of the roll holders. Frank A. Brownell, a local carpenter who had prepared the cases for the experimental roll holders made by Walker and Eastman, agreed to make the cases and assemble the roll holders in his shop. A local metalworking firm produced the metal frames for the holders. Although the regular production of roll holders did not begin in January of 1885, as originally planned, Brownell had production under way in the early spring of 1885. Production of the stripping film, however, proved obdurate. The blistering of the coating was a persistent problem and, at first, could not be solved. As the production of the roll holders began, the need for film became urgent. Eastman turned in desperation to coating long strips of fine-quality photographic paper with emulsion and employing this negative paper as film. Only late in 1885 did the stripping film reach the market.[31]

Initially the leaders of photography reacted favorably to the new system. It won the highest award in photography at the London Invention Exhibition held in the late spring of 1885. By August and September most of the national and British photographic journals had endorsed the system, particularly the roll holder. Leon Warnerke wrote one of the most noteworthy reviews of the Walker-Eastman roll holder, describing it as "of different construction" from his own. He further observed that "details of the mechanism of this apparatus elicit general admiration. It is made on the interchange system, so useful when large numbers of the apparatus of the uniform size is [*sic*] to be produced; it is devised and made like Americans are in the habit of doing to special machinery."[32] Clearly, Warnerke recognized the originality of the Walker-Eastman instrument and that the holder had been designed with the object of mass production and interchangeable parts. Other enthusiastic responses to the system included "ingenious mechanism" and "one of the most perfect pieces of mechanism yet introduced into photography."[33] Yet, the enthusiasm seemed largely confined to the roll holder mechanism.

The inadequate film remained a serious barrier to the general acceptance of the system. The negative paper film first marketed with

[30] Jenkins, chap. 4.
[31] Ibid.
[32] Warnerke, p. 603.
[33] Ibid.

the system, although simple to operate, provided prints which were substantially inferior to prints produced from glass negatives. The stripping film introduced somewhat later, although it provided excellent-quality prints, proved complicated to manipulate, even for professional photographers. The stripping film was so complicated and delicate—requiring development, soaking, separation, squeegee-ing, and varnishing—that after two years of further minor improvements and promotion of the stripping film, professional photographers continued to reject the roll film system as being too complicated. Eastman began to recognize that his roll film system was a failure.

III. *From the Roll Film to the Kodak System*

The development of the roll film system did provide some new sources of sales and profit as a by-product. The experience of producing negative paper film on a continuous basis provided the opportunity to produce factory-sensitized positive print paper on a large scale. Being one of the first companies to produce such paper in the United States, the Eastman firm pioneered in the production of sensitized paper by means of continuous-flow-process machinery. Many professional photographers were at first reluctant to adopt this paper because they were accustomed to sensitizing their own paper. However, because the highly sensitive developing paper of the Eastman company allowed and encouraged production of enlarged prints, the Rochester firm obtained a growing market. As a means of exploiting the enlargement market, the Eastman company also developed a printing and enlarging service whereby customers could mail negatives to the company and regular or enlarged prints would be produced using the Eastman bromide paper. By 1887 the service reached a volume of 5,000–6,000 prints per day and employed mechanized printing techniques.[34] By this time photographic paper represented

TABLE 1

PERCENTAGE SALES, 1887

Equipment	Percentage
Paper:	
Positive	46
Negative	20
Apparatus	23
Dry plates	11

SOURCE.—"Eastman Companies," George Eastman notebook (Rochester, N.Y.: Eastman Kodak Co., Patent Museum).

[34]Eastman to Matthews Northrup and Co., December 18, 1886, GEC.

two-thirds of the company's sales (see table 1). Hence, by 1887 the Eastman company had largely failed with both the production of dry plates and with the roll film system but maintained its financial integrity through its pioneering production of bromide paper and the provision of a developing and printing service.

George Eastman succinctly described the failure of the roll film system and the key conceptual change which he introduced as a consequence: "When we started out our scheme of film photography we expected that everybody that used glass plates would take up films, but we found that the number that did this was relatively small and that in order to make a large business we would have to reach the general public and create a new class of patrons."[35]

This change in conception of who was to practice photography constituted one of the most revolutionary ideas in the history of photography. Drawing on the inventive and production experience he had gained in nearly a decade of work in the field, Eastman delineated three separate functions in photography: (1) producing photosensitive materials, (2) taking a picture by exposure of photosensitive materials in a camera, and (3) developing, fixing, and printing in order to produce finished positive images. The Eastman company had experience in the production of photosensitive materials and in the service function of developing and printing, and it possessed a patent-protected system of photography. Therefore, Eastman conceived of utilizing these resources to transform the roll film system intended for the professional to an amateur system of photography consisting of a simple-to-operate film camera, stripping film, and a factory service for developing and printing the delicate and hard-to-operate film.

The Eastman company already possessed all the requisites for a system of amateur photography based on the roll film system, except for the roll film camera. Although Eastman had worked during the middle 1880s on the design of a roll film detective camera,[36] the difficulties with the stripping film led him to modify that camera to accept glass plates as well. Nevertheless, his initial effort in camera design was not a success.[37] Undaunted, he began in the summer of

[35]Eastman testimony, Goodwin v. EKCo., 1:353.

[36]"Detective" was used to describe a class of small hand cameras introduced with the advent of gelatin plates and employed furtively by certain photographers.

[37]Newhall, "Photographic Inventions," p. 38. Initially the detective camera was not to employ glass plates (Eastman to E. S. Osborne, December 22, 1885, and Eastman to Reickemeyer, March 3, 1886, GEC). During the spring of 1886 awareness of the complexity of operation of stripping film led to abandoning its promotion for beginners (Eastman to H. A. Howlitt, May 5, 1886, GEC). Within a few weeks, a decision was made to equip the detective camera for glass plates (Eastman to J. W. Buel, June 25, 1886, GEC).

1887 to work on a new simple-to-operate roll film camera. He may well have obtained the assistance of Frank Brownell or employees in Brownell's shop, but ultimately he was responsible for the conception of an amateur camera and a system of photography which would place all of the complexities of photography in the hands of the manufacturer and a simple camera in the hands of nearly everyone six years of age or older.

As Eastman continued his design work on the amateur camera through the fall of 1887, he paid attention not only to its convenience for the user but also to its ease of manufacture. At this time he noted, "I think the experience we have had in getting out the detective camera will enable us to avoid most of the difficulties in manufacture. The trouble with the detective is that no matter how successfully it works, it will always be hard to make."[38] In the patent specification for the new amateur roll film camera, Eastman mentioned certain elements of design that were made in consideration of "convenience of manufacture and simplicity of construction."[39]

In December of 1887 Eastman created the name "Kodak" for the new camera.[40] The Kodak camera consisted of a simple camera box and lens (fig. 1), a roll holder at the back of the box (fig. 2), and a case which enclosed both the camera and the roll holder. With dimensions of 6½ × 3¼ × 3¾ inches, it was smaller than the newly popular detective plate cameras. The first Kodak camera included a unique lens and shutter arrangement, with the lens in a cylinder with cutout sides which when rotated also acted as a shutter (fig. 3). A small button on the side of the camera actuated the continuous shutter (fig. 4). The lens shutter, which was made by the Bausch and Lomb Optical Company in Rochester, included a fixed-focus f/8 rectilinear lens with a 2¼-inch focal length.

From January to late June 1888, Brownell made preparations for the production of the Kodak camera and Eastman personally arranged the marketing details, including instruction manuals, advance literature to supply dealers, and advertising in appropriate publications. Late in June the first Kodak cameras were in production, and early in July Eastman exhibited the new camera at the annual photographers' convention in Minneapolis. The panel of judges at the convention awarded it the medal as *the* invention of the year in photography.[41]

[38]Eastman to W. J. Stillman, Scovill Manufacturing Co., October 22, 1887, GEC.

[39]U.S. Patent no. 388, 850 (issued September 4, 1888), p. 2.

[40]He created the name from a distinctive arrangement of letters. He gave special attention to K, which was the first letter of his mother's name (U.S. Patent Office, *Official Gazette*, September 4, 1888, p. 1072).

[41]Eastman Dry Plate and Film Co. to Minnesota Tribune Co., July 23, 1888, GEC.

Fig. 1.—First Kodak camera with time exposure cap. (Courtesy Donald Ryon, Patent Museum, Eastman Kodak Co., Rochester, N.Y.)

Fig. 2.—Roll holder mechanism in first Kodak camera, showing circular focal-plane diaphragm (*Scientific American,* September 15, 1888, p. 164).

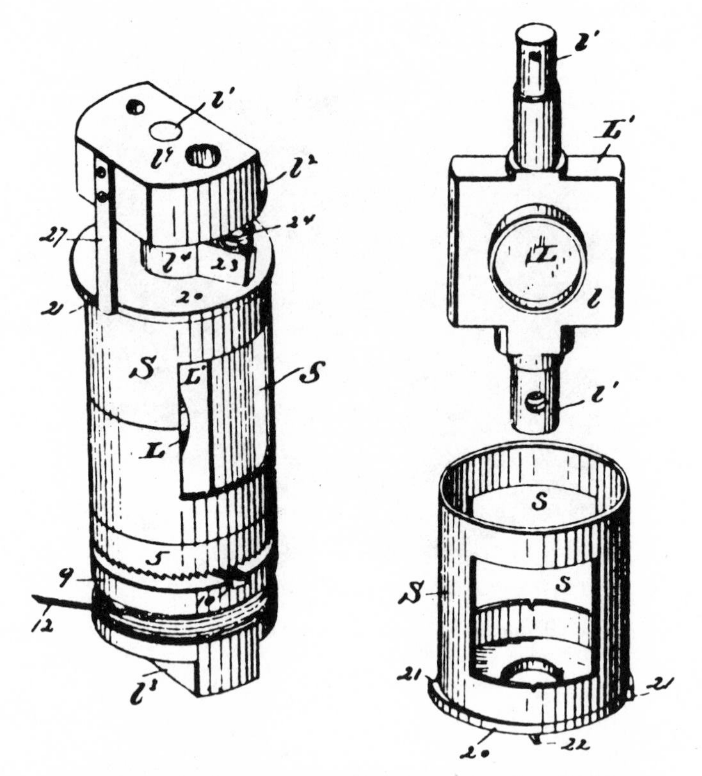

FIG. 3.—Elements of cylindrical lens-shutter system on first Kodak camera (U.S. patent no. 388,850).

Soon the camera, loaded with a 100-exposure roll of film, was on the market for $25. The novice photographer had only to point the camera toward the desired subject and "push the button." When he had exposed the film, he had only to return the camera to the factory where for $10 the film was removed and replaced with a fresh roll of film and the exposed film processed. Eastman captured the simplicity of the system in his often-quoted advertising slogan, "You press the button—we do the rest." By late summer the demand for Kodak cameras and film astounded even the usually unexcitable Eastman. He observed: "From present indications it will be the most popular thing of the kind ever introduced."[42] Most likely the experience of the major stockholder of the Eastman company, Henry A. Strong, reflected that of thousands of people all over the world. Eastman related:

> I gave one of these cameras to Mr. Strong who took it with him on a trip to Tacoma on Puget Sound a few weeks ago. It was the first

[42] Ibid.

FIG. 4.—Cutaway front view of first Kodak camera (*Scientific American,* September 15, 1888, p. 159).

> time he had ever carried a camera, and he was tickled with it as a boy over a top. I never saw anybody so pleased over a lot of pictures before. He apparently had never realized that it was a possible thing to take pictures himself.[43]

As did Strong, the public responded to the opportunity with enthusiasm, and soon the production facilities were burdened with work.

During the next several years the Kodak system won an enthusiastic reception throughout the world and laid the foundation for American leadership in the photographic industry. Meanwhile, despite the patents on the Kodak system, the Eastman company as well as other American firms and individuals outside the Eastman circle intro-

[43]Eastman to W. J. Stillman, July 6, 1888, GEC.

duced major modifications and improvements, including the substitution of nitrocellulose for the paper base for film, the introduction of daylight-loading film, and the reduction in the size of the camera by placement of the film spools in front of rather than behind the focal plane. Production difficulties with the celluloid film contributed to a sharp decline in sales of both film and cameras in 1892–93, but from the middle 1890s the Eastman company recovered from the difficulty, acquired systematically most of the patents on competing camera designs and roll film features, and struggled to increase production capacity quickly enough to keep pace with the rapidly growing demand.[44]

The change from professional to amateur predominance not only transformed the photographic industry from one characterized by decentralized, handicraft modes of production in 1879 to one characterized by centralized, mechanized modes of production in 1899, but, more important, signaled the emergence of a mass market in photography. Despite the quality limitations of the census data, they indicate the enormous growth in sales from 1879 to 1904 (see table 2). The creation of this new mass amateur market and the systematic patent policy pursued by Eastman laid the foundation for large-scale enterprise in the photographic industry and the emergence of American business leadership on an international scale.

* * *

During the decade 1879–89 photography passed through two significant revolutions: the introduction of gelatin emulsions and the creation of mass amateur photography. Both innovations arose within the context of the needs and insights of practicing photographers and the changing technological and economic environment. George Eastman, one of the few nonprofessional photographers involved in

TABLE 2

SALES (IN MILLIONS OF DOLLARS), 1879–1904

	1879	1889	1899	1904
Value of products of American photography materials and apparatus manufacturers*...	0.25	2.75	7.80	13.02
Eastman Kodak sales (U.S.)†	...	0.45	2.28	5.13

*U.S., Department of Commerce, Bureau of the Census, *Abstract of the Census of Manufacture*, 1914 (Washington, D.C.: Government Printing Office, 1917), p. 672.
†U.S. v. Eastman Kodak Co., pp. 2565–69.

[44]Jenkins, chap. 4.

the American photographic business at that time, pioneered in the centralized, factory production of photosensitive materials. He brought to the endeavor the goal of continual sales and profit growth and a strategy of large-scale, mechanized production which would prevent the entry of "miscellaneous competition." However, recognizing his declining market position in the dry-plate business and its implication for his economic goals, Eastman sought late in 1883 and early 1884 to develop a new patent-protected system of photography to replace the gelatin glass plate system. Although he and Walker improved on, patented, and introduced an alternative system, the deficiencies of the film prevented its commercial success with the traditional market—the professional photographer.

Faced with failure once again, Eastman made the key conceptual change in who was to predominate in the practice of photography from the professional to the novice. Along with this, he redefined the market and the requisite design for apparatus and materials. Drawing on the resources and experience of his company in the production of photosensitive materials and in the provision of photofinishing service, Eastman committed himself to the design of a camera for novice use and to the "salvaging" of the unsuccessful roll film system by converting it to the highly popular Kodak system.

Employing modifications of the essential innovation and patent strategies with which he had entered the photographic industry, Eastman finally succeeded in achieving and ultimately far transcending the goals he had originally set for himself and his business. This achievement was a product of his and Walker's thorough understanding of contemporary photographic technology, their combined technical capacities, Eastman's ambitious business goals and his initial failure to achieve those goals within the traditional technological framework, and his intimacy with considerations of the interrelationship between technology and the market. The latter allowed him successively to modify and redirect both his technical and business resources, ultimately leading to the recognition and successful exploitation of the mass amateur market.

A New Role for Professional Scientists in Industry: Industrial Research at General Electric, 1900–1916

GEORGE WISE

Only an attentive stockholder would have noticed the item. It was buried near the bottom of the Report of the Third Vice President of the General Electric Company, in that firm's 1902 Annual Report. After describing the important details of manufacturing and engineering for the year 1901, Vice President Edwin W. Rice noted: "Although our engineers have always been liberally supplied with every facility for the development of new and original designs and improvements of existing standards, it has been deemed wise during the past year to establish a laboratory to be devoted exclusively to original research. It is hoped by this means that many profitable fields may be discovered."[1]

With this, a major American corporation announced to the world that it had embarked on an experiment in the support of scientific research. This marked the formal unveiling of a program that had actually begun on December 15, 1900, when Willis R. Whitney, an assistant professor of chemistry at the Massachusetts Institute of Technology, had arrived in Schenectady, New York, to begin devoting two days a week to research at GE's largest manufacturing works.[2]

He was not the first professional scientist to be employed in American industry—or even in General Electric. Nor was his laboratory the

DR. WISE is a historian employed by the Corporate Research and Development Center of the General Electric Company, Schenectady, New York. The author wishes to thank Jeffrey Sturchio and Leonard Reich for their useful suggestions and for letting him read their unpublished manuscripts bearing on questions discussed in this article. He also would like to acknowledge the many useful insights into the history of General Electric research gained through conversations with Dr. Herman A. Liebhafsky and Dr. Guy Suits, both formerly of General Electric.

[1]General Electric Co., *Annual Report for 1902* (New York, 1903), p. 13.

[2]Whitney's starting date is listed in his personnel file at the General Electric Research and Development Center, Schenectady, N.Y. (hereafter cited as GE RD). The two-days-a-week arrangement is described in Whitney to Ida Schulze, March 9, 1901, on file at the Public Information Unit, GE RD.

first established by that company or its predecessors. That honor must be reserved for Thomas Edison's Menlo Park.[3] But Edison's focus was on invention. Whitney's effort marks a pioneering attempt by American industry to employ scientists in a new role—as "industrial researchers" rather than as inventors, engineers, testers, or calculators. This paper will attempt to identify the nature and emergence of that role in the specific case of GE, and will offer some suggestions highlighted by the GE experience whose possible applicability to other industrial laboratories deserves investigation.

The story of the General Electric Research Laboratory has already been told in two excellent accounts: from the outside, by historian Kendall Birr; and from the inside, by Laurence A. Hawkins, longtime executive engineer of the laboratory. In the quarter-century since these studies were carried out, historians have asked new questions about the relationship of science and technology in the industrial age. And new primary sources have become available, particularly regarding the GE laboratory's origins and early years. A central question can now be addressed more effectively than was previously possible: how did one major American industrial corporation learn to attract and use professional scientists? While industry is a natural habitat for the engineer, with his emphasis on the *use* of knowledge and technique, it is less obviously the home for the seeker of knowledge.[4]

Previous treatments of this question have given oversimplified answers. One view pictures the scientist as being "institutionalized" into the role of an industrial employee, much as a spirited horse might be broken to pull a plow.[5] Another, a view oversimplified in the opposite direction, pictures the research laboratory as protected from indus-

[3]For a perceptive view of the significance of Edison to industrial research, see Thomas P. Hughes, "Edison's Method," *American Patent Law Association Bulletin* (July–August 1977), pp. 433–50.

[4]See Kendall Birr, *Pioneering in Industrial Research* (Washington, D.C., 1957); Laurence A. Hawkins, *Adventures into the Unknown* (New York, 1950). For revision of the "technology as applied science" interpretation that colors the Birr and Hawkins books, see Derek de Solla Price, "On the Historiographic Revolution in the History of Technology," *Technology and Culture* 15 (January 1974): 42–48; and the special issue of *Technology and Culture* (17 [October 1976]), "Interaction of Science and Technology in the Industrial Age."

[5]The institutionalization viewpoint is given, for example, in David F. Noble, "Science and Technology in the Corporate Search for Order" (Ph.D. diss., University of Rochester, 1974). Noble states (p. 172): "Although GE probably laid greater emphasis upon individual initiative, and less on directed projects than most industrial laboratories, science was nevertheless a handmaiden to corporate interests, rather than the other way around. As the laboratory grew in size, the role of the scientist or engineer came more and more to resemble that of a worker on the 'production line.'"

trial reality—a kind of university-in-exile insulated from the demands of the factory.[6]

I see the situation at General Electric from 1900 on as quite different. At GE, the industrial scientist was not molded into a worker on the intellectual assembly line, nor was he shielded from the hard facts of business.

Instead, this laboratory succeeded because it created a *new* role for professional scientists—a blend of research freedom and practical usefulness not available before 1900. General Electric did not merely lure people away from alternative paths, capturing individuals who otherwise would have been independent inventors, or luring potential pure scientists into prostituting their research talents. Instead, its main appeal was to individuals who did not fit easily into university or college teaching positions, yet still wished to gain recognition as professional scientists. It also appealed strongly to individuals with inventive originality who did not wish to take the risks of entrepreneurship. It came to serve as an *alternative,* rescuing certain individuals from the necessity of choosing between other unsatisfactory career paths.

I will support this view by briefly describing some of the roles played by American scientists in industry in the late 19th century, by indicating reasons for the emergence of a new role, and by outlining the specific features of that role as it emerged at GE. For concreteness, I will identity as "professional scientists" those individuals who earned a Ph.D. degree in physical science or mathematics, or who, without that degree, carried out and published original work in those fields. Emphasis will be on three individuals whose characteristics and career paths illustrate both the situation facing certain professional scientists at the turn of the century and the GE response. They are Willis R. Whitney, who gave up a promising academic career to organize and direct the GE lab; William D. Coolidge, who gave up an equally promising start in experimental physical chemistry to become a major contributor to the lab's success; and Irving Langmuir, 1932 Nobel laureate for chemistry, whose career represents a remarkable combination of scientific and technological achievement.

[6]The "university-in-exile" view is given, for example, in John J. Beer and W. David Lewis, "Aspects of the Professionalization of Science," in *The Professions in America,* ed. Kenneth S. Lynn (Boston, 1965), pp. 110–30, esp. p. 115: "At such [industrial] laboratories . . . research work was placed under the direction of men who were purely scientific in their interests and who managed to instill much of the university atmosphere into the industrial situation . . . in laboratories of this type advanced research was clearly distinguished from development."

Professional Scientists in Late 19th-Century U.S. Industry

Science and industry were not independent of one another in the United States before 1900. But their contacts then were more limited in scope, intermittent, and irregular than they are today. Most scientists were teachers in colleges or universities. Many of these academics provided part-time consulting services to industry. Some, such as Columbia's Michael I. Pupin, managed to secure and exploit valuable patents without leaving their teaching positions. A few Ph.D.'s, such as Lee de Forest (physics, Yale, 1895) chose the path of the independent inventor-entrepreneur. Others became employees of established corporations—but not as research scientists.[7]

For example, Charles Dudley (chemistry, Yale, 1874) set up and managed a chemical analysis laboratory for the Pennsylvania Railroad. In 1885, the Bell Telephone Company placed its Mechanical Department, charged with inventing new telephone apparatus, in the hands of Hammond V. Hayes (physics, Harvard, 1877), whose staff included other Ph.D.'s, such as William W. Jacques (physics, Johns Hopkins, 1879). Hayes, however, could only occasionally encourage scientific research. For the most part, his department's mission was closely focused on telephone business needs.[8]

The electrical industry also fitted scientists into testing and engineering jobs. Edward Leamington Nichols (physics, Göttingen, 1879) set up Thomas Edison's first lamp-testing laboratory at Menlo Park. Charles Proteus Steinmetz (who was forced to flee Germany for political reasons before receiving the doctorate in mathematics he had earned at the University of Breslau) began as a draftsman with the small electrical firm of Eickemeyer and Osterheld. He was quickly assigned to applied research work in magnetism and carried out some outstanding original work. When Eickemeyer and Osterheld was purchased by General Electric in 1892, he was placed in GE's calculating department, and later became that company's first (and last) chief consulting engineer.[9] Louis Bell, a Ph.D. physicist (Johns

[7]A good overview is George S. Daniels, ed., *Nineteenth Century American Science: A Reappraisal* (Evanston, Ill., 1972), especially Daniel J. Kevles, "On the Flaws of American Physics," pp. 133–52, and Carroll Pursell, "Science and Industry," pp. 231–49. Of interest on one particular career is Michael I. Pupin, *From Immigrant to Inventor* (New York, 1924).

[8]Leonard Reich, "Radio Electronics and the Development of Industrial Research in the Bell System" (Ph.D. diss., Johns Hopkins University, 1977), gives both an excellent overview of early U.S. industrial research and much specific information about early research efforts at Bell Telephone and Western Electric. M. D. Fagen, ed., *A History of Engineering and Science in the Bell System* (Murray Hill, N.J., 1975), is also very helpful.

[9]On Nichols, see Paul W. Keating, *Lamps for a Brighter America* (New York, 1957), p. 51; the best recent account of Steinmetz's achievements is James E. Brittain, "C. P.

Hopkins, 1888) directed the construction of GE's first major alternating-current power system, while Louis Duncan (physics, Johns Hopkins, 1885) helped design electric locomotives for Westinghouse.

Some of these scientists-turned-engineers (such as Nichols and Duncan) quickly returned to teaching. Other professionally trained scientists, such as Louis Bell, Frederick Perrine (physics, Princeton, 1885), and Eugene F. Roeber (chemistry, Berlin, 1892) became editors of technical periodicals. Almost all who remained outside academia, however, found it impossible to do important work in basic science. Steinmetz, for example, relates in an autobiographical essay that by the early 1890s his engineering responsibilities had occupied much of the time he had hoped to devote to pure mathematics. His jobs at GE involved more engineering "troubleshooting" than scientific research, and his heavy involvement in professional society and engineering education efforts further diluted his efforts.[10]

Before 1900, then, these professional scientists served American industry as part-time consultants, or as full-time occupants of non-research positions. But the changing industrial situation at the turn of the century, coupled with changes in the job market for academic scientists, forced industrial leaders to look for new ways to make use of scientific talent.

On the industrial side, the birth of American industrial research coincides with the period of consolidation that followed the birth or youth of some of America's corporate giants. Certain of these firms—including American Telephone and Telegraph (AT & T), General Electric, and Eastman Kodak—had achieved their positions largely through technological leadership embodied in patents. Expiration of these patents opened the door to new competition. Retaining technological superiority required more attention than part-time consultants, or scientists burdened with engineering responsibilities, could provide.[11]

The supply side of the picture was also changing. As Daniel Kevles

---

Steinmetz and E. F. W. Alexanderson: Creative Engineering in a Corporate Setting," *Proceedings of IEEE* 64 (September 1976): 1413–17.

[10]I have pieced together the career trajectories of Bell, Duncan, Perrine, and Roeber from their entries in *Who Was Who in America,* vol. 1, *1897–1942* (Chicago, 1942). The difficulty of finding time for pure mathematics in his engineering career is described in Charles P. Steinmetz, "Steinmetz's Own Estimate of His Work," originally published in the *GE Monogram,* December 1923, and reprinted in Emil J. Remscheid, *Reminiscences of Steinmetz* (Schenectady, N.Y., 1977), pp. 62–67.

[11]For recent statements of this view, see Reich (n. 8 above) and Reese V. Jenkins, *Images and Enterprise* (Baltimore, 1975), esp. chap. 14, "Eastman Kodak Research and Development," pp. 300–318.

has pointed out, the number of teaching jobs for scientists was rapidly growing, in response to the sharp rise in college enrollments and the growth of professional engineering education that began in the 1890s. The new opportunities emphasized teaching, not research. While industry could not, in 1900, offer positions with the professional prestige of a professorship, it could compete in another way—with research facilities superior to those of almost all American universities except the Johns Hopkins, Chicago, and Harvard. Adequate equipment budgets, and the promise of full time at the bench rather than at the blackboard, could be used by industry as recruiting attractions.[12]

By 1900, then, supply-and-demand pressures began to combine to force a reevaluation of the industrial use of scientists. It was in this context that some GE leaders proposed a new role for the professional scientist in their company.

The Lure of Industry: The Example of Willis R. Whitney

"I often think, though probably wrongly," wrote Willis R. Whitney in 1918, "that men in academic positions ought not to be paid as much as men in industrial positions. On the average, the men in the latter case work harder, overcome more obstacles, worry more, and compromise with their natural desires much oftener."[13] Whitney's experience as director of the GE Research Laboratory from its founding in 1900 until 1932 reflected this need to compromise. He was indeed paid far more than he had been in his academic post. But the money aspect represented only the surface of a complex story.

Technical threats to GE's lighting business triggered the sequence of events that brought Whitney to his new post. In 1900, the electric lighting business of the General Electric Company was based on the high-resistance, carbon filament incandescent lamp pioneered by Joseph Swan and Thomas Edison. Two major lighting innovations threatened this type with obsolescence. The "glower" lamp, invented by German chemist Walther Nernst, achieved superior efficiency through use of a ceramic filament heated to incandescence at a higher (and therefore more light-producing) temperature than carbon could sustain. Unlike the Edison lamp, it did not require an evacuated enclosure. The mercury vapor lamp, being developed by American in-

[12]Daniel Kevles, "The Study of Physics in America, 1865–1916" (Ph.D. diss., Princeton University, 1964), pp. 175–219; and Kevles, *The Physicists* (New York, 1978), esp. chap. 5, "Research and Reform," pp. 60–74.

[13]W. R. Whitney to E. J. Crane, November 14, 1918, W. R. Whitney Collection, Schenectady Archives of Science and Technology, Union College, Schenectady, New York (hereafter cited as Whitney Collection).

ventor Peter Copper Hewitt and others, was even more efficient. Its green-colored light and high voltage blocked immediate commercialization, but these problems appeared soluble (and, as the modern fluorescent light demonstrates, in the long run they were solved).[14]

General Electric had been less active in exploring these new fields than had the leader of its sole major American rival in the electrical industry. George Westinghouse had secured the American rights to Nernst's lamp and was supporting Cooper Hewitt's development work.[15]

Charles Proteus Steinmetz had paid at least one visit to Cooper Hewitt's laboratory by mid-1900. His perception of the lighting threat it represented catalyzed his formal proposal for a laboratory focused on fundamental research, mainly in the field of lighting. His letter of September 21, 1900, to Vice-President Edwin W. Rice, head of GE's manufacturing and engineering, specifically mentions mercury vapor and Nernst-type lamps as principal research targets.[16]

The key point in the proposal was that the laboratory should be kept separate from day-to-day production responsibilities. Two other GE technical leaders concurred in the proposal: patent attorney Albert G. Davis and consultant Elihu Thomson (who, as a founder of one of GE's predecessor companies, a prolific inventor, and Rice's former high school teacher and long-time mentor, had enormous influence with him). Rice quickly secured the approval of company president Charles Coffin. Before the end of October, Rice and Davis were on their way to Boston to discuss the post of director with Whitney. His name had been suggested to them by an MIT colleague, Charles R. Cross, a respected physicist with ties to the electrical industry.

Rice recognized from the beginning that the concept of the laboratory had to be broader than mere support for GE's lighting business if he wished to recruit first-rate researchers. "We all agreed it was to be a real scientific laboratory," he wrote later. And, looking back upon that

[14]On the development of electric lighting technology in the 1890s, see Aaron A. Bright, *The Electric Lamp Industry* (New York, 1949), esp. pp. 220–30.

[15]Henry G. Prout, *A Life of George Westinghouse* (New York, 1921), pp. 234–36. Note that Westinghouse backed these ventures personally, rather than formally including them within the Westinghouse Electric and Manufacturing Company.

[16]On the founding of the GE Research Laboratory (a story that Birr and Hawkins left unclear in their books, perhaps because of unavailability of the primary documents mentioned in this footnote), see items L 2986–92, John Winthrop Hammond File (hereafter cited as Hammond File), Main Library, Building 2, General Electric Company, Schenectady, N.Y. These include summaries of letters from Steinmetz and Albert G. Davis proposing formation of the laboratory, and interviews given in 1926–27 by Whitney, Rice, and Davis on the circumstances of the laboratory's founding.

first employment interview, Albert G. Davis was to recall how anxious he had been to convince Whitney that "industrialists . . . possessed scientific ideals."[17]

Industrialists were also in a position to make a very generous salary offer. In exchange for an annual stipend of $2,400, Whitney would be required to spend only two days a week in Schenectady. This salary figure corresponded to the pay of a full professor, a level Whitney could not expect to reach for several years. He could retain his MIT post. And his position, from the first, would not be merely that of consultant or researcher, but "director of an experimental electrochemical laboratory"—implying that he would soon have assistants, equipment, and a research budget.[18] It is true that in the beginning, this "laboratory" had to share the cramped quarters of a barn behind Steinmetz's Schenectady rooming house. But this detail, the publicity equivalent of a president's being born in a log cabin, was due to temporary space limitations at the fast-expanding Schenectady works, not to a lack of commitment on the part of the laboratory's founders.

Whether by luck or calculation, the GE leaders had made the right offer to the right man at the right time. Whitney saw the offer as an outlet for as yet unfulfilled ambition. "The only thing I want now," he wrote to a friend in the spring of 1901, "is to accomplish some great thing for the 'General Electric.' They are giving me free hand [*sic*]

[17]The quote from Rice is in Edwin W. Rice, Jr., "A Tribute to Willis R. Whitney," *GE Review* 36 (January 1933): 3, while that from Davis is in item L 2991, Hammond File (a 1926 interview with Davis). In keeping with my characterization of the GE Research Laboratory as creating a new blend of scientific and technological opportunities, it is suggestive that two of the four sponsors of the laboratory were men whose scientific aspirations had been at least partially deflected by practical work. Steinmetz was one. The other, Elihu Thomson, began his career as an assistant to Edwin J. Houston, "professor" of chemistry at Philadelphia's Central High School. Thomson carried out some experiments on electromagnetic waves and thermochemistry before concentrating his efforts on electrical technology. From 1878 through 1892—his most productive years—he was occupied in building up the Thomson-Houston Company into a worthy challenger of the Edison (and later Westinghouse) companies. Only after business success was assured, and Thomson-Houston and Edison had been merged to form GE, was he able to return to scientific efforts. Their nature—for example, speculations on volcanism, and telescope making—suggest the serious amateur, rather than the professional scientist. Thus his biography by David Woodbury, *Beloved Scientist* (New York, 1944), seems mistitled. See also Karl T. Compton, *Elihu Thomson, 1853–1937,* National Academy of Sciences Biographical Memoirs, vol. 21, no. 4 (1939); *Professor Elihu Thomson, 1853–1937* (Philadelphia, February 16, 1939); and Harold J. Abrahams and Marion B. Savin, eds., *Selections from the Scientific Correspondence of Elihu Thomson* Cambridge, Mass., 1971).

[18]The salary figure is in John T. Broderick, *Forty Years at GE* (Schenectady, N.Y., 1939), p. 82. Whitney gives his title as "director of an electro chemical laboratory" in Whitney to Ida Schulze, March 9, 1901, GE RD.

here to spend and experiment as well as I am able, and I shall die with a ten-ton shadow on my opinion of Whitney if I don't do some good work here."[19]

The pace of advancement at MIT had not satisfied this drive. Despite a German doctorate from the Leipzig school of the great chemist Wilhelm Ostwald, a promising beginning as a researcher, and an excellent teaching record, Whitney was still an instructor in 1900—ten years after first joining the MIT staff as an assistant instructor in 1890. His promotion to assistant professor came just before he received the GE offer. And, shortly before that, his request for a $75 raise in his annual salary had been flatly turned down by MIT's president.[20]

His first taste of industrial work had been more pleasant. In 1899, he and his senior colleague, MIT physical chemist Arthur A. Noyes, had accepted a request and set up a "little chemical manufactory" for reclaiming valuable solvents at a plant of the American Aristotype Company, a manufacturer of photographic supplies. "I have this summer invested all that I can lay hands on," wrote Whitney that fall, describing the venture. It was an uncharacteristic gamble by this normally financially conservative man. It proved a success; by the end of the next year, Whitney had cleared over $20,000.[21]

But much more than the lure of money was involved in Whitney's decision to accept an industrial post. His teaching experiences at MIT and the tone of some of his early addresses on industrial research suggest a man more interested in inspiring research than in performing it. He preferred to carry out the kind of simple qualitative experiments that stimulate thought, rather than the quantitative work on which important publications could be based. And, as his biographer John Broderick has pointed out, he placed great emphasis on the practical usefulness of scientific work.[22]

Two important personal qualities admirably fitted him for this role of directing research. First, his integrity was ironclad. "Whitney does not need a Code of Ethics, no more than he needs a police code, or any code of laws, or a ready-made religion," wrote a fellow chemist,

[19]Whitney to Ida Schulze, March 9, 1901, GE RD.

[20]Whitney's years of service at MIT, beginning as an assistant instructor in 1890, are listed on his transcript in the files of the Institute Archives, Massachusetts Institute of Technology, Cambridge. The $75 raise story was told by Whitney in his Perkin Medal acceptance speech, reprinted in *Industrial and Engineering Chemistry* 13 (February 1921): 162.

[21]Details of the consulting venture are given in Hawkins (n. 4 above) and Whitney to Ida Schulze, September 28, 1898; March 9, 1901, GE RD.

[22]John T. Broderick's anecdotal account of Whitney's life and opinions is *Willis Rodney Whitney* (Albany, N.Y., 1945).

Leo Baekeland, to another, Arthur D. Little. And, second, his sincere efforts to look for the best in everyone brought him the genuine respect and admiration of virtually all with whom he came in contact.[23]

This outward optimism and positive attitude coexisted, with some strain, with his personal ambition and determination to be the boss of his laboratory. His personal depressions, sometimes manifested in physical symptoms, such as the nervous breakdown that put an end to his tenure as laboratory director in 1932, seem to have been at least partial consequences of this balancing effort.[24]

But all this remained far in the future in 1900–1901, as he spent his first eight months attacking, without success, the problems of the mercury vapor lamp. At the same time, he made known to the GE engineers that he was also willing to tackle problems of less-than-fundamental scientific value, like devising a continuous process for baking resistors.[25]

Those eight months convinced him that the job deserved full-time attention. Accordingly, he requested a one-year leave of absence from MIT. That leave was to stretch into a career. In asking MIT's president for it, he wrote: "I do not mean to let the amount of salary influence me so long as I have an independent income as I at present have." He expressed satisfaction that his GE position carried the independence that had been promised him: "There is no evidence on the part of the officers of the Company, of impatience or a wish to interfere at all in my work." But he held no illusions about his role. "I know I was put here for a purpose," he concluded, "that the Company is not primarily a philanthropic asylum for indigent chemists and I must not let it become one even secondarily."[26]

This realism toward the motives of industry matched Whitney's realistic assessment of his own strengths. His scientific interests were catholic and unfocused. He thoroughly enjoyed teaching, but was also anxious to prove himself through practical achievement. He possessed a rare talent for motivating others. At MIT he had not had the full opportunity to exercise these tastes and talents. As director of an

[23]Baekeland to Little, July 16, 1913, Whitney Collection.

[24]On Whitney's health problems, see Broderick, *Willis Rodney Whitney*, pp. 176–78; and Mary Christie (Whitney's secretary) to W. H. Walker, June 3, 1932; Whitney to C. S. Merrill, February 18, 1916, Whitney Collection. The depression caused a reduction in laboratory staff, but the role of the laboratory was reaffirmed.

[25]Whitney's first laboratory notebook (hereafter cited as NBI), July 9–10, 1901; September 11, 1901, on file at GE RD.

[26]Whitney to Henry S. Pritchett, August 6, 1901, Institute Archives, Massachusetts Institute of Technology, Cambridge. I thank Ms. Eleanor Bartlett for locating this letter.

industrial laboratory, he was accepting a role that had not been defined fully in advance. He could—and did—shape it to meet his own personal needs and aspirations.

The Scientist as Inventor: The Example of William D. Coolidge

In meeting the first challenge, the recruiting of a capable staff, Whitney could not offer the lure that had been used on him—the opportunity to shape an institution. Other attractions were needed. Unquestionably, money was one of the important ones. A review of the starting salaries offered by Whitney to staff candidates with recent Ph.D.'s in physics and chemistry during the years 1901–5 indicates that the standard offer was $1,500 per year. This was 15 percent higher than the median figure being offered to university instructors (the post such individuals typically qualified for) in 1900, although well below the maximum instructor's salary of $1,900.[27]

Nonsalary rewards supplemented this pay differential. For example, physicist Ezekiel Weintraub, the most highly regarded scientist on Whitney's original staff, commanded not only a salary of $3,000 a year and summer vacations in European laboratories, but also a clause in his contract stating that "conditions affecting the position he now holds should not be changed so as to make it exceedingly distasteful or intolerable for him."[28]

But in some cases, even the combination of a generous salary offer and the promise of considerable research freedom failed to land promising candidates. Frank Cottrell, a 1901 Ph.D. in chemistry from Leipzig, turned down an attractive offer from Whitney in order to accept a far less remunerative but purely academic post at the University of California. (His main achievement there proved not to be in pure science, but rather a major invention, the Cottrell precipitator for cleaning up industrial smoke.) A second early candidate from Leipzig, George S. Forbes, also resisted Whitney's blandishments.[29]

For those determined to reserve full freedom to choose research topics, to gain the social prestige that came from a university position, and to fit the established role of the scientific professional, the attractions of industrial science could not compete with those of the uni-

[27]The $1,500 figure recurs in Whitney, NB I. The average salary figure for U.S. instructors in physics is given in Paul Forman, John L. Heilbron, and Spencer Weart, "Physics circa 1900," in *Historical Studies in the Physical Sciences,* ed. R. McCormmach (Princeton, N.J., 1976), 5:42. They list it as 5,500 German marks, which, at a conversion rate of 4.2 marks to the dollar, is just over $1,300.

[28]Whitney, NB II, May 29, 1903; July 23, 1903.

[29]On Cottrell, see Frank Cameron, *Cottrell, Samaritan of Science* (Garden City, N.Y., 1952), p. 102; Whitney, NB I, December 28, 1902. On Forbes, see Whitney, NB III, October 24, 1906.

versities. But even those who could not fully achieve this role might respond to it.

One such response has been described by Edwin Layton. The professional engineer began to take on a role that was the "mirror image" of that of the scientist. "The new technologist," Layton notes, "substituted a college education, a professional organization, and a technical literature patterned on that of science."[30] The industrial research laboratory at GE represented a related type of response. Individuals deeply attracted by the content and methods of science, but less interested in its larger questions or professional perquisites, now had an attractive new choice.

A good example is William D. Coolidge, who received his Ph.D. in physics from Leipzig in 1901. At the physical chemistry laboratory established at the turn of the century at MIT by Arthur A. Noyes, Coolidge proved a capable experimenter. He was coauthor of one of the key papers (on the properties of solutions at high pressures) to emerge from that laboratory.[31]

At the same time, he did not fit the academic mold. He was shy and undemonstrative, and had little interest in lecturing or teaching. His publication rate was low. He evidenced little concern with the theoretical side of physics and chemistry.

He also showed some of the traits of the individual inventor. He was an ingenious equipment designer, with a good sense for the critical technical problem and a persistence at seeking the answer. But he showed neither the resources nor the stomach for risk that made entrepreneurs out of others with similar abilities. He was cautious and thrifty. And, in 1905, he was still in debt from his years of overseas studies.[32]

These attributes and circumstances enhanced the appeal of the job offer at the GE lab from his former teacher and colleague Willis R. Whitney in 1905. Coolidge was presented with the chance to be both a contributing scientist and the laboratory's assistant director (for the first three years unofficially, but from 1908 on, with the title) at a salary of $2,400 per year. This was 50 percent above his MIT pay. In

[30]Edwin Layton, "Mirror Image Twins: The Communities of Science and Technology," in Daniels (n. 7 above), p. 210.

[31]Arthur A. Noyes and W. D. Coolidge, "Electrical Conductivity of Aqueous Solutions," *Proceedings of the American Association for the Advancement of Science* 39 (November 1903): 163–219.

[32]For biographical information on Coolidge, I have relied on John A. Miller, *Yankee Scientist* (Schenectady, N.Y., 1963), and Herman A. Liebhafsky, *William D. Coolidge, a Centenarian and His Work* (New York, 1974), as well as Coolidge's own unpublished and updated "Autobiographical Notes," which are on file both at GE RD and at the American Institute of Physics, New York City.

addition, Whitney recognized Coolidge's "scientific ideals" by allowing him to bring to Schenectady the equipment he had used in his experiments with Noyes. A set portion of Coolidge's time was to be devoted to research on this apparatus.

This proved only a gesture. Coolidge immediately began devoting his full time to the laboratory's main effort of 1905—development of improved lamp filaments made of the heat-resistant metal tungsten. Inventions made in Germany and Austria had already shown that the tungsten incandescent lamp was the answer to the challenge posed by the Nernst and mercury vapor types. Coolidge devoted his first five years at GE to inventing and perfecting a process for making tungsten wire for use in such lamps. The resulting patent (issued in 1913) proved one of the most valuable in the lamp industry since Edison's. Meanwhile, the pure-research apparatus languished and eventually was sent back to MIT.[33]

Nor did Coolidge subsequently achieve much in pure science. His numerous papers deal mainly with inventions (notably his many contributions to X-ray technology) rather than scientific results. The call to industry appears to have involved no sacrifice for him. He was a gifted inventor without the economic or entrepreneurial drives that marked the great money-makers. He was a capable and persistent experimenter without the need to explain, publish, and teach that marks many great scientists. The industrial laboratory proved a middle way, suited to his particular mix of talents.

Teaching versus Industrial Responsibilities: The Example of Irving Langmuir

Whitney and Coolidge shared a less than single-minded devotion to the advancement of knowledge purely for its own sake. As noted, Whitney placed a great emphasis on the usefulness of science, a theme recurring in his many speeches, with their references to Francis Bacon and his views. And Coolidge preferred invention to discovery.[34]

Irving Langmuir, on the other hand, demonstrated that an individual did not have to leave pure science behind when he entered an industrial laboratory. In choice of research topics, publication rate, citation frequency, and informal communication with his academic peers, Langmuir closely resembled the traditional model of the pro-

[33]The apparatus story is told in Rice (n. 17 above), p. 4. The Liebhafsky biography cited above gives a particularly good account of Coolidge's invention of ductile tungsten.

[34]See, e.g., W. R. Whitney, "Dreams of the Future," *General Electric Review* 39 (1936): 360–62, as well as the numerous examples in Broderick, *Willis Rodney Whitney*, e.g., pp. 114–15.

fessional scientist. But he did not turn his laboratory into a university-in-exile. He simultaneously pursued important technological studies, coming up with major inventions in the fields of lighting and electronics.[35]

He received his doctorate at Göttingen under Walther Nernst, and chose a teaching career, partly on the advice of his brother Arthur, an industrial chemist who warned that "for further progress, it will be necessary to have experience in teaching."[36]

His first post, as an instructor at Stevens Tech, was a rough beginning: a low starting salary of only $900 a year and a rebellious and indifferent first crop of students. By 1906, however, conditions had improved. "It is going to be a real pleasure to teach this year because the men seem interested," he wrote his mother in October. His salary rose quickly (reaching $1,350 a year by 1908) and was supplemented by numerous consulting opportunities. In 1907, he felt sufficiently satisfied to turn down an offer to become a research physicist with the U.S. Geological Survey on the grounds that "government positions are insecure and advancement is slow."[37]

His satisfaction did not last. Teaching responsibilities crowded out research. In the winter term of 1907, for example, Langmuir taught a full eight-hour day. He never secured the full-time services of a laboratory assistant. The last straw came in 1909. Upon retirement of the head of the Stevens chemistry department, Langmuir found not only that he was denied the post, but that his request for a salary increase to $1,800 per year was flatly turned down by the new chairman.[38]

He had, however, already prepared an escape route. In January 1909, he had written to Whitney (whose activities he had heard about from a college friend turned GE researcher, Colin G. Fink). Noting that "the opportunities for efficient research work are greater in your laboratory than in any other in the country," Langmuir asked Whitney for a summer job. This was arranged, and on July 16 Langmuir wrote to his mother: "On Monday I go to Schenectady and in all probability will do good enough work so that Whitney will offer me a

[35]Albert Rosenfeld's *The Quintessence of Irving Langmuir (Collected Works of Irving Langmuir,* ed. C. Guy Suits and Harold Way [New York, 1962], 12:36–77), the standard account of Langmuir's early career, is incomplete. I have supplemented it by reliance on the Irving Langmuir Collection at the Library of Congress, Washington, D.C. (hereafter cited as Langmuir Collection).

[36]Langmuir to Mrs. Sadie C. Langmuir, May 16, 1905, Langmuir Collection.

[37]Langmuir to Mrs. S. C. Langmuir, August 3, 1905, August 7, 1906, March 27, 1907; John K. Clement to Irving Langmuir, March 8, 1907; Langmuir to Mrs. S. C. Langmuir, October 5, 1906. All in Langmuir Collection.

[38]Langmuir to Mrs. S. C. Langmuir, February 1, 1907; Langmuir to F. J. Pond, July 2, 1909; Pond to Langmuir, July 14, 1909. All in Langmuir Collection.

salary anywhere from $1,200 to $1,400 for the next year. If I like the work I shall accept this . . . while at Schenectady I will be looking around for a really good position in a university."[39] The prediction was incorrect in one way, and probably in a second, also. Langmuir's subsequent letters indicate that the GE job proved itself to be the "really good position" he had sought. And, if Whitney stuck to his previous policies, the salary offer was probably well over $1,400. (Unfortunately, the GE personnel records needed to check this have not survived.)

Giving up teaching to enter industry—exchanging the responsibilities of class preparation, lecturing, and directing student experiments for the different responsibilities of keeping records, securing patents, and providing consulting services—might be regarded as a sacrifice. The preceding letter indicates that Langmuir saw it at first as a temporary expedient. But it turned out to be the exchange of a burden for an opportunity.

What precisely were the industrial responsibilities that Langmuir accepted in 1909? The foremost was the strict requirement that the individual assign all his property rights to his technical work to the sponsoring company. As already described for Weintraub and Coolidge, there could be special arrangements regarding time and treatment. But there could be none regarding technology. Whitney summed up the policy succinctly in a 1909 address. "Whatever invention results from his [the laboratory staff member's] work becomes the property of the company. I believe that no other way is practicable."[40]

From mid-1901, researchers were instructed to make regular entries in their laboratory notebooks. "Write *something* in your notebook every day," Whitney was later to tell new staff members, "even if it's only 'I didn't do a damn thing today.'"[41] Once each month, a "report of work" summing up the month's notebook entries had to be written, signed, witnessed, and submitted. In addition, patentable ideas were described in special letters addressed to Whitney and, if judged promising, forwarded to company patent attorneys.

Comments in Whitney's own lab notebook indicate that he enforced

[39]The sequence of events leading to Langmuir's employment at GE, making it clear that Langmuir, not Whitney, took the initiative in the matter, is clearly laid out in: Langmuir to Whitney, January 20, 1909; Whitney to Langmuir, February 2, 1909; Langmuir to Whitney, February 22, 1909, March 15, 1909, June 12, 1909; and Langmuir to Mrs. S. C. Langmuir, July 16, 1909, all in Langmuir Collection.

[40]Willis R. Whitney, "Organization of Industrial Research," *Journal of the American Chemical Society* 32 (January 1910): 71–78, esp. p. 74.

[41]The "I didn't do a damn thing today" quote was related to me by Herbert C. Pollock, who heard it from Whitney.

these reporting requirements. In the case of one employee who balked, Whitney noted: "Second talk with Jackson on salary. Told him to wait two weeks to see if I could satisfy myself as to his willingness to do right about reports." In another case, he found it necessary to deliver a verbal reminder to a staff member about the "Co.'s right to all inventions, & c."[42]

Whitney felt a corresponding, if less constant, pressure from his own superiors. The laboratory's Advisory Council—a group that included lab cofounders Rice, Steinmetz, and Thomson, as well as other GE technical leaders—gave explicit advice about desirable directions for further work. They also voiced concern about keeping a record to justify the laboratory's existence, as evidenced in a note Whitney made after a 1903 Advisory Council meeting: "Mr. Rice wants account kept of the 'potboilers' we succeed in doing together with their estimated value to the company."[43]

Could this atmosphere of commercial responsibility appeal to an individual whose expressed desire was to be "free to do research as I wish"? In the case of Irving Langmuir, it could and did.[44]

One reason it did involved the excellent facilities and support available. By early 1910, Langmuir had his own personal assistant, a skilled toolmaker named Samuel Sweetser. He informally directed the work of a growing group of assistants with bachelor's degrees in science and at least limited ability to do research work. And while the experiments had their primary focus on improving the incandescent lamp (in particular, determining the effect of the degree of vacuum on lamp life), Langmuir was free to choose his own route to the goal. The one he chose—investigations of chemical reactions at low pressures—also yielded a stream of publishable results in chemistry and physics.[45]

He drew no line between pure and applied research. In 1912 and 1913, for example, his efforts ranged from electron physics to a weekly trip to Great Barrington, Massachusetts, to help veteran inventor William Stanley design a better electric stove. Characteristically, even the stove project yielded theoretical and experimental pa-

[42]Whitney, NB I, February 25, 1904; NB II, March 29, 1903.

[43]Whitney, NB II, October 5, 1903.

[44]Rosenfeld (n. 35 above), p. 42, gives the "free to do research . . ." quote as Langmuir's answer to a question posed by Prof. R. S. Woodward to him while he was a junior at Columbia's School of Mines.

[45]E.g., "A Chemically Active Modification of Hydrogen," *Journal of the American Chemical Society* 34 (1912): 1310; "The Dissociation of Hydrogen into Atoms," ibid., p. 860; "Chemical Reactions at Very Low Pressures. I.," ibid. 35 (1913): 105; "The Effect of Space Charge and Residual Gases on Thermionic Currents in High Vacuum," *Physical Review* 2 (1913): 450.

pers on heat transfer.[46] He was quick to publish in the scientific and engineering journals, averaging more than five papers a year from 1912 on. But he was just as quick to note the practical applications of his work, call in a patent attorney, and suggest key patent claims.[47]

For Langmuir, patents served as a kind of industrial equivalent to the service activities of the academic scientist, such as textbook writing or lecturing. He valued patenting below publication but recognized it as a necessary contribution to the institution that paid his salary. And his patents alone—especially his 1916 patent on the gas-filled incandescent lamp—made that salary one of GE's best investments.[48]

During the next forty years, his research freedom matched that of any academic scientist. He did not abuse it. He continued to participate in attacks on highly practical problems, such as design of industrial electronic devices, methods of aircraft navigation, and the lubrication of meter bearings. This was not a question of paying a price for scientific privileges. Rather, such projects helped maintain a role that Langmuir found had fitted him far better than had an academic position.

Langmuir was not a typical case. He exemplified the creative limit of industrial research. But even in its atypicality, his situation served as an attraction to other individuals making the choice between industrial and academic work. For example, the physicists Albert W. Hull (later a president of the American Physical Society and a member of the National Academy of Sciences) and Saul Dushman (a leading expert in high-vacuum research) came to GE in 1912 and 1913 from teaching positions. Both credit the influence and example of Langmuir with encouraging them to make a career in industry. Not all recruits came to stay—for example, chemist Colin G. Fink and physicist Wheeler P. Davey went from GE to academic posts. Industry attracted few individuals as creative as Langmuir and retained even fewer of them. But those few served as advertisements to other pro-

[46]Irving Langmuir, E. Q. Adams, and G. S. Meikle, "Flow of Heat thru Furnace Walls: The Shape Factor," *Transactions of the American Electrochemical Society* 24 (1913): 53; Irving Langmuir "Convection and Radiation of Heat," ibid. 23 (1913): 299.

[47]On Langmuir's varied technical interests, see Irving Langmuir, "Reports of Work," January 13 and April 26, 1912, Whitney Library, GE RD; Eliot Q. Adams, "Reports of Work," 1911–12, Whitney Library, GE RD; "Minutes of GE Engineering Council," December 24, 1928, copy in Langmuir Collection; and interviews with Willem A. Westendorp and Harold A. Mott-Smith, March 1, 1974, and February 25, 1977, on file at GE RD.

[48]On the significance of Langmuir's patents, see Bright (n. 14 above), pp. 317–25; and W. Rupert MacLaurin, *Invention and Innovation in the Radio Industry* (New York, 1949), pp. 90–100, 164.

fessional scientists of the possibilities of a true scientific career in industry.[49]

Conclusions: GE's Industrial Research Role and Its Possible Implications

By 1916, with the full industrial adoption of the laboratory's contributions to tungsten lamps, and the beginnings of adoption of its work on radio tubes and X-rays, the General Electric Research Laboratory had achieved permanent institutional status within the General Electric Company. Its staff included a dozen Ph.D.-level scientists; some fifty engineers, skilled assistants, and technicians; and a labor force of over 100 for glassblowing, metalworking, and other supporting tasks. Each of these categories would double in numbers by 1930. The post of executive engineer had been created to handle day-to-day administration, freeing Whitney's time for the jobs he did best: sustaining the enthusiasm of his staff through daily tours of the lab, and sounding out other GE leaders about company technical problems suitable for attack by the lab.[50]

Two Whitney trademarks were well established by 1916: his cheery greeting, "Are you having fun?" and the "Come In, Rain or Shine" sign posted over his always-open door. But it was not his personality that accounted for the laboratory's gaining a permanent status within GE; it was the results that had been achieved. The laboratory had been set up to fend off a threat to GE's lamp business, and it had succeeded in this mission. As Chief Justice William Howard Taft was to note in a 1926 decision of the U.S. Supreme Court:

> The Electric Company is the owner of three patents—one of 1912 to Just and Hanaman, the basic patent for the use of tungsten filaments in the manufacture of electric lamps; the Coolidge patent of 1913, covering a process of manufacturing tungsten filaments by which their tensile strength and endurance are greatly increased; and third, the Langmuir patent of 1916, which is for the use of gas in a bulb by which the intensity of the light is substantially heightened. These three patents cover com-

[49]See Albert W. Hull, "Autobiography," on file at GE RD and at the Center for the History of Physics, American Physical Society, New York City; and Saul Dushman, "An Album of Memories," on file at GE RD.

[50]Statistics of the laboratory, indicating a growth in staff from two in 1904 to forty in 1905, a quadrupling of the size of the staff from 1905 to 1915, a further doubling from 1915 to 1930, and a subsequent collapse in size during the Depression (falling below 200), are on file at GE RD. The numerical reduction during the Depression was carried out by separating radio tube work and releasing some staff. But the role and permanence of the laboratory were reaffirmed (see, e.g., William D. Coolidge, Laboratory Notebook 19, December 19, 1933, GE RD).

pletely the making of the modern electric lights with the tungsten filament, and secure to the Electric Company the monopoly of their making, using, and vending.[51]

Other than this business impact, what more general significance did the establishment of the General Electric Research Laboratory have? How did it supplement the concept of industrial research that had already been put forward by Edison at Menlo Park and West Orange, and by the German chemical industry?

One answer is in the new role it offered to professional scientists. To a degree that had previously been impossible, GE's researchers were able to maintain many of the characteristics of the professional scientist's role, while at the same time serving as full-time industrial employees.

They could publish in scientific journals—although the publications had to be approved by management, a requirement that often meant delay. They met weekly in colloquia to hear a paper delivered by a staff member or a visitor. A look at lab notebooks indicates that they were permitted to interrupt their main lines of work to pursue purely scientific speculations—for example, the ideas about the structure of the atom that occupied the thoughts of some of them during the period 1915–20.[52]

The GE laboratory was not a carbon copy of an academic science department, however. Judged on purely scientific criteria, its productivity was much lower. The recent authoritative study by Forman et al. on academic physics circa 1900 provides a very rough bench mark for comparison. It reports that U.S. physics laboratories produced three scientific papers, on the average, for each man-year of scientific effort. For the GE laboratory, even if one counts only staff members with the Ph.D. degree as potential paper producers and includes publications in engineering journals and the *General Electric Review,* the figure for 1900–1910 was less than one paper per man-year. Figuring on a different basis, the academic physics labs produced one paper for every $3,750 in laboratory expenses; the "cost" of each GE publication

[51]United States v. General Electric Co., et al., 272 U.S. 476 (1926). The Just and Hanaman patents referred to were purchased by GE, while the Coolidge and Langmuir patents were outcomes of Research Laboratory work.

[52]On the colloquium, see Whitney, NB 1, November 15 and 29, 1901; NB II, February 29, 1908. For a good example of the mixture of science and technology in succeeding notebook entries, see Saul Dushman, Laboratory Notebook I, pp. 109–25, Whitney Library, GE RD, in which Dushman shifts gears from the analysis of electric cable insulations, a not very exciting technical problem, to one of the most exciting scientific issues of the day, the arrangement of electrons in the atom.

was over $10,000 in annual laboratory expenses.[53] Of course, these figures are extremely imperfect indicators, not empirical measures of productivity. In both universities and industry, publication was only one of a number of activities covered by laboratory expenses. And publication statistics are often heavily dependent on the work of a few individuals—such as Irving Langmuir at GE. They represent only a crude beginning at measuring research productivity.

Papers were not, after all, GE research's most important product. Freedom to publish was instead part of a mechanism by which professional scientists were induced to enter a new role. Edwin W. Rice had stated in 1901 that the new laboratory would be devoted "exclusively to original research."[54] Distinctions among research, development, and engineering were then, and remain, ambiguous. The term "original research" could mean one thing to the industry executive who wrote it, and another to the professional scientist who read it. The context suggests that Rice thought of original research as a means to new inventions, not as an end in itself. "It is hoped by this means," he had said in that 1901 statement, "that many profitable fields may be discovered." Whatever the aim, however, the statement was a message from an industrial leader that the activities of professional scientists were something that industry ought to support.

The important fact is not that GE scientists published so little, but that they published at all. John J. Beer has pointed out that the pioneering German industrial laboratories of the 19th century "felt that fundamental research was the province of the academic institutions," and "did little in the line of purely speculative inquiry." Edison's unwillingness to be distracted into purely scientific areas is also well documented.[55]

[53]Forman et al. (n. 27 above), p. 127. The GE figures are based on my own count of early reprints on file at the Whitney Library, GE RD.

[54]General Electric Co., *Annual Report for 1902,* p. 13.

[55]John J. Beer, "Coal Tar Dye Manufacture and the Origins of the Modern Industrial Research Laboratory," in *The Development of Western Technology since 1500,* ed. Thomas P. Hughes (New York, 1964), pp. 137–38. On Edison, Matthew Josephson notes in *Edison* (New York, 1959), p. 278, that "he habitually assumed a pose of studied indifference to the work of theoretical scientists and mathematicians," and quotes Edison as saying (1884), "I have never had time to go into the aesthetic side of my work." Edison associate Charles L. Clarke quotes Edison as saying of a member of the Menlo Park staff: "When he happens to note some phenomenon new to him, though easily seen to be of no importance in his apparatus, he gets side-tracked, follows it up, and loses time. We can't be spending time that way. We have to keep working up things of practical value—that is, what this laboratory is for. We can't be like the old German professor who, so long as he can get his black bread and beer, is content to spend his whole life studying the fuzz on a bee" (Francis Jehl, *Menlo Park Reminiscences* [Dearborn,

Rather than looking for a line of descent linking Edison's laboratory, the German research institutions, and the American industrial laboratories of the early 20th century, it seems more accurate to view each as an effective response to its economic and social environment. The specific GE response grew out of the conditions of industrial consolidation and increasing status of physical scientists that marked early 20th-century America. As a successful response, it served as an influence on succeeding waves of scientists and research directors.

Harvard graduate student and chemist James B. Conant, for example, retained a lasting memory of a 1913 lecture by Willis R. Whitney extolling the opportunities for scientists in industry. Princeton graduate student and physicist Arthur H. Compton took away from his 1915 visit to the General Electric Research Lab the conviction that pure science was a highly practical career field, as well as an intellectual challenge. Although both men ultimately chose academic careers, they retained close personal and business ties with industrial laboratories. Similar stories were repeated for many others.[56]

Research directors strongly influenced by the General Electric conception of industrial research included C. E. K. Mees, founder of Eastman Kodak's laboratory, and Charles A. Kettering, General Motors' first research director. A visit by Mees to Schenectady while on his way to set up a research organization for George Eastman helped convince him to temper the regimented organization common in his native Britain with a touch of Willis R. Whitney's emphasis on scientific freedom. Kettering, a frequent visitor to Schenectady from 1912 on, came to echo Whitney's view that a research director could suggest and inspire, but could not "direct."[57]

Whitney was certainly influential in shaping American industrial research. But his personal contributions should not be overrated. Nor should those of the model provided by GE. Each company faced its

---

Mich., 1939], 2:862). Clarke concluded: "Edison has always called himself an inventor and nothing more." It should be noted, however, that Edison drew heavily, wherever possible, on scientific publications, and made extensive use of the most up-to-date scientific instrumentation that he could acquire. On this point, see Hughes, "Edison's Method" (n. 3 above).

[56]James B. Conant, *My Several Lives* (New York, 1970), pp. 25–26; Compton to Whitney, June 5, 1914, Whitney Collection. For an excellent overview of the relations between industrial and academic physics during this period and subsequently, see Spencer R. Weart, "The Rise of 'Prostituted Physics,' " *Nature* (July 1, 1976), pp. 13–17.

[57]On Mees, see Jenkins (n. 11 above), p. 310 (where Whitney is mistakenly referred to as "William" R. Whitney); on Kettering, see T. A. Boyd, ed., *Prophet of Progress* (New York, 1961), pp. 77–101.

own blend of competitive challenges and opportunities. Not all had research champions placed, like GE's Edwin W. Rice, in high corporate positions. Only through detailed examinations of the internal records of such industrial research organizations as the Western Electric, Eastman Kodak, Du Pont, and Corning Glass laboratories can the typicalness of the General Electric experience be fully evaluated. Did these organizations define the role of the industrial researcher in the same way that GE did?

It will also be worth examining some "failures"—cases, for example, where the "university-in-exile" model of research freedom matching that of a university proved impossible to sustain. Charles A. Skinner tried it at Westinghouse in 1917, but after three years, management pressures for more visible results forced a drastic reorganization. Ernest Fox Nichols tried to graft a "pure research laboratory" onto GE's lamp works at Cleveland, Ohio, in the early 1920s, but the same problems occurred. Did specific personal or business circumstances cause these outcomes? Or do they indicate that a too zealous commitment to pure science on the part of an industrial laboratory can be harmful to its survival?[58]

At GE, Willis R. Whitney's policy of responsiveness to business needs, and his making no claim of exclusive devotion to pure research, definitely helped insure institutional survival. His policies were not without their weaknesses. An intensely inward-looking orientation caused the staff to miss some of the broader implications of discoveries. For example, those applying Langmuir's work in electronics to radio thought only in terms of point-to-point communication and overlooked broadcasting. Whitney's mixture of strictness (for example, the requirement that every researcher punch a time clock) and lenience (for example, the lack of a formal organization chart, and a loose policy of accounting for expenditures on equipment and material) could be confusing. The empirical bent of individuals like Whitney and Coolidge caused an undervaluing of theory.

But these weaknesses did not seriously detract from the significance of the role played by the General Electric Research Laboratory in the years 1900 to 1916. It showed that, in at least one company, a new role was available for professional scientists. Those attracted to the content of the physical sciences, the identity of a researcher rather than a tester or engineer, the wish to attack practical problems, and the desire to share in the financial rewards offered by industry could now choose a course that would address all of these motives.

[58]Skinner's experiment is described in *The Cosmos of Arthur Holly Compton,* ed. Marjorie Johnston (New York, 1967), pp. 24, 239–42. The Cleveland lamp works experiment was described to me by A. E. Newkirk and C. Guy Suits.

Custom Design, Engineering Guarantees, and Unpatentable Data: The Air Conditioning Industry, 1902–1935

GAIL COOPER

Factory air conditioning appeared in 1902 as an industry dominated by custom design, and air conditioning remained that way for nearly thirty years. Each installation was unique, specially designed to fit a particular set of circumstances. Thus Carrier Engineering Corporation assured prospective customers in 1921 that "it is customary for engineers who specialize in the design of such equipments to treat each problem individually and develop the most efficient equipment for the specific requirements of the client."[1] This approach to production established the overriding importance of engineering expertise within the industry, and the resultant centrality of technical knowledge to business success conferred on engineers both a shaping and controlling power over the emerging technology that was unbroken until the appearance of mass production.

Custom design has been less well studied than mass production.[2] Perhaps that neglect stems from an association of tailor-made products with a bygone era of craft production. Custom production may seem a historic relic, superseded by mass production. That sense of the new pushing aside the old appears as early as 1832 when Charles Babbage made a distinction between "making" and "manufacturing."[3] In his

Dr. Cooper is associate professor in the Department of History at Lehigh University. This essay is part of a larger history of air conditioning published by The Johns Hopkins University Press. Dr. Cooper is currently working on a history of military contracting and the development of statistical quality control in the U.S. and Japan.

[1] *Weather Vein* 1 (May 1921): 42.

[2] See David Hounshell, *From the American System to Mass Production* (Baltimore, 1984); Otto Mayr and Robert C. Post, eds., *Yankee Enterprise: The Rise of the American System of Manufactures* (Washington, D.C., 1981); and Merritt Roe Smith, *Harpers Ferry Armory and the New Technology: The Challenge of Change* (Ithaca, N.Y., 1977).

[3] Martin Campbell-Kelly, ed., *The Works of Charles Babbage: The Economy of Machinery and Manufactures* (London, 1989), 8: 85–86.

discussion of Henry Maudsley's work for the British navy, he describes a system of "making" that included the production of single-item goods with general-purpose tools to the specifications of the client at a relatively high cost. This description of making, or custom production, and the language to describe it prevailed for nearly a century. In 1914 Frederick Halsey, longtime editor of *American Machinist,* self-consciously used the same terms in his machine shop text to distinguish between the work of the shop and the work of the factory.[4]

When they sat down to write their own book on shop practice in 1935, Joseph Wickham Roe and Charles W. Lytle borrowed heavily from Halsey and reproduced the "making" and "manufacturing" distinction, but this time with a difference.[5] Substituting the term "building" for "making," their characterization of custom production focused less on the methods of production and more on the distinctive economic and social relationships that flowed from its practice. "The two processes of production," Roe and Lytle wrote, "from initial sale to final acceptance follow different courses."[6] Custom design pivoted on a direct relationship between buyer and seller, devolving on the engineer an important role from beginning to end. In this system of production, according to Roe and Lytle, "the sale precedes the building and even much of the designing, and the engineer is intimately concerned in the selling as he must convince the purchaser of the superiority of his design."[7] Their description of custom production in all its commercial trappings vividly depicts the features of a vital system of production rather than the shadowy outline of a dying craft.

Roe and Lytle's assessment is useful not simply for their sensitivity to the economic facets of custom production but, more important, for their reportage of a modern variant of the building tradition. They conceded that "building methods will always have their place and are the only ones possible for a large and unstandardized work which must be made to suit special conditions. Great progress has been made, however, particularly in America, in the partial standardization of such work by standardizing the subassemblies and other details employed."[8] They cited the machinery for materials handling as the perfect example

[4]Frederick A. Halsey, *Methods of Machine Shop Work* (New York, 1914).

[5]Joseph Wickham Roe and Charles W. Lytle, *Factory Equipment* (Scranton, Pa., 1935), p. 1. When they published their book, Roe and Lytle were both professors of industrial engineering at New York University. Roe's varied career had previously included stints at Worthington Pump and Machinery Company, Winchester Repeating Arms Company, Pierce Arrow Motor Car Company, and Yale University. He may be best known to readers as the author of *English and American Tool Builders* (New Haven, Conn., 1916).

[6]Roe and Lytle, p. 3.

[7]Ibid.

[8]Ibid, p. 9.

of standardized parts adapted to individual needs. The combination of the two methods, custom design with manufactured subassemblies, drew their approbation, and they pointed out that it "may often be the wisest and most profitable method of production."[9] This distinctive style might be characterized as a combination of the two classic systems of making and manufacturing, but it is more accurate to define it as a variant of custom production since the economic relationships and commercial realities of this method of production differ little from the classic system of "building." It is the place of the engineer in this distinctive system of production that is of special interest.

While superficially linked to an older artisanal tradition, modern custom production is aggressively engineering oriented. Artisanal production may have privileged the consumer, but modern custom production counterbalances the traditional power of the buyer with technical expertise and the prestige of science, tipping the scales in favor of the designer. It is this strong position of the engineer relative to both the technology and the buyer that makes custom production such an interesting arena.

Many custom-designed goods are site-specific; that is, they are unique to the extent that they are adapted to the geography of the place they occupy. Most civil engineering falls into this category, as does architectural design at its best. In contrast, some scientific instruments and machine tools are produced to the particularities of the clients' needs rather than the demands of location. Yet whether adapted to specific locations or special needs, custom-designed products are sold primarily on the basis of their qualities or performance characteristics. The importance of that process lies in the fact that the manufacture of machinery to specific needs and conditions inevitably requires the talents of an engineer or technologist. Custom production relies on engineering talent in the same way that mass production relies on specialized machine tools and flexible production employs skilled labor. This method of production, with its attendant economic relationships, places primary emphasis on the engineer as designer. Thus it is a revealing place to examine the culture of engineering.

From the beginning, air conditioning conformed to Roe and Lytle's pattern of building custom-designed systems out of standardized parts. Thus the importance of the industry is not in its unique patterns but in its illustrative qualities. Air conditioning may be notable only because it developed into a mass-produced consumer product as well, throwing into high relief the distinguishing features of the early industry.

[9]Ibid., p. 1.

The Early Industry

"Air conditioning was born in industry," declared one engineer.[10] Indeed, the commercial development of air conditioning rested heavily on the need of factories to control the levels of humidity for uniform production. Contrary to popular notions, then, air conditioning began with a concern to control the humidity levels of interior spaces, not the temperature. The two are, of course, inevitably linked; hot air has a larger capacity to hold moisture than does cold air. Yet in 1902 there were as many engineers interested in increasing humidity levels for factory processing as there were those interested in decreasing humidity for human comfort. In fact, there were probably a great many more. That smaller cadre of engineers engaged in comfort cooling is well represented by Alfred Wolff, a New York ventilation engineer, who in 1902 designed an air conditioning system for the New York Stock Exchange that provided both temperature and humidity control.[11] His clients included Cornelius Vanderbilt, Andrew Carnegie, the Metropolitan Museum of Art, and a handful of other wealthy customers; and, while this exclusive clientele supported a thriving engineering practice for Wolff, it was not broad enough to sustain an industry.

A more important market for air conditioning emerged in the form of factories that processed hygroscopic materials, that is, materials that absorbed moisture from the air. Textiles, tobacco, pasta, sausage, black powder, chewing gum, candy, chocolate, and flour were only a few of the industries engaged in that kind of processing. Changing humidity levels within the factory meant that materials changed their size, appearance, and handling characteristics; cigarette machinery jammed, chocolates turned gray, and cotton threads broke in adverse weather conditions. The kind of system designed to prevent problems in factory production is called "process air conditioning." Thus, in the same year that Wolff designed the comfort system for the New York Stock Exchange, Willis Carrier installed perhaps the first process air conditioning system in the

[10]Phillip L. Davidson, "Air Conditioning: Its Highlights and Newer Aspects," *Refrigerating Engineering* 29 (March 1935): 19.

[11]Bernard A. Nagengast, "Alfred Wolff: HVAC Pioneer," *American Society of Heating, Refrigerating and Air-conditioning Engineers (ASHRAE) Journal* 32 (January 1990): S66–S80; "Alfred R. Wolff," *American Society of Mechanical Engineers (ASME) Transactions* 31 (1909): 1055–57; "Death of Alfred Wolff," *Heating and Ventilating Magazine* 6 (January 1909): 26; "Heating, Ventilating and Air Cooling at the New York Stock Exchange," *Engineering Record* 51 (April 1905): 413–14, 436–37, 464–67, 490–99; Donald A. Kepler, "Air Conditioning the New York Stock Exchange," *Heating, Piping and Air Conditioning* 19 (April 1947): 69–73; Richard G. Ohmes and Arthur K. Ohmes, "Early Comfort Cooling Plants," *Heating, Piping and Air Conditioning* 8 (June 1936): 310–12.

Sackett-Wilhelms printing plant in Brooklyn to stabilize the registry of multicolor prints. From a virtually even start then, process air conditioning quickly eclipsed the commercial importance of comfort air conditioning.

Industrial air conditioning systems were all tailor-made. They were designed specifically for each factory building and for a particular industrial process. These installations were both site-specific and need-sensitive. That tradition of custom design put a great deal of importance on engineering expertise. Its centrality was heightened even more by the industry's practice of offering engineering guarantees rather than machinery guarantees. Perhaps the best example of the pivotal importance of engineering design to business leadership is the career of Willis Carrier and the Carrier Engineering Corporation.

Carrier is often called the "father of air conditioning."[12] Despite his prominence, he shares the honor of pioneering the field with other capable engineers, notably Alfred Wolff and Stuart Cramer, a southern textile engineer who coined the term "air conditioning" in 1906. Carrier's preeminence in the field is perhaps best understood in terms of his gift of combining theory and practice, the tenacity of his commitment to this emerging field of engineering, and his business success.

He was a Cornell University graduate with a degree in mechanical engineering, whose first job in 1901 with the Buffalo Forge Company set his professional identity as a "centrifugal fan man," a specialist in fans and ventilation.[13] The company first tried to make a salesman out of him, but the man assigned to supervise him, J. Irvine Lyle, sent Carrier back to the home office in 1902 with the recommendation that he be given research and development work.[14] This early assessment of Carrier's talents proved extremely accurate. Later associates told numerous stories of an absentminded theorist—of the occasion, for instance, when he volunteered to make dinner reservations in the hotel dining room but had forgotten his mission by the time he arrived there and instead

[12]Margaret Ingels, *Willis Haviland Carrier, Father of Air Conditioning* (Garden City, N.J., 1952). Ingels held a bachelor of mechanical engineering (1916) and a master of engineering (1920) from the University of Kentucky. She first worked for the Chicago Telephone Company and then for the Carrier Engineering Corporation until 1921 when she embarked on a successful eight-year stint as a laboratory researcher. She rejoined Carrier Engineering Corporation in 1929. While there, she wrote an "engineering biography" of Carrier that was substantially rewritten by a company publicist and published as above. The original draft of her book exists in the Carrier Papers, Accession no. 2511, Cornell University Archives, Ithaca, New York, hereafter referred to as "working manuscript."

[13]Interview of Willis Carrier by Margaret Ingels, June 22, 1948, Carrier Papers, no. 2511, Box 29, Cornell University Archives.

[14]*Weather Vein* 1 (June 1921): 9.

sat down to eat by himself, leaving his party of friends waiting endlessly in the hotel lounge. Carrier's absentmindedness was the flip side of his talent for analytical thinking; thus, in an industry that rewarded engineering skill with business success, Carrier's idiosyncracies were markers of the personal strengths he brought to the development of this new technology. It was his engineering skill and business success, two inseparable qualities, that account for his subsequent recognition.

Carrier attributed his belief in the importance of first principles to his mother, who explained fractions to him by cutting up apples in the family's kitchen and telling him, "Think to the bottom of things. . . . It's better to learn 'why' of one thing than to see the surface of a dozen things." However, the limits of theory must have been forcibly borne in on Carrier in 1906 when he stood in the Chronicle Cotton Mills in Belmont, North Carolina, to inspect his first installation in a textile mill. In cotton mills, high humidities were necessary to keep the fibers elastic, especially during spinning and weaving. Dry air caused thread breakage and a consequent machine shutdown. Traditional methods of humidification could not cope with the increased heat produced by faster machinery speeds, and a faster schedule of production in textile manufacturing. Carrier's Chronicle Cotton Mills installation was meant to relieve such humidification problems, but the intense heat of the machinery was more than he bargained for. He had designed the system at the head office in Buffalo, New York, without ever having been in a cotton mill, and as he felt the incredible heat the machinery produced—still hot enough to raise a burn several minutes after it had been shut down—he knew the system would never work as designed.[15]

In that moment, Carrier learned the importance of matching the capacity of an air conditioning system to the activities within a building and not just to the building itself. This awareness of the role of heat generated within the factory was new to the humidification industry. Traditionally, companies that handled industrial humidification made design estimates on the basis of building size. *Kent's Mechanical Engineers' Pocket Book,* which one engineer called "the Bible of heating engineers," advocated a rule of thumb based on total volume of the building.[16] As one manufacturer of humidifiers remembered common practice in 1906, "Textile manufacturers had been content to buy—and humidifier manufacturers had been content to sell humidifiers on a 'perhaps' basis. So many humidifiers to about so many cubic feet."[17] Heating and

[15]Interview of Willis Carrier by Margaret Ingels, August 8, 1949, Carrier Papers, no. 2511, Box 27, Cornell University Archives.

[16]Ingels, working manuscript, pp. 53–54, Carrier Papers, no. 2511, Box 28, Cornell University Archives.

[17]Advertisement, Parks-Cramer Company, *Southern Textile Bulletin* 17 (August 14, 1919): 4.

ventilating engineers had refined that formula from the simple cubic contents of a room to a method for considering the exposure of the walls and the role of outside climate. Yet even the improved methods for estimation produced occasional but spectacular failures. In 1918, *Heating and Ventilating* magazine noted that "the science on this subject did not develop very rapidly. . . . Some very simple facts and theories were literally dragged forth from obscurity by repeated failures in obtaining uniform results, because of the fact that physical laws had been overlooked."[18]

With a new awareness of the role of heat generated within the factory, Carrier began to estimate 2,520 BTU per hour for each horsepower used to drive the machinery; Cramer used the value of 2,550 BTU per hour.[19] Another early air conditioning pioneer called this value "one of the real keys to the calculation of the size of equipment."[20] The construction of the mill and its location were important, but so, too, was the heat generated within the factory.

The importance of accurately estimating the heat load and consequently the size of the machinery was all the more crucial because of the nature of guarantees that companies offered their clients. Often these were atmospheric guarantees rather than the more common guarantee on machinery. Indeed, contracts specified a range of temperature and humidity levels that could be maintained in the factory depending on the season. Air conditioning engineers quite simply sold air.

A Carrier engineer recalled the power of that kind of guarantee to clinch a business deal. A company representative reported that he faced a lower bid from rival B. F. Sturtevant Company and persuaded the client to read the Sturtevant proposal aloud so that the two could be compared word-for-word. As he remembered, the exercise revealed that "both quoted upon the same basic essentials, that is, air washer, heater, fan, ducts, and the controls. The situation got worse and worse until they finally came to the guarantees and that let the cat out of the bag.

[18]"Developments in Heating and Ventilating during the Past Five Decades," *Heating and Ventilating* 15 (August 1918): 18.

[19]Carrier's value was calculated from data published in W. H. Carrier, "A New Departure in Cooling and Humidifying Textile Mills," *Textile World Record* 33, suppl. (May 1907): 363–69. Cramer's value is cited in an L. L. Lewis memo, "Competition in the Early Days," December 13, 1956, Carrier Papers, no. 2511, Box 4, Cornell University Archives. Lewis took Cramer's value from Stuart Cramer, *Useful Information for Cotton Manufacturers* (Charlotte, N.C., 1904). The presence of this data in public print leads the historian to conclude that they were not secret, yet abundant testimony from engineers on both sides—those privileged and those excluded—claims that this information and a complex of values like it that were necessary for accurately determining heat loads were not generally known.

[20]L. L. Lewis, memo, "Competition in the Early Days," December 13, 1956, Carrier Papers, no. 2511, Box 6, Cornell University Archives.

Sturtevant had guaranteed the volume of air which the system outlets would handle and that all of it would be delivered from the outlets at a relative humidity of 70%; Carrier guaranteed that the system would maintain 70% in the mill."[21]

Guaranteeing atmospheric conditions throughout the mill rather than at the outlet was a strong selling point but a more difficult engineering proposition. Atmospheric conditions were established by a dynamic relationship between the air conditioning system and factory conditions. Designing a system that could respond to changing conditions within the factory required just such a precise estimate of heat generated by lights, people, machinery, and industrial processes that Carrier became adept at providing. These kinds of guarantees reinforced both the need for custom design of systems and the primacy of engineering expertise.

Thus air conditioning companies could successfully meet atmospheric guarantees only if the designer were familiar with the requirements of specific industrial processes and designed the system accordingly. The New York consulting engineer Walter Fleisher believed that the need for industry-specific engineering expertise was one of the factors that kept the number of air conditioning companies small. He explained that "to cope with industrial air conditioning, one had to be familiar with the industry and as the need for an air conditioning installation was sporadic in any one vicinity only those companies who were nationally involved were able to maintain an air conditioning engineering force able to cope with the widely spread requirements for installations or engineering works."[22] This reliance on expertise led to the appearance of a small coterie of air conditioning firms that dominated the industry.

The importance of engineering expertise does not mean that patents played no role in the industry. Patenting activity in this period centered around control instruments for regulating temperature and humidity. There were at least three significant groups of patents among the pioneers: Cramer patented a control device in 1904; Carrier patented his control instrument in 1906; and William G. R. Braemer patented a similar one in 1907.[23] In 1936, when Carrier recalled the pioneering air conditioning companies, he named the S. W. Cramer Company, the Carrier Air Conditioning Company of America, and Warren Webster &

[21]Ibid.

[22]Walter Fleisher, "Air Conditioning—Past, Present, and Future," speech presented at New York University, 1947, Carrier Papers, no. 2511, Box 28, Cornell University Archives.

[23]Stuart Cramer, U.S. Patent 811,383, filed December 28, 1904, issued January 30, 1906; Willis Carrier, U.S. Patent 854,270, filed July 16, 1906, issued May 21, 1907; William Braemer, U.S. Patent 885,173, filed September 11, 1907, issued April 21, 1908.

Company—companies that installed systems based on these three patents.[24] This taxonomy of the early industry reveals Carrier's own preference for defining the industry in terms of inventions and patents. As he was the holder of one of the three pivotal patents, this organizing principle ensured for Carrier a prominent place in the pantheon of pioneers, a prominence reinforced by the fact that some regarded the patented control device of his onetime coworker, Braemer, as essentially derivative of Carrier's own invention despite its separate patent rights.

In addition to these three patent holders, at least three instrument companies manufactured temperature and humidity controls: the Taylor Instrument Company, the Powers Regulator Company, and the Johnson Service Company. As early as 1902, Wolff installed a Johnson Service Company humidistat in the Andrew Carnegie residence for the automatic regulation of humidity, and the American Blower Corporation apparently did all its air conditioning from 1906 on with controls from these three companies.[25] Thus it seems clear that control instruments were a major element in the competition of the new industry, but they were not decisive. These companies were pioneers but did not necessarily retain their dominance. The practice of offering customers atmospheric guarantees shifted importance away from machinery and emphasized instead engineering expertise. In contrast to Carrier's list of industry leaders, consulting engineer Walter Fleisher estimated that there were probably only three air conditioning companies active on the national level in the early years: the Carrier Engineering Corporation, the American Blower Corporation, and his own firm, W. L. Fleisher & Company. Each of these companies rose to leadership through engineering skill. Indeed, the convergence of custom design and atmospheric guarantees led to the emergence of an oligopoly of engineering firms.

Those oligopolistic conditions made it easier for the industry leaders to control the information so essential for success. Estimating heat loads accurately and sizing the equipment accordingly through an acquaintance with the industry and a knowledge of engineering principles were essential skills for meeting atmospheric guarantees. Among the ventilation fraternity in 1906, this varied kind of information was valuable and not widely known. At a meeting of heating and ventilating engineers, Edward Berry of Philadelphia broached the subject of rules for estimating the size of equipment: "I think there must be some of the older men

[24] W. H. Carrier, "Progress in Air Conditioning in the Last Quarter Century," *Heating, Piping and Air Conditioning* 8 (August 1936): 447–59.

[25] [A. P. Trautwein], "Record of the Principal Work of Alfred R. Wolff," *Stevens Institute of Technology Indicator* 26, no. 1 (January 1909): 14; "American Blower Corp.," Carrier Papers, no. 2511, Box 28, Cornell University Archives.

of the Society who have adopted rules which are accurate enough. There seems to be no way of getting track of such trade secrets," he lamented.[26]

Indeed, the commercial importance of such data encouraged secrecy. Monte Calvert has argued that, among 19th-century machine shops that produced specialized machines for industrial customers, commercial relationships were grounded in personal relationships and trust, not innovative techniques. As a consequence, technical information flowed freely from one shop to the next.[27] The air conditioning industry did not conform to that pattern. Unlike the clients of these machine shops, the majority of air conditioning customers were buying their first system and had no prior experience with the firm with which they contracted. In this new industry, technical competence was not only the key to fulfilling guarantees but the rock on which commercial competitiveness was honed. Not surprisingly, any technical advantage that could not be protected by a patent was maintained as proprietary knowledge. As engineer Phillip L. Davidson remembered, "All engineering knowledge was very closely guarded as trade secrets, even such first grade principles as the fact that a machine requiring 1 h.p. to operate would liberate 42.5 Btu. per minute."[28]

Such principles of engineering design have been identified by Edwin Layton as engineering science, and he has persuasively argued for the role of engineering science in establishing engineering as an independent "mirror-image twin" of the scientific community.[29] Yet, little has been said of the role of engineering science in the commercial realm. That knowledge can be regarded as basic engineering science, or, alternately, it can be seen as proprietary knowledge. The engineering community saw it both ways.

Air conditioning developed alongside the American Society of Heating and Ventilating Engineers (ASH&VE), founded in 1904. In his 1917 presidential address before the society, Harry M. Hart imagined that engineers would turn to the society to gain expertise from other engineers. He linked the "increasing demand for suitable equipment to meet new requirements in heating and air conditioning" to an increase in membership. He imagined that these new members were motivated

[26]"Topical Discussions: Advantages and Disadvantages Attending the Use of the Thumb Rules," *ASH&VE Transactions* 12 (1906): 207.

[27]Monte Calvert, *The Mechanical Engineer in America, 1830–1910: Professional Cultures in Conflict* (Baltimore, 1967), pp. 6–7.

[28]Phillip L. Davidson, "Air Conditioning Reminiscences," *Refrigerating Engineering* 37 (1939): 236.

[29]Edwin Layton, "Mirror-Image Twins: The Communities of Science and Technology in 19th-Century America," *Technology and Culture* 12 (October 1971): 562–80.

by "the need of a closer co-operation among engineers in order to be better able to meet the new demands."[30] That ideal of professional cooperation had suffered, however, when the society organized a session on unusual air conditioning and drying installations in factories and experienced difficulty persuading active companies to contribute papers. "A seeming reluctance by some of our members to present papers of this character," Hart scolded, "would indicate a lack of accurate data or a mistaken idea that the giving out of such data would result in the loss of some advantage, presumably over competitors."[31] He continued, "This subtle reluctance that seems to grip so many of our members has been one of the greatest drawbacks to the advancement of the science of heating and ventilating."[32]

Within the air conditioning industry, manufacturers and engineering firms were split on their willingness to promote the free flow of technical information. Generally, manufacturers of component parts like fans, air washers, and refrigeration equipment were active in disseminating information that would help their customers use their products most successfully. Like early automobiles, air conditioning equipment was generally assembled from the parts supplied by a variety of manufacturers. Thus it utilized standard components common in the ventilation industry, but each system was assembled on demand according to the needs of the client. Manufacturers' catalogs often contained "textbook" sections that included simple tabular data but also the basic engineering principles behind the problems customers faced. One engineer declared that "the manufacturer of heating apparatus is the greatest disseminator of knowledge concerning apparatus there is," while another marveled that any blower manufacturer "will send any one any information that he may ask, even to the extent of a whole library."[33]

The Buffalo Forge Company was one such manufacturer. Its catalog noted that the company "has always taken the stand that engineering data and developments should not be hoarded as hidden treasures but should be made available for the use and edification of the engineering profession in general."[34] Carrier's psychrometric table was one example. While Carrier used the Department of Agriculture psychrometric tables

[30]"Proceedings, Twenty-third Annual Meeting, 1917: The Presidential Address," *ASH&VE Journal* 23, no. 3 (April 1917): 348.

[31]Ibid.

[32]Ibid., p. 349.

[33]J. D. Hoffman, "Reasons Why the Science of Heating and Ventilating Engineering Should Be Observed," *ASH&VE Transactions* 14 (1908): 89; "Topical Discussions: No. 2. Reluctance to Divulge Alleged Secrets," *ASH&VE Transactions* 17 (1911): 309.

[34]Buffalo Forge Company, "Buffalo Fan System of Heating, Ventilating and Humidifying," Catalog no. 700 (New York, [1920]).

published in 1900, they did not provide the precision that he needed for air conditioning. With support from Buffalo Forge, he derived his own values and published the resulting table in the textbook section of a 1906 company catalog.

Engineering companies, in contrast, were less likely than manufacturers to reveal the engineering data behind their successful installations. Carrier personally made the transition from a manufacturing company to an engineering firm with a consequent change of perspective. From 1907 to 1914, he worked at the Buffalo Forge Company subsidiary, Carrier Air Conditioning Company of America. He cherished the ambition, conceived in 1905, of publishing a handbook on air conditioning, and his professional papers were part of that greater plan. As Margaret Ingels described it, "When no paper was read, and therefore none published, on a subject he wanted to use as a reference in his handbook, he or [Frank L.] Busey would write the paper, present it, and when it was published he'd have his reference."[35] The handbook, *Fan Engineering,* appeared in 1914 while Carrier was still at Buffalo Forge.

As war loomed in Europe, however, the senior partner at Buffalo Forge, William Wendt, worried about business stability. Primarily a fan manufacturer, the company felt it could no longer continue to compete with its own customers by maintaining an engineering subsidiary.[36] Faced with this crisis, seven Buffalo Forge employees headed by Carrier and Lyle established a completely independent company, Carrier Engineering Corporation (CEC). It was clearly an engineering company in which expertise was the principal commodity. With the exception of control instruments, it did no manufacturing, instead buying fans from Buffalo Forge and refrigeration equipment from the York Manufacturing Company.[37] The creation of air conditioning systems out of standard parts supplied by a variety of manufacturers was essentially the same process in both the old and new company, but a patent agreement between Carrier and Buffalo Forge, which gave the latter part of the rights to all of Carrier's patents, extended beyond the term of his employment with the company. Until the expiration of the patent agreement, Carrier could not compete on the basis of innovative new hardware without yielding half the advantage to his old company. Thus CEC began life as an engineering firm. As CEC told its customers, "We

[35]Ingels, working manuscript, p. 188, Carrier Papers, no. 2511, Box 28, Cornell University Archives.

[36]Ibid., p. 192.

[37]L. Logan Lewis, "Memo Relating to Carrier Engineering Corporation and Its Agreement with Buffalo Forge Company," November 13, 1958; "Fate Plays a Hand: Another Month and We Would Never Have Organized Carrier Corporation," Carrier Papers, no. 2511, Box 12, Cornell University Archives.

are not interested primarily in the manufacture and sale of equipment for this is an organization of engineering specialists."[38]

As an engineering firm, the new company pursued an explicit policy of maintaining a measure of secrecy about its methods. "For many years," recalled L. L. Lewis, a CEC engineer, "we refused to divulge the air-handling capacity of the system and revealed only its requirements for floor space, power, water, and steam."[39] In particular, the company was not forthcoming about the details of how it estimated its designs. Lewis called CEC's competitors "troublesome rather than effective," noting that "no consulting engineer had then developed the skill that would enable him to specify air conditioning."[40] Carrier's 1906 insight at Belmont that machinery, people, lights, and processes all produced a substantial amount of heat that affected air conditioning performance turned the company's attention to the mill interior rather than its exterior. Indeed, it was Carrier's ability to put a quantitative figure on many of the activities that occurred within the factory that gave the new company a competitive edge. As one company member recalled, "Up to about 1925 Carrier people had, to all practical purposes, a monopoly on the brains and the know-how of air conditioning. Consequently its policy was to educate its own people but at the same time to take extreme measures to prevent this knowledge and practical experience from getting into the hands of outsiders—especially competitors but including consulting engineers and contractors in pretty much the same class."[41] In this new circumstance, Carrier Engineering Corporation's British subsidiary now complained that Carrier's *Fan Engineering* handbook was too informative, and the company agreed to refrain from distributing it in Britain.[42]

Because air conditioning was still developing, an engineer's "experience" quickly became "expertise." Each installation presented new problems that company engineers solved on the job, and those solutions

[38]"Carrier Research Service Available to Manufacturers," *Weather Vein* 5, no. 4 (1925): 45.

[39]L. L. Lewis, "The Romance of Air Conditioning," pamphlet (Syracuse, N.Y., n.d.), pp. 7, 9.

[40]L. L. Lewis, "Competition in the Early Days," memo (December 13, 1956), p. 1, Carrier Papers, no. 2511, Box 6, Cornell University Archives.

[41]"Architects & Engineers Manual Historical Notes," June 20, 1958, Carrier Papers, no. 2511, Box 2, Cornell University Archives. Use of the phrases "we" and "our competitors" marks this unidentified memo as the product of a company member. Although it does not carry an attribution, it matches the style of many other memos that L. L. Lewis included in this collection and matches also the sentiments expressed in a similar document of known authorship.

[42]A. M. Sanderson to J. I. Lyle; E. T. Murphy to A. E. Stacey, Jr., December 6, 1927, Stacey Correspondence, Box C-3, Carrier Collection, United Technologies Archives, East Hartford, Connecticut.

informed subsequent practice. The continuing importance of practical experience was made clear by John R. Allen, professor of mechanical engineering at the University of Michigan, in the preface to his 1905 textbook on ventilation. He noted that "the design of heating and ventilating systems has not been reduced to an exact science. . . . One reason for this is the lack of exact experimental data governing some of the most important factors entering into these calculations. This lack must be filled from the designer's experience."[43]

Since a separate research department was not established at CEC until 1919, the factory became a laboratory for the company engineers. Industrial applications became the main avenue for advancements in the art. This phenomenon posed a dilemma for the company: how could one base a firm's competitiveness on this highly individualistic engineering skill? The corporation's answer was to convert individual engineering skill into corporate expertise. In early 1916, each engineer recorded his experience and deposited it in company files called "Confidential Data" files.[44] This formed the heart of CEC's "unpatentable" engineering knowledge. Organized by industry, each report was authored by the company's specialist and represented a summation of his experience.

The extent to which the firm could effectively expropriate individual skill was problematic. In a new field in which so much was unknown, the Confidential Data files record a style of engineering that was often highly intuitive. The "art" of air conditioning, which depended on an individual's intuition informed by experience, proved difficult to reduce to paper. One engineer wrote frankly in a document deposited in the files, "There are so many independent variables in this part of the proposition that it is impossible to carry the calculations in logical order and proper sequence. The writer had a fairly accurate knowledge of many of the sizes from previous experience, and consequently was able to assume certain factors quite accurately. In the case of others, a method of cut and try, which is not indicated here, was followed."[45] This contrast between the art of engineering, based on individual experience and intuition, and the science of engineering, expressed in unambiguous mathematical formulas, mirrored the conflict between shop culture and school culture in the larger realm of mechanical engineering. Clearly, business displayed a greater affinity to "science" than "art," to empowering the corporation rather than the individual. Despite the

[43]John R. Allen, *Notes on Heating and Ventilation* (Chicago, 1905), p. 3.

[44]"Architects & Engineers Manual Historical Notes," June 20, 1958.

[45]L. L. Lewis, "Theatre Cooling," September 7, 1921, Confidential Data no. 24, Carrier Collection, United Technologies Archives.

difficulties of recording this kind of design, the Confidential Data files were a valuable resource for CEC.

Thus unpatentable data was the third leg, along with custom design and engineering guarantees, which supported a business oligopoly in the early air conditioning industry. The number of air conditioning firms was small and remained that way until the middle of the 1930s. While the number and identity of the principal firms varied over this thirty-year span, the group was never large. Refrigeration engineers, whose interest in the new technology during the 1930s increased in proportion to the deepening economic crisis, were vocal about the tight circle that controlled the industry. "The very crux of air conditioning is air movement, and practically all the crucial knowledge of it is in the hands of a few experienced men," a frustrated refrigeration engineer wrote in 1934.[46] As late as 1936, a heated debate broke out at a large symposium where the chairman claimed that there were not four air conditioning companies in the United States competent to estimate a job accurately.[47] David Fiske, editor of *Refrigerating Engineering*, acknowledged "that the art of air conditioning originated in the art of ventilation, which was an art understood by a very small number of men and manufacturers. The art has continued to be 'closely held.' "[48]

The Ideal of Artificial Climate

That combination of power, both technical and commercial, allowed the engineering community to shape the new technology. While Carrier was inclined to give greatest credit to those individuals whose inventive imagination created its first machines, the importance of engineering expertise in re-creating each system—in designing a site-specific, need-sensitive installation—meant that it is more accurate to say that the technology was defined by the small coterie of engineering firms that created the new installations.

Cramer coined the term "air conditioning" in 1906 to include "humidifying and air cleansing, and heating and ventilation."[49] Yet behind the term "air conditioning" was a shifting set of machinery, and industry leaders like CEC were anxious to standardize the technology.

[46]Everett R. Ryan, "How Will Air Conditioning Be Marketed?" *Refrigerating Engineering* 28 (August 1934): 65.

[47]"Editorially Speaking," *Refrigerating Engineering* 32 (September 1936): 137.

[48]David L. Fiske, "Air Conditioning and Engineering," *Refrigerating Engineering* 30 (July 1935): 25.

[49]Stuart Cramer, *Useful Information for Cotton Manufacturers*, 2d ed. (Charlotte, N.C., 1906), 3: 1208. This is an "extract with slight modifications" of Cramer's paper before the tenth annual convention of the American Cotton Manufacturers Association, Asheville, N.C., May 16–17, 1906.

In 1911 Carrier similarly defined air conditioning as "control of desired atmospheric conditions . . . with respect to moisture, temperature and purity."[50] These characteristics were not chosen randomly. "Temperature" and "humidity" were inseparable elements that logically needed to be addressed together. "Cleanliness" or "purity," however, was a bonus provided by the technology's use of water sprays to control humidity—water sprays that washed lint, dust, and other impurities from the air. "Control" was provided by several newly invented control instruments on which both Cramer and Carrier had taken out patents. Thus, this definition—control of temperature, humidity, and purity—was one that, in a sense, started with the machinery and described its capabilities.

Increasingly, however, air conditioning was conceptualized by engineering companies not simply as a range of functions or machinery but also as artificial climate. Carrier Engineering expressed this grand vision of the technology when it adopted the term "manufactured weather" to describe its systems. This conceptualization differed from the simpler functional definition in that it suggested an ideal that shaped rather than described the technical components of the system. In comfort installations of the 1920s, for example, the ideal of artificial climate led to experimentation with ionization, ozone, variable-speed fans, and ultraviolet lamps as methods for reproducing the full range of natural climate indoors. But above all, artificial climate contained the idea that mechanical ventilation would replace rather than augment natural ventilation. It contained the ideal of an atmospherically controlled environment. A 1921 CEC publication expressed this clearly: "Obviously, therefore, the *weather* inside a modern Textile Mill must be manufactured—the temperature and Relative Humidity must be accurately controlled."[51] The conceptualization of air conditioning as manufactured weather expressed not only the potential of the new technology in a way that might appeal to manufacturers but also the brash self-confidence of the air conditioning engineers. The implementation of this mechanical ideal not only relied on their knowledge and skills but also on the willingness of manufacturers and workers to yield atmospheric control of the factory to a technical elite.

Like atmospheric guarantees, the concept of artificial climate placed the emphasis on air conditions rather than machinery. Similarly, its successful realization was dependent on correctly judging the dynamic relationship between machinery, the building, and its occupants. And in ways that perhaps were not quite anticipated, the successful installation

[50]Willis H. Carrier and Frank L. Busey, "Air-Conditioning Apparatus; Principles Governing Its Application and Operation," *ASME Transactions* 33 (1911): 1055.

[51]*Weather Vein* 1 (March 1921): 17.

of any system of artificial climate required that the air conditioning company control all three elements. Thus engineers attempted to fulfill both their guarantees and the promise of the technology by sealing windows, closing doors, and freezing work patterns and production processes. To achieve environmental control, air conditioning engineers insisted on control first over the factory building and then increasingly over the activities within the building.

The Tripartite Control over Atmospheric Conditions

While designing a system to meet atmospheric guarantees was a matter of engineering expertise, maintaining atmospheric control over the building led to a persistent involvement in production processes themselves. A larger role for air conditioning engineers in the work of the factory challenged and circumscribed both the decisions of management and the prerogatives of workers. In fact, clients who operated either the air conditioning machinery or their production processes in unanticipated patterns were a constant irritant to an engineering company hostage to its atmospheric guarantees. The conflict over changing patterns of production is illustrated by CEC's exchange with Atlas Powder Company's Wolf Lake plant.

Wolf Lake was using a CEC installation to dry black powder when the company began to complain to CEC about excessive moisture in the final product.[52] Faced with a balky system, Wolf Lake personnel "attempted to help themselves out of their trouble by several methods of baffling," but CEC engineers concluded grumpily "that some of the baffles are undoubtedly interfering with proper circulation in the room."[53] The corporation sent H. B. Forbes to investigate; he was perhaps aware of his boss's assessment that "it will be helpful to all of us if we can sell the local people on the idea of this dryer. They are apparently a bit old-fashioned and this type of dryer was purchased over their objections."[54] Due to an explosion a few days before his arrival, Forbes had to wait to get half a dryerful of powder but his ultimate conclusion was that the company had been using a batch dryer as a continuous dryer, adding a new load of wet powder before removing the first load.[55] Carrier Engineering insisted on batch

[52]I. K. O'Brien to P. F. Pie, November 29, 1933; A. P. Shanklin to A. E. Stacey, Jr., November 29, 1933, Stacey Correspondence, Box C-1, Carrier Collection, United Technologies Archives.

[53]A. P. Shanklin to Chicago Office, November 29, 1933, A. E. Stacey Correspondence, Box C-1, Carrier Collection, United Technologies Archives.

[54]A. P. Shanklin to A. E. Stacey, Jr., December 4, 1933, Stacey Correspondence, Box C-1, Carrier Collection, United Technologies Archives.

[55]H. B. Forbes to A. E. Stacey, Jr., n.d., Stacey Correspondence, Box C-1, Carrier Collection, United Technologies Archives.

drying even though it might slow production. Atmospheric guarantees meant that the company had a continuing interest in how their product was used and quite a bit of power to reorder the factory around the needs of the air conditioning system.

Carrier engineers, sent to resolve complaints by clients about system performance, were likely to complain in turn about the independent actions of factory officials. Factory personnel closed dampers, built baffles, and ran the machinery contrary to design assumptions.[56] A year later the Wolf Lake superintendent was alarmed to find black powder in the return air duct and began running the machines entirely on outside air, even though the system had been designed to handle a given percentage of recirculated air.[57] Efforts to make good on guarantees often pitted engineers against their clients, with engineers insisting on a larger role in factory production methods or schedules.

Indeed, the most comfortable position for the air conditioning company trying to make good on an atmospheric guarantee was to freeze the production process around an industry standard. If, for instance, all rayon plants adopted the same processes, they would present the same engineering problems for the air conditioning engineer. Carrier Engineering counted on some standardization within the industry to make the heavy initial investment in engineering work pay off in subsequent installations. For that reason they declined to take a job for a company that dried golf clubs because they felt the potential number of installations was too small to warrant the time involved. In general, diversity in production methods created uncertainty in new installations; improvements in processing overset design parameters. In one instance, CEC engineer A. E. Stacey noted with exasperation that the problem was not with the company's air conditioning system but with the client, Amoskeag Manufacturing Company. Amoskeag had a "lack of knowledge of the business," Stacey maintained.[58] Using his general experience in the industry to suggest that it ran one process too long, he charged that the company "did not know how to make rayon."[59] The focus on atmospheric conditions as the product and artificial climate as the ideal set engineers on a collision course with both management and labor as they tried to transform the factory into an atmospherically controlled environment.

[56]H. B. Forbes to A. E. Stacey, August 27, 1920, Stacey Correspondence, Box C-1, Carrier Collection, United Technologies Archives.

[57]P. L. Davidson to A. E. Stacey, August 31, 1934, Stacey Correspondence, Box C-1, Carrier Collection, United Technologies Archives.

[58]A. E. Stacey to C. G. Norton and E. T. Lyle, March 5, 1931, Stacey Correspondence, Box C-1, Carrier Collection, United Technologies Archives.

[59]Ibid.

No less a conflict evolved with workers. For workers in hygroscopic industries—industries that worked moisture-sensitive materials—part of their skill was based on an understanding of the complexity of natural material and its response to weather and to mechanized processing. Those manufacturers who processed hygroscopic materials watched the weather carefully. Factories were affected by changes in climate, season, and weather, and adverse environmental conditions could bring mechanical production to a halt. Production could be problematic when unchanging production machinery encountered variable natural materials. Weather-sensitive materials strongly linked the factory to the natural world, a world that air conditioning engineers hoped to replace with manufactured weather.[60]

In the textile industry, for example, spinners and weavers regulated windows to change environmental conditions. While managers generally operated the mechanical equipment, windows were within easy reach of most workers. Windows became a contested ground, for managers were never sure that workers were trying to achieve the best production conditions. While closing windows could increase humidity within certain limits, that increase was almost always achieved at the expense of ventilation and personal comfort. Because of the conflict between the needs of production and the conditions of comfort, the manner in which workers chose to regulate the windows under their immediate control was not easy to predict. Those who were paid by piecework rates might decide to keep the windows closed in warm weather to preserve high humidity levels and increase their productivity despite the personal discomfort, or they might sacrifice wages for better working conditions.

That conflict between process and comfort undermined the authority of factory operatives who were often experienced judges of proper environmental conditions in an inexact craft. The proper levels of humidification depended on the type and size of the yarn or cloth being produced, the length of time the fibers were exposed to the conditioned air of the workroom, and the type of mechanical processes being used, among other variables. Carrier Engineering's textile expert, E. P. Heckel, admitted in the Confidential Data files that it would be "quite a proposition" to itemize the different humidities required for the different departments and different grades of cloth. Not only did variety make this task difficult, so did difference of opinion. "I have known

[60]For a discussion of complex natural materials and the engineering perspective, see Charles M. Haines, "The Industrialization of Wood: The Transformation of a Material" (Ph.D. diss., University of Delaware, 1990). Sigfried Giedion in *Mechanization Takes Command: A Contribution to Anonymous History* (New York, 1948) discusses the attempts to make organic material mesh with mechanical production machinery.

some spinners that would be delighted if you maintained for them relative humidities of 50%," he wrote, "and other spinners who would jump at you with all fours if you tried to give them less than 65%." Carrier Engineering routinely supplied 60 percent relative humidity in cotton spinning and 75 percent in cotton weaving.[61] Nevertheless, it is clear that, while these humidity levels were produced with great precision and consistency, the standards themselves were averages; they did not represent any greater insight into the response of natural materials to weather fluctuations than that supplied by experienced spinners, weavers, or overseers. What an air conditioning system did supply was greater capacity, automatic controls, and an end to conflict between labor and management over the manner in which windows were regulated.

Despite the considerable knowledge that factory workers possessed about environmental conditions, engineers insisted on total control over the environment. Artificial climate displaced natural climate and the skills that went along with it. A series of air conditioning advertisements by the Parks-Cramer Company makes the underlying issues quite graphic (figs. 1 and 2). At least eleven advertisements drove home the point that open windows in the textile mill were bad for production and profits.[62] One proclaimed that "spinners can't be trusted with a window"; another advised owners to break the "window-lust of your spinners."[63] The text that followed these provocative statements impugned not the motivation of workers but their skill at atmospheric control. "Asking spinners to create uniform conditions in the spinning and weaving rooms by adjusting windows is like asking untrained weather observers to give an accurate weather report," the ad maintained.[64] While the cutout doll figures of female spinners and male supervisors seem to depict the conflict between labor and management, it is the unseen figure of the engineer who is really the chief actor in this drama, for the struggle is, in essence, one between craft knowledge and engineering science.

The rivalry between Carrier air conditioning engineers and skilled workers can be seen more clearly in other industries. Perhaps the best documented example of the impact of the new controlled factory

[61]E. P. Heckel, "Textile Humidities," December 28, 1920, Confidential Data no. 18, Carrier Collection, United Technologies Archives.

[62]Parks-Cramer Company advertisement, *Textile Bulletin* 48 (June 27, 1935): 32; 48 (July 25, 1935): 2; 49 (October 31, 1935): 28; 50 (April 16, 1936): 48; 50 (May 28, 1936): 12; 50 (June 25, 1936): 52; 50 (July 23, 1936): 13; 50 (August 20, 1936): 16; 51 (November 19, 1936): 17; 51 (December 17, 1936): 44; 51 (January 21, 1937): 44.

[63]*Textile Bulletin* 50 (May 28, 1936): 12; 48 (June 27, 1935): 32.

[64]*Textile Bulletin* 48 (June 27, 1935): 32.

FIG. 1.—One of a series of Parks-Cramer Company advertisements that illustrates the conflict over open windows in textile mills. (*Textile Bulletin* 48 [June 27, 1935]: 32.)

environment on labor can be found in the macaroni industry. Macaroni factories displayed a pattern of intermittent production typical of industries that processed hygroscopic materials. As one journal explained, "Macaroni cannot be made every day in the year or month or week. It must be a dry day, or the substance from which it is made will not bind properly."[65] Particularly difficult was the drying of pastes into the finished pasta, a complex operation that depended on the product's shape, the factory conditions, and the weather. Excessive humidity promoted the souring or spoiling of macaroni through bacterial growth while excessive heat dried the product only on the top layer, sealing in the moisture below and leading to checking, cracking, and breaking. As one trade journal noted, "The length of time required to cure or dry macaroni and spaghetti varies according to the process employed and how much the process is affected by the atmospheric conditions outside

[65]"Philadelphia-made Macaroni Sent Abroad," *Confectioners and Bakers Gazette* 25 (February 1904): 26.

Fig. 2.—This ad suggests that spinners could not regulate windows and humidity levels within the factory as well as the new air conditioning systems. (*Textile Bulletin* 50 [May 28, 1936]: 12.)

the factory and the standards of quality maintained. Some manufacturers complete the process in three days; others require five or six days."[66] Part of the success of drying macaroni was dependent on the skill of the workers. A report on the numerous Philadelphia manufacturers singled out one: "The factory of Frank Cuneo is scrupulously clean and the workmen employed unusually skilled at their trade. They have the best luck in successfully 'curing' the large quantity of pastes made by the

[66]"Production of Macaroni," *Macaroni and Noodle Manufacturers Journal* 9 (August 1911): 7.

factory."[67] By 1907, however, one Bowery manufacturer confided to the journal's correspondent, "So far as making macaroni is concerned, it seems to me there is room, in the way of machinery for decided progress and improvement."[68]

Carrier Engineering soon sensed a new market for its skills, applying humidity control to the problem of industrial drying. After initial tests in the Bellanca Macaroni Company's Buffalo plant in 1916, CEC signed contracts in 1917–18 with three macaroni manufacturers, Skinner Macaroni Company, Crescent Macaroni and Cracker Company, and Foulds Milling Company.[69] It is clear that engineers saw the advantage to manufacturers of replacing skilled male workers with less expensive, less troublesome, and less knowledgeable workers. The Confidential Data files record how CEC engineers viewed the mechanization of macaroni drying and, perhaps, also the basis on which the installations were sold to manufacturers. Carrier Engineering macaroni specialist Russell Tree wrote:

> Some manufacturers have an idea that only the Italian understand the drying of macaroni. That may have been true at one time, but today it is absolute rot. When I say it does not require macaroni skill, I mean that it does not require a man familiar with macaroni to dry it with our dryer. All a man needs is a regular amount of human intelligence. You may wonder what that may mean to the manufacturer. It means that instead of a small field of men to hire from, he can chose [*sic*] from a large field. Then too, if a man quits, there is no great loss, for a new one can be taught in less than a week to do with our dryer what in the past men have taken years to learn.[70]

Quite plainly, this was technology in the service of management. For the manufacturer, part of the appeal of the new technology was to break the hold of skilled workers on production schedules and techniques and give this control to management.

[67]"Philadelphia Trade Conditions," *Macaroni and Noodle Manufacturers Journal* 5 (March 1907): 13.

[68]"Greater New York and Brooklyn," *Macaroni and Noodle Manufacturers Journal* 5 (August 1907): 7.

[69]Ingels, working manuscript, p. 213, Carrier Papers, no. 2511, Box 28, Cornell University Archives. In Ingels's manuscript, the third company is identified only as "Foulds"; research strongly suggests the firm was Foulds Milling Company, Inc. See *Moody's Manual of Railroads and Corporation Securities,* vol. 3, Nineteenth Annual Number, Industrial Section (New York, 1918), pp. 600–601.

[70]R. T. Tree and P. L. Davidson, "Macaroni Drying with Ejector Unit Kiln," October 30, 1920, p. 6, Confidential Data, Box C-12, Carrier Collection, United Technologies Archives.

Yet the technology that CEC developed did more than replace skilled workers. In a move that set him in direct rivalry with workers, Carrier believed that he could improve on traditional drying techniques, not simply duplicate them. Rather than give these macaroni companies atmospheric guarantees, the company was emboldened to guarantee production rates. Preliminary tests at the Bellanca factory suggested that drying rates could be gradually increased rather than held constant over the drying period, and he installed in the Skinner plant a new control instrument, a thermotyne invented by H. Y. Norwood of Taylor Instrument Company, which gradually lowered the dewpoint setting with a cam.[71] Carrier predicted, and CEC guaranteed, a twenty-hour drying time.

The results were disastrous. Before the macaroni was completely dry, it cracked and fell off the drying racks onto the floor. Tree and R. Leslie Jones filled the drying racks again, and within twelve hours another 10,000 pounds of macaroni had collapsed onto the floor. Carrier Engineering had no formal research department until 1919, so Carrier, Tree, Jones, and a Taylor instrument representative turned the Skinner plant into an experimental laboratory. As Tree recalled: "We ruined a lot of macaroni before arriving at the minimum and safe drying period, and we paid for the ruined macaroni at the rate of five cents a pound. We walked miles during the tests, from one drying room to another. A man from Taylor Instruments companies (who inspected controls) ended his first day in his sock feet."[72]

None of the macaroni installations met its guarantees. The company refused to finish payment on the system; two years later they sued for the return of their initial payment of $13,936. The courts decided against Skinner, but in retaliation—whether by management or workers—CEC's erecting man, R. B. Winfrey, got beaten up by two unidentified men.[73] Tree evaluated the experience for the company's Confidential Data files, noting that "we have never met with any serious difficulty in producing real high grade macaroni; all of our drying plants have done this but our troubles have been due to the time limits of our—let's be frank—foolish guarantees."[74]

The failure of CEC to replace traditional craft skills with mechanical systems appears to be largely due to its failure to recognize the complexity of the relationship between natural materials and environ-

[71]Ingels, working manuscript, p. 213, Carrier Papers, no. 2511, Box 28, Cornell University Archives.

[72]Ingels, working manuscript, p. 214.

[73]Ingels, working manuscript, p. 215.

[74]Tree and Davidson (n. 70 above).

mental conditions. Marveling over that complexity, one engineer wrote that "the commercial problem with respect to drying hinges upon the fact that no two macaronis have exactly the same drying characteristics."[75]

This was not the last time that the company underestimated the complexity of natural materials. In 1919 CEC formed Tobacco Treating Inc., and large-scale experiments were begun in drying tobacco leaves. The most prized leaves were those that emerged from drying with a rich, gold color and with excellent handling qualities; these were used as the "wrappers" for cigars and brought a good price. CEC's drying process, achieved through mechanical environmental controls, yielded leaves all of which had that rich gold color. Yet further experiments revealed that these machine-dried leaves had none of the superior working qualities of high-grade tobacco produced under craft methods. The president of Porto Rican–American Tobacco Company advised CEC that "the cigar makers have refused to work anymore of the wrappers cured through your process on account of the same being entirely too tender."[76] To Carrier he explained, "The cigar-makers and strippers refuse to work it on the ground that it breaks in their hands."[77] Not only had engineers failed to produce wrappers with good handling characteristics, but the new curing process eliminated the color difference that allowed cigar makers to distinguish between fair- and good-quality leaves. The tobacco curing company collapsed.

In both cases, engineers had before them a model of the process of drying. Had they been interested simply in aiding management's attempt to break the monopoly of skilled workers by duplicating their skills, that model would have been the easiest to follow. But CEC engineers chose to try to radically improve traditional methods, greatly accelerating the process or dramatically changing the product. They pursued a different path, and from it we can extrapolate a different goal. This drive for dramatic technical change was fueled by the recognition that skilled workers were their greatest rival to technical expertise in the factory; a substantial change in technology would put engineers unquestionably in control.[78]

In both macaroni and tobacco drying, CEC had not been able to improve on the product of traditional craft practice. Engineers clearly

[75]L. L. Lewis, April 11, 1949, Carrier Papers, no. 2511, Box 26, Cornell University Archives.

[76]L. Toro to Carrier Engineering Corporation, July 17, 1920, Stacey Correspondence, Box C-1, Carrier Collection, United Technologies Archives.

[77]L. Toro to Willis Carrier, June 17, 1920, Stacey Correspondence, Box C-1, Carrier Collection, United Technologies Archives.

[78]This shift in factory production is comparable to Frederick W. Taylor's introduction of high-speed steel, which he used, not simply to increase the power of management, but also to increase the influence of engineers in the factory.

knew less about the complexities of natural materials than skilled workers did. Part of the difficulty they experienced was a clash between the quantitative approach of engineering and the sensual approach of craft production. In yet another example of this, the consulting engineer Walter Fleisher lamented his own failure in drying. He had calculated that to dry skins for a leather company his equipment should be able to remove 10 percent of the water by weight. Yet even after the removal of 17–26 percent, the client claimed the skins were still not dry. Only the removal of an additional 1.5 percent satisfied the company. "Their method of testing a skin to see whether it was dry," Fleisher recounted, "was entirely by feeling. They seemed to know what they considered was a dry skin, and cared very little for percentages of moisture removed."[79] Once again, engineering approaches to the processing of natural materials failed to replicate older methods, and craftsmen were often the judges who set the standards. In the eyes of these workmen at least, quantification led to oversimplification, not to precision.

Carrier's response to Fleisher's dilemma and his own was the call for more science, not less. He argued that the appropriate level of residual moisture in a properly dried material could be experimentally determined and quantified. He called on scientific laboratories to produce such data for a range of materials and for engineers to thus arm themselves.[80] The air conditioning engineer's dilemma was that he was in direct rivalry with skilled workers for control of production processes, with adversaries whose claim to authority was experiential knowledge. Laboratory constants and quantitative standards were essential advantages to engineers who lost the first round in the match between craft and engineering.

In a way that was consonant with the popular enthusiasm for "science" and "efficiency," air conditioning engineers thus fell back on science and quantification as a way to bolster the uncertain performance of environmental engineering in their conflict with skilled workers over factory production. In 1917, CEC's J. I. Lyle, in his presidential address before the ASH&VE, called for the establishment of a research laboratory funded by the society to determine more precise values for the standard factors in their field. As a result of his leadership, the society established a lab in Pittsburgh in 1918 at the newly built Bureau of Mines facility, becoming the only major engineering society to support such a facility. The ASH&VE laboratory and the increasing role of the science of engineering (as opposed to the art of engineering) performed several different functions for the profession; importantly,

[79]"Discussion of Papers on Drying," *ASH&VE Transactions* 23 (1917): 279.
[80]Ibid., p. 280.

the rhetoric and prestige of science was a crucial element in strengthening the engineer's influence in a regulatory battle then raging in the United States over ventilation standards. Yet less public but just as important was the role of laboratory-defined quantitative standards in securing for air conditioning engineers a more effective and more authoritative role in factory production.

For a large body of workmen at the turn of the century—those employed in hygroscopic factories in particular—"skill" could be more accurately characterized as the skills necessary to match natural materials to industrial production. Workmen were essentially knowledgeable about the natural world, and factories were inevitably connected to it. That reality made industries that processed natural materials more chaotic and organic than rational and mechanical. It was this connection to nature that air conditioning engineers sought to break through the adoption of new technology and new standards. However, if a close look at the practice of engineering documents an unsurprising hostility toward labor, it also reveals an ambivalence toward management. Greater authority for engineers was achieved by nibbling away at the traditional prerogatives of both management and labor. The engineer's struggle for control over factory production was allied to that of management but not identical to it. As long as each system was the product of engineering design, custom-made for each factory, the division of power over factory production became a three-way split among workers, engineers, and managers. With that larger voice in the direction of the factory, engineers promoted a technology that had embedded in it a quantitative approach to life and one that tried to reorder the factory around those values.

This mechanical ideal of artificial indoor climate was dominant as long as the industry remained a custom-designed industry in the hands of a small group of engineering firms. The editor David L. Fiske, reviewing American Blower's new engineering text, conceded, "It may all prove to be different in the future, but right now one can but say: Here is an outfit that knows its business—its definition of air conditioning is one that others must accept."[81]

Mass Production and the Redefinition of Air Conditioning

The industry did change, however. By the 1930s it was beginning to shift to a new emphasis on mass production. While engineering companies continued to custom design industrial systems with an ever stronger understanding of engineering science and the needs of

[81]Fiske (n. 48 above).

individual industries, CEC itself gradually shifted the basis of its competitiveness. Freed from his patent agreement with Buffalo Forge, Willis Carrier returned to invention and patenting, and CEC acquired factory space for the manufacture of air conditioning equipment. In 1928 the firm began to de-emphasize its traditional base in industrial air conditioning and turned with enthusiasm to the prospect of tapping into the enormous market represented by comfort air conditioning; thus began a period that one company engineer later ruefully described as its "comfort-cooling drunk."[82]

The different relationship between buyers and sellers in the comfort air conditioning market changed the competitive structure of the industry. Large electrical manufacturers like General Electric and Frigidaire challenged the dominance of older but much smaller engineering firms like CEC. For these established manufacturing companies, air conditioning was an appliance rather than an activity. For them, engineering was confined to the invention process itself, an engineering effort that could be secured through patents. The window air conditioner was developed in this decade and was the antithesis of "manufactured weather." Designed and sold without any certain knowledge of the conditions under which it would be used, it came with no guarantees about the atmospheric conditions it would produce when plugged in. It was sold with the expectation that it would mitigate existing conditions, not create an ideal climate. Carrier's company tended to think of the room cooler as a transitional technology, with all the comfort installations gradually yielding to central station systems. Thus it imagined that, while the manufacture of air conditioning equipment would become more standardized and readily available from one company, the need for some degree of skill to match the equipment to each locality would remain an element of the business, along with precise standards of atmospheric control. But the loss of commercial dominance meant that the company was no longer able to define the technology. The influential position of firms engaged in custom-designed systems yielded to a more diverse mix of companies. Carrier's firm remained an industry leader but yielded its exclusive dominance. Economic power, market expertise, and manufacturing ability all rivaled engineering expertise as factors that would predict business success.

[82]L. L. Lewis, "Defense Sales in World War II," December 12, 1950, Sales Conference, p. 4, Carrier Papers, no. 2511, Box 12, Cornell University Archives. Lewis dates the shift in emphasis to comfort air conditioning to 1922 when the company began theater air conditioning. By 1928 it began to manufacture comfort cooling for the home as well.

The Place of Custom Design

The centrality of engineering expertise to custom design marked an era in air conditioning development when a small engineering community was influential in defining the technology. Placing the responsibility for technical development at the door of the engineering community is now an old-fashioned idea. Our efforts to widen the circle of technological determinants, first through the social context of technology, and then through the theoretical lens of social constructionism, have brought an increasingly complex understanding of technological development. Today, few scholars will argue with the proposition, stated in simple terms, that technology embodies both technical constraints and social decisions. However, the distinguishing characteristic of early air conditioning development is the compression of the competing social groups, such as businessmen and engineers, into a small controlling body. These early air conditioning engineers conceded some battles over the control of the factory environment to managers and skilled workers, reinforcing our understanding that technology is the sum of diverse interests. However, their hold on the technology, which came from the necessity for their skills not only in the creation, but in the constant re-creation, of the technology, gave them immense control. Within the framework of multicausal theory of technological development, we must still grapple with the issue of disproportionate power, for not all actors are equal. Custom production provides a greater role for technical expertise, the technologist, and technical values.

The relative neglect of custom production is in sharp contrast to the attention garnered by mass production. As epitomized by the Ford Motor Company, mass production has captured both popular and scholarly interest, in part because of the belief that technology is an important source of national prosperity. For many, technical innovation enriches the individual or business sponsoring it, gives jobs to those who produce it, and puts new goods on the market. This view of technology creating "national" wealth by empowering these three important groups—business, labor, and consumers—perhaps helps account for the strength of the Fordist ideal, as it blends the values of democratic politics, free enterprise, and social mobility. The rags-to-riches story of Henry Ford's rise from farmer to industrialist, the bonanza of the Five Dollar Day for assembly-line workers, and the affordability of the Tin Lizzie are all important elements in mass production's domination of the American imagination. The caveats to this rosy picture—the public humiliation of Ford in a famous court case that decided that "ignorant" was a factual description of the man and not a libelous statement; the tedium, routine, and dehumanizing power of assembly-line work; and the seemingly stagnant technology of the Model T—all seemed simply a confirmation of the checks and balances afforded by this tripartite sharing of industrial prosperity. No one group gleaned all the advantages.

Only the deployment of flexible production in the Japanese automobile industry succeeded in toppling the reputation of mass production from its place at the apex of modern industrial development.[83] Historians pointed out that mass production did not show a straight-line development from interchangeable parts to the assembly line. David Hounshell's history of Ford's mass production and its constituent elements—interchangeable parts, specialized machine tools, quantity production, and the assembly line—showed a disparate use of these techniques before they came together at Highland Park.[84] Scholars read the historical record with an increasing interest in the successful use of different modes of production. Particularly notable in the effort to document the historical complexity of industrial production is Philip Scranton's detailed examination of flexible production in Philadelphia's textile industry. Calling for a more holistic approach to the history of industry, his study encompasses the culture and the practice of flexible production—both necessary, Scranton argues, to understand the business decisions of Philadelphia proprietors.[85]

The reemergence of a vigorous system of flexible production makes it impossible to write a simple linear industrial history with mass production as the teleological end, despite the appeal of its apparent balance of social interests. Clearly, these are alternative systems layered in historically changing patterns, to which custom production should be added. To see mass production and flexible production as two competing systems is to substitute one confining paradigm for another. But while we reject the notion of mass production as a system that contains both the effective production of wealth and its own logic of distribution, it is important to describe any production system in terms of its tangled relationships among management, labor, and consumers. Clearly, that complexity is one of the strengths of Scranton's work. He links

[83]James P. Womak, Daniel T. Jones, and Daniel Roos, *The Machine that Changed the World* (New York, 1990); Michael A. Cusamano, *The Japanese Automobile Industry: Technology and Management at Nissan and Toyota* (Cambridge, Mass., 1985); Taiichi Ono, *Toyota Production System: Beyond Large-Scale Production* (Cambridge, Mass., 1988); Japanese Management Association, ed., *Kanban Just-in-Time at Toyota: Management Begins at the Workplace* (Stamford, Conn., 1986).

[84]Hounshell (n. 2 above).

[85]Philip Scranton, *Figured Tapestry: Production, Markets, and Power in Philadelphia Textiles, 1885–1941* (New York, 1989), and "Diversity in Diversity: Flexible Production and American Industrialization, 1880–1930," *Business History Review* 65 (Spring 1991): 27–90.

production techniques to larger economic relationships and thus provides a broader depiction of the process of production. An explicit linkage between technology and economic relations might recast the way production systems in general are defined: mass production could be characterized as a system that produces quantity goods that are meant to compete on the basis of price, flexible production as a system that produces a variety of goods that offers consumer choice, and custom production as the manufacture of single goods bought on the basis of performance. Such an integration of consumption and production would highlight the differences between "armory practice" and "mass production," a promising starting point for any analysis of American industry in the Cold War era when massive defense spending so shaped industrial structure and production practices.

It is, of course, our interest in technology's role as an agent of prosperity that had focused our attention on mass production rather than custom design. But an enhanced concern with understanding the technical community and its formative influence on technology might lead to a reconsideration of the latter. Certainly, within the air conditioning industry, the conjunction of custom design, engineering guarantees, and unpatentable data placed engineers at the heart of an emerging technology.

"Touch Someone": The Telephone Industry Discovers Sociability

CLAUDE S. FISCHER

The familiar refrain, "Reach out, reach out and touch someone," has been part of American Telephone and Telegraph's (AT&T's) campaign urging use of the telephone for personal conversations. Yet, the telephone industry did not always promote such sociability; for decades it was more likely to discourage it. The industry's "discovery" of sociability illustrates how structural and cultural constraints interact with public demand to shape the diffusion of a technology. While historians have corrected simplistic notions of "autonomous technology" in showing how technologies are produced, we know much less about how consumers use technologies. We too often

DR. FISCHER is professor of sociology at the University of California, Berkeley. Some material presented here was initially delivered to the Social Science History Association, Washington, D.C., October 1983. The research was supported by the National Endowment for the Humanities (grant RO-20612), the National Science Foundation (grant SES83-09301), the Russell Sage Foundation, and the Committee on Research, University of California, Berkeley. Further work was conducted as a Fellow at the Center for Advanced Study in the Behavioral Sciences, Stanford, California, with financial support from the Andrew W. Mellon Foundation. Archival research was facilitated by the generous assistance of people in the telephone industry: at AT&T, Robert Lewis, Robert Garnet, and Mildred Ettlinger; at the San Francisco Pioneer Telephone Museum, Don Thrall, Ken Rolin, and Norm Hawker; at the Museum of Independent Telephony, Peggy Chronister; at Pacific Bell, Robert Deward; at Bell Canada Historical, Stephanie Sykes and Nina Bederian-Gardner; at Illinois Bell, Rita Lapka; John A. Fleckner at the National Museum of American History also provided assistance. Thanks to those interviewed for the project: Tom Winburn, Stan Damkroger, George Hawk Hurst, C. Duncan Hutton, Fred Johnson, Charles Morrish, and Frank Pamphilon. Several research assistants contributed to the work: Melanie Archer, John Chan (who conducted the interviews), Steve Derné, Keith Dierkx, Molly Haggard, Barbara Loomis, and Mary Waters. And several readers provided useful comments on prior versions, including Victoria Bonnell, Paul Burstein, Glenn Carroll, Bernard Finn, Robert Garnet, Roland Marchand, Michael Schudson, John Staudenmaier, S.J., Ann Swidler, Joel Tarr, Langdon Winner, and auditors of presentations. None of these colleagues, of course, is responsible for remaining errors.

take those uses (especially of consumer products) for granted, as if they were straightforwardly derived from the nature of the technology or dictated by its creators.[1]

In the case of the telephone, the initial uses suggested by its promoters were determined by—in addition to technical and economic considerations—its cultural heritage: specifically, practical uses in common with the telegraph. Subscribers nevertheless persisted in using the telephone for "trivial gossip." In the 1920s, the telephone industry shifted from resisting to endorsing such sociability, responding, at least partly, to consumers' insistent and innovative uses of the technology for personal conversation. After summarizing telephone history to 1940, this article will describe the changes in the uses that telephone promoters advertised and the changes in their attitudes toward sociability; it will then explore explanations for these changes.[2]

[1]See C. S. Fischer, "Studying Technology and Social Life," pp. 284–301 in *High Technology, Space, and Society: Emerging Trends*, ed. M. Castells (Beverly Hills, Calif., 1985). For a recent example of a study looking at consumers and sales, see M. Rose, "Urban Environments and Technological Innovation: Energy Choices in Denver and Kansas City, 1900–1940," *Technology and Culture* 25 (July 1984): 503–39.

[2]The primary sources used here include telephone and advertising industry journals; internal telephone company reports, correspondence, collections of advertisements, and other documents, primarily from AT&T and Pacific Telephone (PT&T); privately published memoirs and corporate histories; government censuses, investigations, and research studies; and several interviews, conducted by John Chan, with retired telephone company employees who had worked in marketing. The archives used most are the AT&T Historical Archives, New York (abbreviated hereafter as AT&T ARCH), and the Pioneer Telephone Museum, San Francisco (SF PION MU), with some material from the Museum of Independent Telephony, Abilene (MU IND TEL); Bell Canada Historical, Montreal (BELL CAN HIST); Illinois Bell Information Center, Chicago (ILL BELL INFO); and the N. W. Ayer Collection of Advertisements and the Warshaw Collection of Business Americana, National Museum of American History, Smithsonian Institution, Washington, D.C. A bibliography on the social history of the telephone is unusually short, especially in comparison with those on later technologies such as the automobile and television. There are industrial and corporate histories, but the consumer side is largely untouched. For some basic sources, see J. W. Stehman, *The Financial History of the American Telephone and Telegraph Company* (Boston, 1925); A. N. Holcombe, *Public Ownership of Telephones on the Continent of Europe* (Cambridge, Mass., 1911); H. B. MacMeal, *The Story of Independent Telephony* (Chicago: Independent Pioneer Telephone Association, 1934); J. L. Walsh, *Connecticut Pioneers in Telephony* (New Haven, Conn.: Morris F. Tyler Chapter of the Telephone Pioneers of America, 1950); J. Brooks, *Telephone: The First Hundred Years* (New York, 1976); A. Hibbard, *Hello-Goodbye: My Story of Telephone Pioneering* (Chicago, 1941); Robert Collins, *A Voice from Afar: The History of Telecommunications in Canada* (Toronto, 1977); R. L. Mahon, "The Telephone in Chicago," ILL BELL INFO, MS, ca. 1955; J. C. Rippey, *Goodbye, Central; Hello, World: A Centennial History of North-*

A Brief History of the Telephone

Within about two years of A. G. Bell's patent award in 1876, there were roughly 10,000 Bell telephones in the United States and fierce patent disputes over them, battles from which the Bell Company (later to be AT&T) emerged a victorious monopoly. Its local franchisees' subscriber lists grew rapidly and the number of telephones tripled between 1880 and 1884. Growth slowed during the next several years, but the number of instruments totaled 266,000 by 1893.[3] (See table 1.)

As long-distance communication, telephony quickly threatened telegraphy. Indeed, in settling its early patent battle with Western Union, Bell gave financial concessions to Western Union as compensation for loss of business. As local communication, telephony quickly overwhelmed nascent efforts to establish signaling exchange systems (except for stock tickers).

During Bell's monopoly, before 1894, telephone service consisted basically of an individual line for which a customer paid an annual flat fee allowing unlimited calls within the exchange area. Fees varied widely, particularly by size of exchange. Bell rates dropped in the mid-1890s, perhaps in anticipation of forthcoming competition. In 1895, Bell's average residential rate was $4.66 a month (13 percent of an average worker's monthly wages). Rates remained high, especially in the larger cities (the 1894 Manhattan rate for a two-party line was $10.41 a month).[4]

On expiration of the original patents in 1893–94, thousands of new telephone vendors, ranging from commercial operations to

---

western Bell (Omaha, Nebr.: Northwestern Bell, 1975); G. W. Brock, *The Telecommunications Industry: The Dynamics of Market Structure* (Cambridge, Mass, 1981); I. de S. Pool, *Forecasting the Telephone* (Norwood, N.J., 1983); R. W. Garnet, *The Telephone Enterprise: The Evolution of the Bell System's Horizontal Structure, 1876–1909* (Baltimore, 1985); R. A. Atwood, "Telephony and Its Cultural Meanings in Southeastern Iowa, 1900–1917" (Ph.D. diss., University of Iowa, 1984); Lana Fay Rakow, "Gender, Communication, and the Technology: A Case Study of Women and the Telephone" (Ph.D. diss., University of Illinois at Urbana-Champaign, 1987); and I. de S. Pool, ed., *The Social Impact of the Telephone* (Cambridge, Mass., 1977). (Note that AT&T, Bell, and similar corporate names refer, of course, to these companies—or their direct ancestors—up to the U.S. industry reorganization of January 1, 1984.)

[3]Statistics from AT&T, *Events in Telecommunications History* (New York: AT&T, 1979), p. 6; U.S. Bureau of the Census (BOC), *Historical Statistics of the United States*, Bicentennial Ed., pt. 2 (Washington, D.C., 1975), pp. 783–84.

[4]Rates are reported in scattered places. For these figures, see BOC, *Telephones and Telegraphs 1902*, Special Reports, Department of Commerce and Labor (Washington, D.C., 1906), p. 53; and *1909 Annual Report of AT&T* (New York, 1910), p. 28. Wage data are from *Historical Statistics* (n. 3 above), tables D735–38.

TABLE 1

TELEPHONE DEVELOPMENT, 1880–1940

	Number of Telephones	Telephones per 1,000 People	Percentage in Bell System	Percentage Independent, Connected to Bell	Percentage Residential, Connected to Bell
1880	54,000	1	100	0	...
1885	156,000	3	100	0	...
1890	228,000	4	100	0	...
1895	340,000	5	91	0	...
1900	1,356,000	18	62	1	...
1905	4,127,000	49	55	6	...
1910	7,635,000	82	52	26	...
1915	10,524,000	104	57	30	...
1920	13,273,000	123	66	29	68
1925	16,875,000	145	75	24	67
1930	20,103,000	163	80	20	65
1935	17,424,000	136	82	18	63
1940	21,928,000	165	84	16	65
1980	180,000,000	790	81	19	74

SOURCES.—U.S. Bureau of the Census, *Historical Statistics of the United States*, Bicentennial Ed., pt. 2 (Washington, D.C., 1975), pp. 783–84; and U.S. Bureau of the Census, *Statistical Abstract of the United States 1982–83* (Washington, D.C., 1984), p. 557.

small cooperative systems, sprang up. Although they typically served areas that Bell had ignored, occasional head-to-head competition drove costs down and spurred rapid diffusion: almost a ninefold increase in telephones per capita between 1893 and 1902, as compared to less than a twofold increase in the prior nine years.[5]

Bell responded fiercely to the competition, engaging in price wars, political confrontations, and other aggressive tactics. It also tried to reach less affluent customers with cheaper party lines, coin-box telephones, and "measured service" (charging by the call). Still, Bell lost at least half the market by 1907. Then, a new management under Theodore N. Vail, the most influential figure in telephone history, changed strategies. Instead of reckless, preemptive expansion and price competition, AT&T bought out competitors where it could and ceded territories where it was losing. With tighter fiscal con-

[5]BOC, *Telephones, 1902* (n. 4 above); Federal Communications Commission (FCC), *Proposed Report: Telephone Investigation* (Washington, D.C., 1938), p. 147. AT&T has always officially challenged this interpretation; see, e.g., *1909 Annual Report of AT&T*, pp. 26–28.

trol, and facing capital uncertainties as well, AT&T's rate of expansion declined.[6] Meanwhile, the "independents" could not expand much beyond their small-town bases, partly because they were unable to build their own long-distance lines and were cut off from Bell-controlled New York City. Many were not competitive because they were poorly financed and provided poor service. Others accepted or even solicited buyouts from AT&T or its allies. By 1912, the Bell System had regained an additional 6 percent of the market.

During this competitive era, the industry offered residential customers a variety of economical party-line plans. Bell's average residential rate in 1909 was just under two dollars a month (about 4 percent of average wages).[7] How much territory the local exchange covered and what services were provided—for example, nighttime operators—varied greatly, but costs dropped and subscriber lists grew considerably. These basic rates changed little until World War II (although long-distance charges dropped).

In the face of impending federal antitrust moves, AT&T agreed in late 1913 to formalize its budding accommodation with the independents. Over several years, local telephone service was divided into regulated geographic monopolies. The modern U.S. telephone system—predominantly Bell local service and exclusively Bell long-distance service—was essentially fixed from the early 1920s to 1984.

The astronomical growth in the number of telephones during the pre-Vail era (a compound annual rate of 23 percent per capita from 1893 to 1907) became simply healthy growth (4 percent between 1907 and 1929). The system was consolidated and technically improved, and, by 1929, 42 percent of all households had telephones. That figure shrank during the Depression to 31 percent in 1933 but rebounded to 37 percent of all households in 1940.

Sales Strategies

The telephone industry believed, as President Vail testified in 1909, that the "public had to be educated . . . to the necessity and ad-

[6]See, e.g., *Annual Report of AT&T*, 1907–10; and FCC, *Proposed Report* (n. 5 above), pp. 153–154. On making deals with competitors, see, e.g., Rippey (n. 2 above), pp. 143ff.

[7]*1909 Annual Report of AT&T*, p. 28. Charges for minimal, urban, four-party lines ranged from $3.00 a month in New York (about 6 percent of the average manufacturing employee's monthly wages) to $1.50 in Los Angeles (about 3 percent of wages) and much less in small places with mutual systems; see BOC, *Telephones and Telegraphs and Municipal Electric Fire-Alarm and Police-Patrol Signaling Systems, 1912* (Washington, D.C., 1915); and *Historical Statistics* (n. 3 above), table D740.

vantage of the telephone."[8] And Bell saluted itself on its success in an advertisement entitled "Blazing the Way": Bell "had to invent the business uses of the telephone and convince people that they were uses. . . . [Bell] built up the telephone habit in cities like New York and Chicago. . . . It has from the start created the need of the telephone and then supplied it."[9]

"Educating the public" typically meant advertising, face-to-face solicitations, and public relations. In the early years, these efforts included informational campaigns, such as publicizing the existence of the telephone, showing people how to use it, and encouraging courteous conversation on the line.[10] Once the threat of nationalization became serious, "institutional" advertising and publicity encouraged voters to feel warmly toward the industry.[11]

As to getting paying customers, the first question vendors had to ask was, Of what use is this machine? The answer was not self-evident.

For roughly the first twenty-five years, sales campaigns largely employed flyers, simple informational notices in newspapers, "news" stories supplied to friendly editors (many of whom received free service or were partners in telephony), public demonstrations, and personal solicitations of businessmen. As to uses, salesmen typically

[8]Testimony on December 9, 1909, in State of New York, *Report of the Committee of the Senate and Assembly Appointed to Investigate Telephone and Telegraph Companies* (Albany, 1910), p. 398.

[9]Ayer Collection of AT&T Advertisements, Collection of Business Americana, National Museum of American History, Smithsonian Institution.

[10]See, e.g., *Pacific Telephone Magazine* (PT&T employee magazine, hereafter PAC TEL MAG), 1907–40, passim; 1914 advertisements in SF PION MU folder labeled "Advertising"; MU IND TEL "Scrapbook" of Southern Indiana Telephone Company clippings; advertisements in directories of the day; "Educating the Public to the Proper Use of the Telephone," *Telephony* 64 (June 21, 1913): 32–33; "Swearing over the Telephone," *Telephony* 9 (1905): 418; and "Advertising and Publicity—1906 –1910," box 1317, AT&T ARCH.

[11]On AT&T's institutional advertising, see R. Marchand, "Creating the Corporate Soul: The Origins of Corporate Image Advertising in America" (paper presented to the Organization of American Historians, 1980), and N. L. Griese, "AT&T: 1908 Origins of the Nation's Oldest Continuous Institutional Advertising Campaign," *Journal of Advertising* 6 (Summer 1977): 18–24. FCC, *Proposed Report* (n. 5 above), has a chapter on "Public Relations"; see also N. R. Danielian, *AT&T: The Story of Industrial Conquest* (New York, 1939), chap. 13. For a defense of AT&T public relations, see A. W. Page, *The Bell Telephone System* (New York, 1941). Among the publicity efforts along these lines were "free" stories, subsidies of the press, and courting of reporters and politicians (documented in AT&T ARCH). In one comical case, AT&T frantically and apparently unsuccessfully tried in 1920 to pressure Hal Roach to cut out from a Harold Lloyd film he was producing a burlesque scene of central exchange hysteria (see folder "Correspondence—E. S. Wilson, V.P., AT&T," SF PION MU).

stressed those that extended applications of telegraph signaling. For example, an 1878 circular in New Haven—where the first exchange was set up—stated that "your wife may order your dinner, a hack, your family physician, etc., all by Telephone without leaving the house or trusting servants or messengers to do it." (It got almost no response.)[12] In these uses, the telephone directly competed with—and decisively defeated—attempts to create telegraph exchanges that enabled subscribers to signal for services and also efforts to employ printing telegraphs as a sort of "electronic mail" system.[13]

In this era and for some years later, the telephone marketers sought new uses to add to these telegraphic applications. They offered special services over the telephone, such as weather reports, concerts, sports results, and train arrivals. For decades, vendors cast about for novel applications: broadcasting news, sports, and music, night watchman call-in services, and the like. Industry magazines eagerly printed stories about the telephone being used to sell products, alert firefighters about forest blazes, lullaby a baby to sleep, and get out voters on election day. And yet, industry men often attributed weak demand to not having taught the customer "what to do with his telephone."[14]

In the first two decades of the 20th century, telephone advertising became more professionally "modern."[15] AT&T employed a Bos-

[12]Walsh (n. 2 above), p. 47.

[13]S. Schmidt, "The Telephone Comes to Pittsburgh" (master's thesis, University of Pittsburgh, 1948); Pool, *Forecasting* (n. 2 above), p. 30; D. Goodman, "Early Electrical Communications and the City: Applications of the Telegraph in Nineteenth-Century Urban America" (unpub. paper, Department of Social Sciences, Carnegie-Mellon University, n.d., courtesy of Joel Tarr); and "Telephone History of Dundee, Ontario," City File, BELL CAN HIST.

[14]On special services and broadcasting, see Walsh (n. 2 above), p. 206; S. H. Aronson, "Bell's Electrical Toy: What's the Use? The Sociology of Early Telephone Usage," pp. 15–39, and I. de S. Pool et al., "Foresight and Hindsight: The Case of the Telephone," pp. 127–58, both in Pool, ed., *Social Impact* (n. 2 above); "Broadening the Possible Market," *Printers' Ink* 74 (March 9, 1911): 20; G. O. Steel, "Advertising the Telephone," *Printers' Ink* 51 (April 12, 1905): 14–17; and F. P. Valentine, "Some Phases of the Commercial Job," *Bell Telephone Quarterly* 5 (January 1926): 34–43. For illustrations of uses, see, e.g., PAC TEL MAG (October 1907), p. 6, (January 1910), p. 9, (December 1912), p. 23, and (October 1920), p. 44; and the independent magazine, *Telephony*. E.g., the index to vol. 71 (1916) of *Telephony* lists the following under "Telephone, novel uses of": "degree conferred by telephone, dispatching tugs in harbor service, gauging water by telephone, telephoning in an aeroplane." On complaints about not having taught the public, see the quotation from H. B. Young, ca. 1929, pp. 91, 100 in "Publicity Conferences—Bell System—1921–34," box 1310, AT&T ARCH, but similar comments appear in earlier years, as well as positive claims, such as Vail's in 1909.

[15]The following discussion draws largely from examination of advertisement collections at the archives listed in n. 2. Space does not permit more than a few examples

ton agency to dispense "free publicity" and later brought its chief, J. D. Ellsworth, into the company. It began national advertising campaigns and supplied local Bell companies with copy for their regional presses. Some of the advertising was implicitly competitive (e.g., stressing that Bell had long-distance service), and much of it was institutional, directed toward shaping a favorable public opinion about the Bell System. Advertisements for selling service employed drawings, slogans, and texts designed to make the uses of the telephone—not just the technology—attractive. (The amount and kind of advertising fluctuated, especially in the Bell System, in response to competition, available supplies, and political concerns.)[16]

From roughly 1900 to World War I, Bell's publicity agency advertised uses of the telephone by planting newspaper "stories" on telephones in farm life, in the church, in hotels, and the like.[17] The national advertisements, beginning around 1910, addressed mostly businessmen. They stressed that the telephone was impressive to customers and saved time, both at work and at home, and often noted the telephone's convenience for planning and for keeping in touch with the office during vacations.

A second major theme was household management. A 1910 series, for example, presented detailed suggestions: Subscribers could telephone dressmakers, florists, theaters, inns, rental agents, coal dealers, schools, and the like. Other uses were suggested, too, such as conveying messages of moderate urgency (a businessman calling home to say that he will be late, calling a plumber), and conveying invitations (to an impromptu party, for a fourth at bridge).

Sociability themes ("visiting" kin by telephone, calling home from a business trip, and keeping "In Touch with Friends and Relatives")

---

of hundreds of advertisements in the sources. See esp. at AT&T ARCH, files labeled "Advertising and Publicity"; at SF PION MU, folders labeled "Advertising" and "Publicity Bureau"; at BELL CAN HIST, "Scrapbooks"; at ILL BELL INFO, "AT&T Advertising" and microfilm 384B, "Adver."; and at the Ayer Collection (n. 9 above), the AT&T series.

[16]For explicit discussions, see Mahon (n. 2 above), e.g., pp. 79, 89; Publicity Vice-President A. W. Page's comments in "Bell System General Commercial Conference, 1930," microfilm 368B, ILL BELL INFO; and comments by Commercial Engineer K. S. McHugh in "Bell System General Commercial Conference on Sales Matters, 1931," microfilm 368B, ILL BELL INFO. On the origins of in-house advertising, see N. L. Griese, "1908 Origins" (n. 11 above).

[17]See correspondence in "Advertising and Publicity—Bell System—1906–1910, Folder 1," box 1317, AT&T ARCH. Some reports claimed that thousands of stories were placed in hundreds of publications. Apparently no national advertising campaigns were conducted prior to these years; Bell marketing strategy seemed largely confined to price and service competition. See N. C. Kingsbury, "Results from the American Telephone's National Campaign," *Printers' Ink* (June 29, 1916): 182–84.

appeared, but they were relatively rare and almost always suggested sending a message such as an invitation or news of safe arrival rather than having a conversation. A few advertisements also pointed out the modernity of the telephone ("It's up to the times!"). But the major uses suggested in early telephone advertising were for business and household management; sociability was rarely advised.[18]

With the decline of competition and the increase in regulation during the 1910s, Bell stressed public relations even more and pressed local companies to follow suit. AT&T increasingly left advertising basic services and uses to its subsidiaries, although much of the copy still originated in New York, and the volume of such advertising declined. Material from Pacific Telephone and Telegraph (PT&T), apparently a major advertiser among the Bell companies, indicates the substance of "use" advertising during that era.[19]

PT&T advertisements for 1914 and 1915 include, aside from informational notices and general paeans to the telephone, a few suggestions for businessmen (e.g., "You fishermen who feel these warm days of Spring luring you to your favorite stream. . . . You can adjust affairs before leaving, ascertain the condition of streams, secure accommodations, and always be in touch with business and home"). Several advertisements mention the home or women, such as those suggesting that extension telephones add to safety and those encouraging shopping by telephone. Just one advertisement in this set explicitly suggests an amiable conversation: A grandmotherly woman is speaking on the telephone, a country vista visible through the window behind her, and says: "My! How sweet and clear my daughter's voice sounds! She seems to be right here with me!" The text reads: "Let us suggest a long distance visit home today." But this sort of advertisement was unusual.

During and immediately after World War I, there was no occasion to promote telephone use, since the industry struggled to meet demand pent up by wartime diversions. Much publicity tried to ease customer irritation at delays.

Only in the mid-1920s did AT&T and the Bell companies refo-

[18]In addition to the advertising collections, see A. P. Reynolds, "Selling a Telephone" (to a businessman), *Telephony* 12 (1906): 280–81; id., "The Telephone in Retail Business," *Printers' Ink* 61 (November 27, 1907): 3–8; and "Bell Encourages Shopping by Telephone," ibid., vol. 70 (January 19, 1910).

[19]Letter from AT&T Vice-President Reagan to PT&T President H. D. Pillsbury, March 4, 1929, in "Advertising," SF PION MU; W. J. Phillips, "The How, What, When and Why of Telephone Advertising," talk given July 7, 1926, in ibid.; and "Advertising Conference—Bell System—1916," box 1310, AT&T ARCH, p. 44.

cus their attention, for the first time in years, to sales efforts.[20] The system was a major advertiser, and Bell leaders actively discussed advertising during the 1920s. Copy focused on high-profit services, such as long distance and extension sets; modern "psychology," so to speak, influenced advertising themes; and Bell leaders became more sensitive to the competition from other consumer goods. Sociability suggestions increased, largely in the context of long-distance marketing.

In the United States, long-distance advertisements still overwhelmingly targeted business uses, but "visiting" with kin now appeared as a frequent suggestion. Bell Canada, for some reason, stressed family ties much more. Typical of the next two decades of Bell Canada's long-distance advertisements are these, both from 1921: "Why night calls are popular. How good it would sound to hear mother's voice tonight, he thought—for there were times when he was lonely—mighty lonely in the big city"; and "it's a weekly affair now, those fond intimate talks. Distance rolls away and for a few minutes every Thursday night the familiar voices tell the little family gossip that both are so eager to hear." Sales pointers to employees during this era often suggested providing customers with lists of their out-of-town contacts' telephone numbers.

In the 1920s, the advertising industry developed "atmosphere" techniques, focusing less on the product and more on its consequences for the consumer.[21] A similar shift may have begun in Bell's advertising, as well: "The Southwestern Bell Telephone Company has decided [in 1923] that it is selling something more vital than distance, speed or accuracy. . . . [T]he telephone . . . almost brings [people] face to face. It is the next best thing to personal contact. So the fundamental purpose of the current advertising is to sell the company's subscribers their voices at their true worth—to help them realize that 'Your Voice is You.'. . . to make subscribers think of the telephone whenever they think of distant friends or relatives. . . . "[22] This attitude was apparently only a harbinger, because during most of the 1920s the sociability theme was largely re-

[20]See n. 16 above.

[21]D. Pope, *The Making of Modern Advertising* (New York, 1983); S. Fox, *The Mirror Makers: A History of American Advertising and Its Creators* (New York, 1984); M. Schudson, *Advertising: The Uneasy Persuasion* (New York, 1985), pp. 60ff; R. Marchand, *Advertising the American Dream: Making Way for Modernity, 1920–1940* (Berkeley, Calif., 1985); and R. Pollay, "The Subsiding Sizzle: A Descriptive History of Print Advertising, 1900–1980," *Journal of Marketing* 49 (Summer 1985): 24–37.

[22]W. B. Edwards, "Tearing Down Old Copy Gods," *Printers' Ink* 123 (April 26, 1923): 65–66.

stricted to long distance and did not appear in many basic service advertisements.

Bell System salesmen spent the 1920s largely selling ancillary services, such as extension telephones, upgrading from party lines, and long distance, to current subscribers, rather than finding new customers. Basic residential rates averaged two to three dollars a month (about 2 percent of average manufacturing wages), not much different from a decade earlier, and Bell leaders did not consider seeking new subscribers to be sufficiently profitable to pursue seriously.[23] The limited new subscriber advertising continued the largely practical themes of earlier years. PT&T contended that residential telephones, especially extensions, were useful for emergencies, for social convenience (don't miss a call about an invitation, call your wife to set an extra place for dinner), and for avoiding the embarrassment of borrowing a telephone, as well as for its familiar business uses. A 1928 Bell Canada sales manual stressed household practicality first and social invitations second as tactics for selling basic service.[24]

Then, in the late 1920s, Bell System leaders—prodded perhaps by the embarrassment that, for the first time, more American families owned automobiles, gas service, and electrical appliances than subscribed to telephones—pressed a more aggressive strategy. They built up a full-fledged sales force. And they sought to market the telephone as a "comfort and convenience"—that is, as more than a practical device—drawing somewhat on the psychological, sensualist themes in automobile advertising. They focused not only on upgrading the service of current subscribers but also on reaching those car owners and electricity users who lacked telephones. And the *social* character of the telephone was to be a key ingredient in the new sales strategies.[25]

Before "comfort and convenience" could go far, however, the Depression drew the industry's attention to basic service once again. Subscribers were disconnecting. Bell companies mounted campaigns to

[23]On rates, see W. F. Gray, "Typical Schedules for Rates of Exchange Service," and related discussion, in "Bell System General Commercial Engineers' Conference, 1924," microfilm 364B, ILL BELL INFO.

[24]Bell Telephone Company of Canada, "Selling Service on the Job," ca. 1928, cat. 12223, BELL CAN HIST.

[25]Comments, esp. by AT&T vice-presidents Page and Gherardi, during "General Commercial Conference, 1928," and "Bell System General Commercial Conference, 1930," both microfilm 368B, ILL BELL INFO, expressed a view that telephones should be part of consumers' "life-styles," not simply their practical instruments. One hears many echoes of "comfort and convenience" at lower Bell levels during this period.

save residential connections by mobilizing *all* employees to sell or save telephone hookups on their own time (a program that had started before the Crash), expanding sales forces, advertising to current subscribers, and mounting door-to-door "save" and "nonuser" campaigns in some communities.[26] The "pitches" PT&T suggested to its employees included convenience (e.g., saving a trip to market), avoiding the humiliation of borrowing a neighbor's telephone, and simply being "modern." Salesmen actually seemed to rely more on pointing out the emergency uses of the telephone—an appeal especially telling to parents of young children—and suggesting that job offers might come via the telephone. Having a telephone so as to be available to friends and relatives was a lesser sales point. By now, a half-century since A. G. Bell's invention, salespeople did not have to sell telephone service itself but had to convince potential customers that they needed a telephone in their own homes.[27]

During the Depression, long-distance advertising continued, employing both business themes and the themes of family and friendship. But basic service advertising, addressed to both nonusers and would-be disconnectors, became much more common than it had been for twenty years.

The first line of argument in print ads for basic service was practicality—emergency uses, in particular—but suggestions for sociable conversations were more prominent than they had been before. A 1932 advertisement shows four people sitting around a woman who is speaking on the telephone. "Do Come Over!" the text reads, "Friends who are linked by telephone have good times." A 1934 Bell Canada advertisement features a couple who have just resubscribed and who testify, "We got out of touch with all of our friends and missed the good times we have now." A 1935 advertisement asks, "Have you ever watched a person telephoning to a friend? Have you noticed how readily the lips part into smiles . . . ?" And 1939 copy states, "Some one thinks of some one, reaches for the telephone, and all is well." A 1937 AT&T advertisement reminds us that "the telephone is vital in emergencies, but that is not the whole of its service. . . . Friendship's path often follows the trail of the telephone wire." These family-and-friend mo-

[26]See A. Fancher, "Every Employee Is a Salesman for American Telephone and Telegraph," *Sales Management* 28 (February 26, 1931): 45–51, 472; "Bell Conferences," 1928 and 1930 (n. 25 above), esp. L. J. Billingsley, "Presention of Disconnections," in 1930 conference; *Pacemaker*, a sales magazine for PT&T, ca. 1928–31, SF PION MU; and *Telephony*, passim, 1931–36.

[27]PT&T *Pacemaker;* interviews by John Chan with retired industry executives in northern California; see also J. E. Harrel, "Residential Exchange Sales in New England Southern Area," in "Bell Conference, 1931" (n. 16 above), pp. 67ff.

tifs, more frequent and frank in the 1930s, forecast the jingles of today, such as ". . . a friendly voice, like chicken soup/is good for your health/Reach out, reach out and touch someone."[28]

This brief chronology draws largely from prepared copy in industry archives, not from actual printed advertisements. A systematic survey, however, of two newspapers in northern California confirms the impression of increasing sociability themes. Aside from one 1911 advertisement referring to farm wives' isolation, the first sociability message in the *Antioch Ledger* appeared in 1929, addressed to parents: "No girl wants to be a wallflower." It was followed in the 1930s with notices for basic service such as "Give your friends straight access to your home," and "Call the folks now!" In 1911, advertisements in the *Marin* (County) *Journal* stressed the convenience of the telephone for automotive tourists. Sociability became prominent in both basic and long-distance advertisements in the late 1920s and the 1930s with suggestions that people "broaden the circle of friendly contact" (1927), "Voice visit with friends in nearby cities" (1930), and call grandmother (1935), and with the line, "I got my telephone for convenience. I never thought it would be such fun!" (1940).[29]

The emergence of sociability also appears in guides to telephone salesmen. A 1904 instruction booklet for sales representatives presents many selling points, but only one paragraph addresses residential service. That paragraph describes ways that the telephone saves time and labor, makes the household run smoothly, and rescues users in emergencies, but the only barely social use it notes is that the telephone "invites one's friends, asks them to stay away, asks them to hurry and enables them to invite in return." Conversation—telephone "visiting"—per se is not mentioned.

A 1931 memorandum to sales representatives, entitled "Your Tele-

[28]There is some variation among the advertising collections I examined. Illinois Bell's basic service advertisements used during the Depression are, for the most part, similar to basic service ads used a generation earlier. The Pacific Bell and Bell Canada advertisements feature sociable conversations much more. On the other hand, the Bell Canada ads are distinctive in that sociability is almost exclusively a family matter. Friendship, featured in U.S. ads all along, emerges clearly in the Canadian ads only in the 1930s. The 1932 ad cited in the text appears in the August 17 issue of the *Antioch* (Calif.) *Ledger*. The "chicken soup" jingle, sung by Roger Miller, was a Bell System ad in 1981. On the "Touch Someone" campaigns, see M. J. Arlen, *Thirty Seconds* (New York, 1980). See also "New Pitch to Spur Phone Use," *New York Times*, October 23, 1985, p. 44.

[29]These particular newspapers were examined as part of a larger study on the social history of the telephone that will include case studies of three northern California communities from 1890 to 1940.

phone," is, on the other hand, full of tips on selling residential service and encouraging its use. Its first and longest subsection begins: "*Fosters friendships*. Your telephone will keep your personal friendships alive and active. Real friendships are too rare and valuable to be broken when you or your friends move out of town. Correspondence will help for a time, but friendships do not flourish for long on letters alone. When you can't visit in person, telephone periodically. Telephone calls will keep up the whole intimacy remarkably well. There is no need for newly-made friends to drop out of your life when they return to distant homes." A 1935 manual puts practicality and emergency uses first as sales arguments but explicitly discusses the telephone's "social importance," such as saving users from being "left high and dry by friends who can't reach [them] conveniently."[30]

This account, so far, covers the advertising of the Bell System. There is less known and perhaps less to know about the independent companies' advertising. Independents' appeals seem much like those of the Bell System, stressing business, emergencies, and practicality, except perhaps for showing an earlier sensitivity to sociability among their rural clientele.[31]

In sum, the variety of sales materials portray a similar shift. From the beginning to roughly the mid-1920s, the industry sold service as a practical business and household tool, with only occasional mention of social uses and those largely consisting of brief messages. Later sales arguments, for both long-distance and basic service, featured social uses prominently, including the suggestion that the telephone be used for converations ("voice visiting") among

[30]Central Union Telephone Company Contracts Department, *Instructions and Information for Solicitors,* 1904, ILL BELL INFO. Note that Central Union had been, at least through 1903, one of Bell's most aggressive solicitors of business. Illinois Bell Commercial Department, *Sales Manual* 1931, microfilm, ILL BELL INFO. Ohio Bell Telephone Company, "How You Can Sell Telephones," 1935, file "Salesmanship," BELL CAN HIST.

[31]Until 1894, independent companies did not exist. For years afterward, they largely tried to meet unfilled demand in the small cities and towns Bell had underserved. In other places, they advertised competitively against Bell. Nevertheless, advertising men often exhorted the independents to use "salesmanship in print" to encourage basic service and extensive use. See, e.g., J. A. Schoell, "Advertising and Other Thoughts of the Small Town Man," *Telephony* 70 (June 10, 1916): 40–41; R. D. Mock, "Fundamental Principles of the Telephone Business: Part V, Telephone Advertising," series in ibid., vol. 71 (July 22–November 21, 1916); D. Hughes, "Right Now Is the Time to Sell Service," ibid., 104 (June 10, 1933): 14–15; and L. M. Berry, "Helpful Hints for Selling Service," ibid., 108 (February 2, 1935): 7–10. See also Kellogg Company, "A New Business Campaign for ———" (Chicago: Kellogg, 1929), MU IND TEL.

friends and family. While it would be helpful to confirm this impressionistic account with firm statistics, for various reasons it is difficult to draw an accurate sample of advertising copy and salesmen's pitches for over sixty years. (For one, we have no easily defined "universe" of advertisements. Are the appropriate units specific printed ads, or ad campaigns? How are duplicates to be handled? Or ads in neighboring towns? Do they include planted stories, inserts in telephone bills, billboards, and the like? Should locally generated ads be included? And what of nationally prepared ads not used by the locals? For another, we have no clear "population" of ads. The available collections are fragmentary, often preselected for various reasons.) An effort in that direction appears, however, in table 2, in which the numbers of "social" advertisements show a clear increase, both absolutely and relatively.

TABLE 2

Counts of Dominant Advertising Themes by Period

Sources and Types of Advertisements	Prewar	1919–29	1930–40
Antioch (Calif.) *Ledger*:			
Social, sociability	1 (1)	1 (1)	6 (4)
Business, businessmen	6 (5)	1 (1)	2 (1)
Household, convenience, etc.	5 (5)	3 (3)	4 (3)
Public relations, other	0 (0)	4 (3)	1 (1)
Total	12 (11)	9 (8)	13 (9)
Approximate ratio of social to others	1:11 (1:10)	1:8 (1:7)	1:1 (1:1)
Marin (Calif.) *Journal*:			
Social, sociability	1 (1)	5 (2)	43 (20)
Business, businessmen	2 (2)	8 (2)	10 (3)
Household, convenience, etc.	12 (12)	3 (3)	20 (20)
Public relations, other	0 (0)	19 (13)	25 (16)
Total	15 (15)	35 (20)	98 (59)
Approximate ratio of social to others	1:14 (1:14)	1:6 (1:9)	1:1 (1:2)
Bell Canada:			
Social, sociability	5 (2)	25 (1)	59* (9)
Business, businessmen	20* (20)	15 (2)	24* (4)
Household, convenience, etc.	28 (28)	3 (3)	23* (6)
Public relations, other	30* (30)	25 (40)	2 (2)
Total	83* (80)	68 (46)	108* (21)
Approximate ratio of social to others	1:16 (1:39)	1:2 (1:45)	1:1 (1:1)

TABLE 2 (*continued*)

Sources and Types of Advertisements	Prewar	1919–29	1930–40
Pacific Telephone, 1914–15:			
Social, sociability	2 (1)	...	...
Business, businessmen	7 (6)	...	...
Household, convenience, etc.	18 (16)	...	...
Public relations, other	16 (9)	...	...
Total	43 (32)	...	...
Approximate ratio of social to others	1:21 (1:31)	...	...
Assorted Bell ads, 1906–10:			
Social, sociability	4 (4)	...	...
Business, businessmen	13 (12)	...	...
Household, convenience, etc.	11 (11)	...	...
Public relations, other	9 (9)	...	...
Total	37 (36)	...	...
Approximate ratio of social to others	1:8 (1:8)	...	...

SOURCES.—Advertisements in the *Antioch Ledger* were sampled from 1906 to 1940 by Barbara Loomis; those in the *Marin Journal* were sampled from 1900 to 1940 by John Chan. The Bell Canada collection appears in scrapbooks at Bell Canada Historical; the Pacific collection is in the San Francisco Pioneer Telephone Museum. The AT&T advertisements are from AT&T ARCH, box 1317. Other, spotty collections were used for the study but not counted here because they were not as systematic. All coding was done by the author.

NOTE.—Counts in parentheses exclude explicitly long-distance advertisements. Usually each ad had one dominant theme. When more than one seemed equal in weight, the ad was counted in both categories. "Social, sociability" refers to the use of the telephone for personal contact, including season's greetings, invitations, and conversation between friends and family. (Note that the inclusion of brief messages in this category makes the analysis a conservative test of the argument that there was a shift toward sociability themes.) "Business, businessmen" refers to the explicit use of the telephone for business purposes or general appeals to businessmen—e.g., that the telephone will make one a more forceful entrepreneur. "Household, convenience, etc." includes the use of the telephone for household management, personal convenience (e.g., don't get wet, order play tickets), and for emergencies, such as illness or burglary. "Public relations, other" includes general institutional advertising, informational notices (such as how to use the telephone), and other miscellaneous. Perhaps the most conservative index is the ratio of non-long-distance social ads to non-long-distance household ads. (Business ads move to speciality magazines over the years; public information ads fluctuate with political events; and long-distance ads may be "inherently" social.) In the *Antioch Ledger*, this ratio changes from 1:5 to 4:3; in the *Marin Journal*, from 1:12 to 1:1; and in Bell Canada's ads, from 1:14 to 1.5:1. Even these ratios understate the shift, for several reasons. One, I was much more alert to social than to other ads and was more thorough with early social ads than any other category. Two, the household category is increased in the later years by numerous ads for extension telephones. Three, the nature of the social ads counted here changes. The earlier ones overwhelmingly suggest using the telephone for greetings and invitations, not conversation. With rare exception, only the later ones discuss friendliness and "warm human relationships" and suggest chats.

*Estimated.

Industry Attitudes toward Sociability

This change in advertising themes apparently reflected a change in the actual beliefs industry men held about the telephone. Alexander Graham Bell himself forecast social chitchats using his invention. He predicted that eventually Mrs. Smith would spend an hour on the telephone with Mrs. Brown "very enjoyably . . . cutting up Mrs. Robinson."[32] But for decades few of his successors saw it that way.

Instead, the early telephone vendors often battled their residential customers over social conversations, labeling such calls "frivolous" and "unnecessary." For example, an 1881 announcement complained, "The fact that subscribers have been free to use the wires as they pleased without incurring additional expense [i.e., flat rates] has led to the transmission of large numbers of communications of the most trivial character."[33] In 1909, a local telephone manager in Seattle listened in on a sample of conversations coming through a residential exchange and determined that 20 percent of the calls were orders to stores and other businesses, 20 percent were from subscribers' homes to their own businesses, 15 percent were social invitations, and 30 percent were "purely idle gossip"—a rate that he claimed was matched in other cities. The manager's concern was to reduce this last, "unnecessary use." One tactic for doing so, in addition to "education" campaigns on proper use of the telephone, was to place time limits on calls (in his survey the average call had lasted over seven minutes). Time limits were often an explicit effort to stop people who insisted on chatting when there was "business" to be conducted.[34]

[32]Quoted in Aronson, "Electrical Toy" (n. 14 above).

[33]Proposed announcement by National Capitol Telephone Company, in letter to Bell headquarters, January 20, 1881, box 1213, AT&T ARCH. In a similar vein, the president of Bell Canada confessed, ca. 1890, to being unable to stop "trivial conversations"; see Collins, *A Voice* (n. 2 above), p. 124. The French authorities were also exasperated by nonserious uses; see C. Bertho, *Télégraphes et téléphones* (Paris, 1980), pp. 244–45.

[34]C. H. Judson, "Unprofitable Traffic—What Shall Be Done with It?" *Telephony* 18 (December 11, 1909): 644–47, and PAC TEL MAG 3 (January, 1910): 7. He also writes, "the telephone is going beyond its original design, and it is a positive fact that a large percentage of telephones in use today on a flat rental basis are used more in entertainment, diversion, social intercourse and accommodation to others, than in actual cases of business or household necessity" (p. 645). MacMeal, *Independent* (n. 2 above), p. 240, reports on a successful campaign in 1922 to discourage gossipers through letters and advertisements. Typically, calls were—at least officially—limited to five minutes in many places, although it is unclear how well limits were enforced.

An exceptional few in the industry, believing in a more "populist" telephony, did, however, try to encourage such uses. E. J. Hall, Yale-educated and originally manager of his family's firebrick business, initiated the first "measured service" in Buffalo in 1880 and later became an AT&T vice-president. A pleader for lower rates, Hall also defended "trivial" calls, arguing that they added to the total use-value of the system. But the evident isolation of men like Hall underlines the dominant antisociability view of the pre–World War I era.[35]

Official AT&T opinions came closer to Hall's in the later 1920s when executives announced that, whereas the industry had previously thought of telephone service as a practical necessity, they now realized that it was more: it was a "convenience, comfort, luxury"; its value included its "trivial" social uses. In 1928, Publicity Vice-President A. W. Page, who had entered AT&T from the publishing industry the year before, was most explicit when he criticized earlier views: "There had also been the point of view [in the Bell System and among the public] about not using the telephone for frivolous conversation. This is about as commercial as if the automobile people should advertise. 'Please do not take out this car unless you are going on a serious errand. . . .' We are faced, I think, with a state of public consciousness that the telephone is a necessity and not to be trifled with, certainly in the home." Bell sales officials were told to sell telephone service as a "comfort and convenience," including as a conversational tool.[36]

Although this change in opinion is most visible for the Bell System, similar trends can be seen in the pages of the journal of the independent companies, *Telephony,* especially in regard to rural customers. Indeed, early conflict about telephone sociability was most acute in rural areas. During the monopoly era, Bell companies largely neglected rural demand. The depth and breadth of

[35]Hall's philosophy is evident in the correspondence over measured service before 1900, box 1127, AT&T ARCH. Decades later, he pushed it in a letter to E. M. Burgess, Colorado Telephone Company, March 30, 1905, box 1309, AT&T ARCH, even arguing that operators should stop turning away calls made by children and should instead encourage such "trivial uses." The biographical information comes from an obituary in AT&T ARCH. Another, more extreme populist was John L. Sabin, of PT&T and the Chicago Telephone Co.; see Mahon (n. 2 above), pp. 29ff.

[36]A. W. Page, "Public Relations and Sales," "General Commercial Conference, 1928," p. 5, microfilm 368B, ILL BELL INFO. See also comments by Vice-President Gherardi and others in same conference and related ones of the period. On Page and the changes he instituted, see G. J. Griswold, "How AT&T Public Relations Policies Developed," *Public Relations Quarterly* 12 (Fall 1967): 7–16; and Marchand, *Advertising* (n. 21 above), pp. 117–20.

that demand became evident in the first two decades of this century, when proportionally more farm than urban households obtained telephones, the former largely from small commercial or cooperative local companies. Sociability both spurred telephone subscription and irritated the largely non-Bell vendors.

The 1907 Census of Telephones argued that in areas of isolated farmhouses "a sense of community life is impossible without this ready means of communication. . . . The sense of loneliness and insecurity felt by farmers' wives under former conditions disappears, and an approach is made toward the solidarity of a small country town." Other official investigations bore similar witness.[37] Rural telephone men also dwelt on sociability. One independent company official stated: "When we started the farmers thought they could get along without telephones. . . . Now you couldn't take them out. The women wouldn't let you even if the men would. Socially, they have been a godsend. The women of the county keep in touch with each other, and with their social duties, which are largely in the nature of church work."[38]

Although the episodic sales campaigns to farmers stressed the practical advantages of the telephone, such as receiving market prices, weather reports, and emergency aid, the industry addressed the social theme more often to them than to the general public. A PT&T series in 1911, for example, focused on the telephone in emergencies, staying informed, and saving money. But one additional advertisement said it was: "A Blessing to the Farmer's Wife. . . . It relieves the monotony of life. She CANNOT be lonesome with the Bell Service. . . ."[39] For all that, telephone professionals who dealt with farm-

[37]BOC, *Special Reports: Telephones: 1907* (Washington, D.C., 1910), pp. 77–78; see also U.S. Congress, Senate, Country Life Commission, 60th Cong., 2d sess., 1909, S. Doc. 705; and F. E. Ward, *The Farm Woman's Problems*, USDA Circular 148 (Washington, D.C., 1920). See also C. S. Fischer, "The Revolution in Rural Telephony," *Journal of Social History* (in press).

[38]Quoted in R. F. Kemp, "Telephones in Country Homes," *Telephony* 9 (June 1905): 433. A 1909 article claims that "[t]he principle use of farm line telephones has been their social use. . . . The telephones are more often and for longer times held for neighborly conversations than for any other purpose." It goes on to stress that subscribers valued conversation with anyone on the line; see G. R. Johnston, "Some Aspects of Rural Telephony," *Telephony* 17 (May 8, 1909): 542. See also R. L. Tomblen, "Recent Changes in Agriculture as Revealed by the Census," *Bell Telephone Quarterly* 9 (October 1932): 334–50; and J. West (C. Withers), *Plainville, U.S.A.* (New York, 1945), p. 10.

[39]The PT&T series appeared in the *Antioch* (Calif.) *Ledger* in 1911. For some examples and discussions of sales strategies to farmers, see Western Electric, "How to Build Rural Lines," n.d., "Rural Telephone Service, 1944–46," box 1310, AT&T ARCH; Stromberg-Carlson Telephone Manufacturing Company, *Telephone Facts for*

ers often fought the use of the line for nonbusiness conversations, at least in the early years. The pages of *Telephony* overflow with complaints about farmers on many grounds, not the least that they tied up the lines for chats.

More explicit appreciation of the value of telephone sociability to farmers emerged later. A 1931 account of Bell's rural advertising activities stressed business uses, but noted that "only within recent years [has] emphasis been given to [the telephone's] usefulness in everyday activities . . . the commonplaces of rural life." A 1932 article in the *Bell Telephone Quarterly* notes that "telephone usage for social purposes in rural areas is fundamentally important." Ironically, in 1938, an independent telephone man claimed that the social theme *had been* but was *no longer* an effective sales point because the automobile and other technologies had already reduced farmers' isolation![40]

As some passages suggest, the issue of sociability was also tied up with gender. When telephone vendors before World War I addressed women's needs for the telephone, they usually meant household management, security, and emergencies. There is evidence, however, that urban, as well as rural, women found the telephone to be useful for sociability.[41] When industry men criticized chatting

---

Farmers (Rochester, N.Y., 1903), Warshaw Collection, Smithsonian Institution; "Facts regarding the Rural Telephone," *Telephony* 9 (April 1905): 303. In *Printers' Ink*, "The Western Electric," 65 (December 23, 1908): 3–7; F. X. Cleary, "Selling to the Rural District," 70 (February 23, 1910): 11–12; "Western Electric Getting Farmers to Install Phones," 76 (July 27, 1911): 20–25; and H. C. Slemin, "Papers to Meet 'Trust' Competition," 78 (January 18, 1912): 28.

[40] R. T. Barrett, "Selling Telephones to Farmers by Talking about Tomatoes," *Printers' Ink* (November 5, 1931): 49–50; Tomblen (n. 38 above); and J. D. Holland, "Telephone Service Essential to Progressive Farm Home," *Telephony* 114 (February 19, 1938): 17–20. See also C. S. Fischer, "Technology's Retreat: The Decline of Rural Telephones, 1920–1940," *Social Science History* (in press).

[41] A 1925 survey of women's attitudes toward home appliances by the General Federation of Women's Clubs showed that respondents preferred automobiles and telephones above indoor plumbing; see M. Sherman, "What Women Want in Their Homes," *Woman's Home Companion* 52 (November 1925): 28, 97–98. A census survey of 500,000 homes in the mid-1920s reportedly found that the telephone was considered a primary household appliance because it, with the automobile and radio, "offer[s] the homemaker the escape from monotony which drove many of her predecessors insane"; reported in *Voice Telephone Magazine*, in-house organ of United Communications, December 1925, p. 3, MU IND TEL. One of our interviewees who conducted door-to-door telephone sales in the 1930s said that women were attracted to the service first in order to talk to kin and friends, second for appointments and shopping, and third for emergencies, while, for men, employment and business reasons ranked first. See also Rakow, "Gender" (n. 2 above), and C. S. Fischer, "Women and the Telephone, 1890–1940," paper presented to the American Sociological Association, 1987.

on the telephone, they almost always referred to the speaker as "she." Later, in the 1930s, the explicit appeals to sociability also emphasized women; the figures in such advertisements, for example, were overwhelmingly women.

In rough parallel with the shift in manifest advertising appeals toward sociability, there was a shift in industry attitudes from irritation with to approval of sociable conversations as part of the telephone's "comfort, convenience, and luxury."

Economic Explanations

Why were the telephone companies late and reluctant to suggest sociable conversations as a use? There are several, not mutually exclusive, possible answers. The clearest is that there was no profit in sociability at first but profit in it later.

Telephone companies, especially Bell, argued that residential service had been a marginal or losing proposition, as measured by the revenues and expenses accounted to each instrument, and that business service had subsidized local residential service. Whether this argument is valid remains a matter of debate. Nevertheless, the belief that residential customers were unprofitable was common, especially among line workers, and no doubt discouraged intensive sales efforts to householders.[42] At times, Bell lacked the capital to construct lines needed to meet residential demand. These constraints seemed to motivate occasional orders from New York not to advertise basic service or to do so only to people near existing and unsaturated lines.[43] And, at times, there was a technical incompatibility

[42]See, e.g., J. W. Sichter, "Separations Procedures in the Telephone Industry," paper P-77-2, Harvard University Program on Information Resources (Cambridge, Mass., 1977); *Public Utilities Digest*, 1930s–1940s, passim; "Will Your Phone Rates Double?" *Consumer Reports* (March 1984): 154–56. Chan's industry interviewees believed this cross subsidy to be true, as, apparently, did AT&T's commercial engineers; see various "Conferences" cited above, AT&T ARCH and ILL BELL INFO.

[43]E.g., commercial engineer C. P. Morrill wrote in 1914 that "we are not actively seeking new subscribers except in a few places where active competition makes this necessary. Active selling is impossible due to rapid growth on the Pacific Coast." He encouraged sales of party lines in congested areas, individual lines in place of party lines elsewhere, extensions, more calling, directory advertisements, etc., rather than expanding basic service into new territories; see PAC TEL MAG 7 (1914): 13–16. And, in 1924, the Bell System's commercial managers decided to avoid canvassing in areas that would require plant expansion and to stress instead long-distance calls and services, especially for large business users; see correspondence from B. Gherardi, vice-president, AT&T, to G. E. McFarland president, PT&T, July 14, 1924, and November 26, 1924, folder "282—Conferences," SF PION MU, and exchage with McFarland, May 10 and May 20, 1924, folder "Correspondence—B. Gherardi," SF PION MU.

between the quality of service Bell had accustomed its business subscribers to expect and the quality residential customers were willing to pay for. Given these considerations, Bell preferred to focus on the business class, who paid higher rates, bought additional equipment, and made long-distance calls.[44]

Still, when they did address residential customers, why did telephone vendors not employ the sociability theme until the 1920s, relying for so long only on practical uses? Perhaps social calls were an untouched and elastic market of consumer demand. Having sold the service to those who might respond to practical appeals—and perhaps by World War I everyone knew those practical uses—vendors might have thought that further expansion depended on selling "new" social uses of the telephone.[45] Similarly, vendors may have thought they had already enrolled all the subscribers they could—42 percent of American households in 1930—and shifted attention to encouraging use, especially of toll lines. We have seen how sales efforts for intercity calls invoked friends and family. But this explanation does not suffice. It leaves as a puzzle why the sociability themes continued in the Depression when the industry focused again on simply ensuring subscribers and also why the industry's internal attitudes shifted as well.

Perhaps the answer is in the rate structures. Initially, telephone companies charged a flat rate for unlimited local use of the service. In such a system, extra calls and lengthy calls cost users nothing but are unprofitable to providers because they take operator time and, by occupying lines, antagonize other would-be callers. Some industry men explicitly blamed "trivial" calls on flat rates.[46] Discouraging "visiting" on the telephone then made sense.

Although flat-rate charges continued in many telephone exchanges, especially smaller ones, throughout the period, Bell and others instituted "measured service" in full or in part—charging additionally per call—in most large places during the era of competition. In St. Louis in 1898, for example, a four-party telephone cost forty-five dollars a year for 600 calls a year, plus eight cents a call in excess.[47] This system allowed companies to reduce basic subscrip-

[44]The story of the Chicago exchange under John L. Sabin illustrates the point. See R. Garnet, "The Central Union Telephone Company," box 1080, AT&T ARCH.

[45]This point was suggested by John Chan from the interviews.

[46]See n. 33, 34. This is also the logic of a recent New York Telephone Co. campaign to encourage social calls: The advertising will not run in upstate New York "since the upstaters tend to have flat rates and there would be no profit in having them make unnecessary calls" (see "New Pitch," n. 28 above).

[47]Letter to AT&T President Hudson, December 27, 1898, box 1284, AT&T ARCH. On measured service in general, see "Measured Service Rates," boxes 1127,

tion fees and thus attract customers who wanted the service only for occasional use.

Company officials had conflicting motives for pressing measured service. Some saw it simply as economically rational, charging according to use. Others saw it as a means of reducing "trivial" calls and the borrowing of telephones by nonsubscribers. A few others, such as E. J. Hall, saw it as a vehicle for bringing in masses of small users.

The industry might have welcomed social conversations, if it could charge enough to make up for uncompleted calls and for the frustrated subscribers busy lines produced. In principle, under measured service, it could. (As it could with long distance, where each minute was charged.) Although mechanical time metering was apparently not available for most or all of this period, rough time charges for local calls existed in principle, since "messages" were typically defined as five minutes long or any fraction thereof. Thus, "visiting" for twenty minutes should have cost callers four "messages." In such systems, the companies would have earned income from sociability and might have encouraged it.[48]

However, changes from flat rates to measured rates do *not* seem to explain the shift toward sociability around the 1920s. Determining the extent that measured service was actually used for urban residential customers is difficult because rate schedules varied widely from town to town even within the same states. But the timing does not fit. The big exchanges with measured residential rates had them early on. For example, in 1904, 96 percent of Denver's residential subscribers were on at least a partial measured system, and, in 1905, 90 percent of those in Brooklyn, New York, were as well. (Yet, Los Angeles residential customers continued to have flat

---

1213, 1287, 1309, AT&T ARCH; F. H. Bethell, "The Message Rate," repr. 1913, AT&T ARCH; H. B. Stroud, "Measured Telephone Service," *Telephony* 6 (September 1903): 153–56, and (October 1903): 236–38; and J. E. Kingsbury, *The Telephone and Telephone Exchanges* (London, 1915), pp. 469–80.

[48]Theodore Vail claimed in 1909 that mechanical time metering was impossible (in testimony to a New York State commission, see n. 8 above, p. 470). See also Judson (n. 34 above), p. 647. In 1928, an operating engineer suggested overtime charges on five-minute calls and stated that equipment for monitoring overtime was now available; see L. B. Wilson, "Report on Commercial Operations, 1927," in "General Commercial Conference, 1928," p. 28, microfilm 368B, ILL BELL INFO. On the five-minute limit, see "Measured Service," box 1127, AT&T ARCH, passim; and Bell Canada, *The First Century of Service* (Montreal, 1980), p. 4. There is no confirmation on how strict operators in fact were in charging overtime. The Bell System, at least, was never known for its laxness in such matters.

rates.)[49] There is little sign that these rate systems altered significantly in the next twenty-five years while sociability themes emerged.

Conversely, flat rates persisted in small exchanges beyond the 1930s. Moreover, sociability themes appeared more often in rural sales campaigns than in urban ones, despite the fact the rural areas remained on flat-rate schedules.

Although concern that long social calls occupied lines and operators—with financial losses to the companies—no doubt contributed to the industry's resistance to sociability, it is not a sufficient explanation of those attitudes or, especially, of the timing of their change.

Technical Explanations

Industry spokesmen early in the era would probably have claimed that technical considerations limited "visiting" by telephone. Extended conversations monopolized party lines. That is why companies, often claiming customer pressure, encouraged, set—or sought legal permission to set—time limits on calls. Yet, this would not explain the shift toward explicit sociability, because as late as 1930, 40–50 percent of Bell's main telephones in almost all major cities were still on party lines, a proportion not much changed from 1915.[50]

A related problem was the tying up of toll lines among exchanges, especially those among villages and small towns. Rural cooperatives complained that the commercial companies provided them with only single lines between towns. The companies resisted setting up more, claiming they were underpaid for that service. This

[49]Denver: letter from E. J. Hall to E. W. Burgess, 1905, box 1309, AT&T ARCH; Brooklyn: BOC, *Telephones, 1902* (n. 4 above); Los Angeles: "Telephone on the Pacific Coast, 1878–1923," box 1045, AT&T ARCH.

[50]On company claims, see, e.g., "Limiting Party Line Conversations," *Telephony* 66 (May 2, 1914): 21; and MacMeal (n. 2 above), p. 224. On party-line data, compare the statistics in the letter from J. P. Davis to A. Cochrane, April 2, 1901, box 1312, AT&T ARCH, to those in B. Gherardi and F. B. Jewett, "Telephone Communications System of the United States," *Bell System Technical Journal* 1 (January 1930): 1–100. The former show, e.g., that, in 1901, in the five cities with the most subscribers, an average of 31 percent of telephones were on party lines. For those five cities in 1929, the percentage was 36. Smaller exchanges tended to have even higher proportions. See also "Supplemental Telephone Statistics, PT&T," "Correspondence—Du Bois," SF PION MU. The case of Bell Canada also fails to support a party-line explanation. Virtually all telephones in Montreal and Toronto were on individual lines until 1920.

single-line connection would create an incentive to suppress social conversations, at least in rural areas. But this does not explain the shift toward sociability either. The bottleneck was resolved much later than the sales shift when it became possible to have several calls on a single line.[51]

The development of long distance might also explain increased sociability selling. Over the period covered here, the technology improved rapidly, AT&T's long-distance charges dropped, and its costs dropped even more. The major motive for residential subscribers to use long distance was to greet kin or friends. Additionally, overtime was well monitored and charged. Again, while probably contributing to the overall frequency of the sociability theme, long-distance development seems insufficient to explain the change. Toll calls as a proportion of all calls increased from 2.5 percent in 1900 to 3.2 percent in 1920 and 4.1 percent in 1930, then dropped to 3.3 percent in 1940. They did not reach even 5 percent of all calls until the 1960s.[52] More important, the shift toward sociability appears in campaigns to sell basic service and to encourage local use, as well as in long-distance ads. (See table 2.)

Cultural Explanations

While both economic and technical considerations no doubt framed the industry's attitude toward sociability, neither seems sufficient to explain the historical change. Part of the explanation probably lies in the cultural "mind-set" of the telephone men.

In many ways, the telephone industry descended directly from the telegraph industry. The instruments are functionally very similar; technical developments sometimes applied to both. The people who developed, built, and marketed telephone systems were predominantly telegraph men. Theodore Vail himself came from a family involved in telegraphy and started his career as a telegrapher. (In contrast, E. J. Hall and A. W. Page, among the supporters of "triviality," had no connections to telegraphy. J. L. Sabin, a man of the

[51]"Carrier currents" allowed multiple conversations on the same line. The first one was developed in 1918, but for many years they were limited to use on long-distance trunk lines, not local toll lines. See, e.g., R. Coe, "Some Distinguishing Characteristics of the Telephone Business," *Bell Telephone Quarterly* 6 (January 1927): 47–51, esp. pp. 49–50; and R. C. Boyd, J. D. Howard, Jr., and L. Pederson, "A New Carrier System for Rural Service," *Bell System Technical Journal* 26 (March 1957): 349–90. The first long-distance carrier line was established in Canada in 1928, after the long-distance sociability theme had emerged; see Bell Canada, *First Century*, no. 46, p. 28.

[52]BOC, *Historical Statistics* (n. 3 above), p. 783.

same bent, did have roots in telegraphy.) Many telephone companies had started as telegraph operations. Indeed, in 1880, Western Union almost displaced Bell as the telephone company. And the organization of Western Union served in some ways as a model for Bell. Telephone use often directly substituted for telegraph use. Even the language used to talk about the telephone revealed its ancestry. For example, an early advertisement claimed that the telephone system was the "cheapest telegraph service ever." Telephone calls were long referred to as "messages." American telegraphy, finally, was rarely used even for brief social messages.[53]

No wonder, then, that the uses proposed first and for decades to follow largely replicated those of a printing telegraph: business communiqués, orders, alarms, and calls for services. In this context, industry men reasonably considered telephone "visiting" to be an abuse or trivialization of the service. Internal documents suggest that most telephone leaders typically saw the technology as a business instrument and a convenience for the middle class, claimed that people had to be sold vigorously on these marginal advantages, and believed that people had no "natural" need for the telephone—indeed, that most (the rural and working class) would never need it. Customers would have to be "educated" to it.[54] AT&T Vice-

[53]On the telegraph background of early telephone leaders, see, e.g., A. B. Paine, *Theodore N. Vail* (New York, 1929); Rippey (n. 2 above); and W. Patten, *Pioneering the Telephone in Canada* (Montreal: Telephone Pioneers, 1926). Interestingly, this was true of Bell and the major operations. But the leaders of small-town companies were typically businessmen and farmers; see, e.g., *On the Line* (Madison: Wisconsin State Telephone Association, 1985). On Western Union and Bell, see G. D. Smith, *The Anatomy of a Business Strategy: Bell, Western Electric, and the Origins of the American Telephone Industry* (Baltimore, 1985). The "cheapest telegraph" appears in a Buffalo flier of November 13, 1880, box 1127, AT&T ARCH. On the infrequent use of the telegraph for social messages, see R. B. DuBoff, "Business Demand and Development of the Telegraph in the United States, 1844–1860," *Business History Review* 54 (Winter 1980): 459–79.

[54]In the very earliest days, Vail had expected that the highest level of development would be one telephone per 100 people; by 1880, development had reached four per 100 in some places; see Garnet (n. 2 above), p. 133, n. 3. It reached one per 100 Americans before 1900 (see table 1). In 1905, a Bell estimate assumed that twenty telephones per 100 Americans was the saturation point and even that "may appear beyond reason"; see "Estimated Telephone Development, 1905–1920," letter from S. H. Mildram, AT&T, to W. S. Allen, AT&T, May 22, 1905, box 1364, AT&T ARCH. The saturation date was forecast for 1920. This estimate was optimistic in its projected *rate* of diffusion—twenty per 100 was reached only in 1945—but very pessimistic in its projected *level* of diffusion. That level was doubled by 1960 and tripled by 1980. One reads in Bell documents of the late 1920s of concern that the automobile and other new technologies were far outstripping telephone diffusion. Yet, even then, there seemed to be no assumption that the telephone would reach the near universality in American homes of, say, electricity or the radio.

President Page was reacting precisely against this telegraphy perspective in his 1928 defense of "frivolous" conversation. At the same conference, he also decried the psychological effect of telephone advertisements that explicitly compared the instrument to the telegraph.[55]

Industry leaders long ignored or repressed telephone sociability—for the most part, I suggest, because such conversations did not fit their understandings of what the technology was supposed to be for. Only after decades of customer insistence on making such calls—and perhaps prodded by the popularity of competing technologies, such as the automobile and radio—did the industry come to adopt sociability as a means of exploiting the technology.

This argument posits a generation-long lag, a mismatch, between how subscribers used the telephone and how industry men thought it would be used. A variant of the argument (posed by several auditors of this article) suggests that there was no mismatch, that the industry's attitudes and advertising accurately reflected public practice. Sales strategies changed toward sociability around the mid-1920s because, in fact, people began using the telephone that way more. This increase in telephone visiting occurred for perhaps one or more reasons—a drop in real costs, an increase in the number of subscribers available to call, clearer voice transmission, more comfortable instruments (from wall sets to the "French" handsets), measured rates, increased privacy with the coming of automatic dial switching, and so on—and the industry's marketing followed usage.

To address this argument fully would require detailed evidence on the use of the telephone over time, which we do not have. Recollections by some elderly people suggest that they visited by telephone less often and more quickly in the "old days," but they cannot specify exact rates or in what era practices changed.[56] On the other hand, anecdotes, comments by contemporaries, and fragments of numerical data (e.g., the 1909 Seattle "study") suggest that residential users regularly visited by telephone before the mid-1920s, whatever the etiquette was supposed to be, and that such calls at least equaled calls regarding household management. Yet, telephone advertising in the period overwhelmingly stressed practical use and ignored or suppressed sociability use.

Changes in customers' practices may have helped spur a change

[55]Page 53 in L. B. Wilson (chair), "Promoting Greater Toll Service," "General Commercial Conference, 1928," microfilm 368B, ILL BELL INFO.

[56]This comment is based on the oral histories reported by Rakow (n. 2 above) and by several interviews conducted in San Rafael, Calif., by John Chan for this project. See also Fischer, "Women" (n. 41).

in advertising—although there is no direct evidence of this in the industry archives—but some sort of mismatch existed for a long time between actual use and marketing. Its source appears to be, in large measure, cultural.

This explanation gains additional plausibility from the parallel case of the automobile, about which space permits only brief mention. The early producers of automobiles were commonly former bicycle manufacturers who learned their production techniques and marketing strategies (e.g., the dealership system, annual models) during the bicycle craze of the 1890s. As the bicycle was then, so was the automobile initially a plaything of the wealthy. The early sales campaigns touted the automobile as a leisure device for touring, joyriding, and racing. One advertising man wondered as late as 1906 whether "the automobile is to prove a fad like the bicycle or a lasting factor in the industry of the country."[57]

That the automobile had practical uses dawned on the industry quickly. Especially after the success of the Ford Model T, advertisements began stressing themes such as utility and sociability—in particular, that families could be strengthened by touring together. Publicists and independent observers alike praised the automobile's role in breaking isolation and increasing community life.[58] As with the telephone, automobile vendors largely followed a market-

[57]Among the basic sources on the history of the automobile drawn from are: J. B. Rae, *The American Automobile: A Brief History* (Chicago, 1965); id., *The Road and Car in American Life* (Cambridge, Mass., 1971); J. J. Flink, *America Adopts the Automobile, 1895–1910* (Cambridge, Mass., 1970); id., *The Car Culture* (Cambridge, Mass., 1976); and J.-P. Bardou, J.-J. Chanaron, P. Fridenson, and J. M. Laux, *The Automobile Revolution,* trans. J. M. Laux (Chapel Hill, N.C., 1982). The advertising man was J. H. Newmark, "Have Automobiles Been Wrongly Advertised?" *Printers' Ink* 86 (February 5, 1914): 70–72. See also id., "The Line of Progress in Automobile Advertising," ibid., 105 (December 26, 1918): 97–102.

[58]G. L. Sullivan, "Forces That Are Reshaping a Big Market," *Printers' Ink* 92 (July 29, 1915): 26–28. Newmark (n. 57 above, p. 97) wrote in 1918 that it "has taken a quarter century for manufacturers to discover that they are making a utility." A 1930s study suggested that 80 percent of household automobile expenditures was for "family living"; see D. Monroe et al., *Family Income and Expenditures. Five Regions,* Part 2. *Family Expenditures,* Consumer Purchases Study, Farm Series, Bureau of Home Economics, Misc. Pub. 465 (Washington, D.C., 1941), pp. 34–36. Recall the 1925 survey of women's attitudes toward appliances (n. 41 above). The author of the report, Federation President Mary Sherman, concluded that "Before toilets are installed or washbasins put into homes, automobiles are purchased and telephones are connected . . . [b]ecause the housewife for generations has sought escape from the monotony rather than the drudgery of her lot" (p. 98). See also *Country Life* and Ward (n. 37 above); E. de S. Brunner and J. H. Kolb, *Rural Social Trends* (New York, 1933); and F. R. Allen, "The Automobile," pp. 107–32 in F. R. Allen et al., *Technology and Social Change* (New York, 1957).

ing strategy based on the experience of their "parent" technology; they stressed a limited and familiar set of uses; and they had to be awakened, it seems, to wider and more popular uses. The automobile producers learned faster.

No doubt other social changes also contributed to what I have called the discovery of sociability, and other explanations can be offered. An important one concerns shifts in advertising. Advertising tactics, as noted earlier, moved toward "softer" themes, with greater emphasis on emotional appeals and on pleasurable rather than practical uses of the product. They also focused increasingly on women as primary consumers, and women were later associated with telephone sociability.[59] AT&T executives may have been late to adopt these new tactics, in part because their advertising agency, N. W. Ayer, was particularly conservative. But in this analysis, telephone advertising eventually followed general advertising, perhaps in part because AT&T executives attributed the success of the automobile and other technologies to this form of marketing.[60]

Still, there is circumstantial and direct evidence to suggest that the key change was the loosening, under the influence of public practices with the telephone, of the telegraph tradition's hold on the telephone industry.

Conclusion

Today, most residential calls are made to friends and family, often for sociable conversations. That may well have been true two or three generations ago, too.[61] Today, the telephone industry encourages such calls; seventy-five years ago it did not. Telephone salesmen then claimed the residential telephone was good for emergencies; that function is now taken for granted. Telephone salesmen then claimed the telephone was good for marketing; that function

[59]Recall that, early on, women were associated in telephone advertising with emergencies, security, and shopping.

[60]On changes in advertising, see sources cited in n. 21 above. The comment on N. W. Ayer's conservatism comes from Roland Marchand (personal communication).

[61]It is difficult to establish for what purpose people actually use the telephone. A few studies suggest that most calls by far are made for social reasons, to friends and family. (This does not mean, however, that people subscribe to telephone service for such purposes.) See Field Research Corporation, *Residence Customer Usage and Demographic Characteristics Study: Summary*, conducted for Pacific Bell, 1985 (courtesy R. Somer, Pacific Bell); B. D. Singer, *Social Functions of the Telephone* (Palo Alto, Calif.: R&E Associates, 1981), esp. p. 20; M. Mayer, "The Telephone and the Uses of Time," in Pool, *Social Impact* (n. 2 above), pp. 225–45; and A. H. Wurtzel and C. Turner, "Latent Functions of the Telephone," ibid., pp. 246–61.

persists ("Let your fingers do the walking. . . . ") but never seemed to be too important to residential subscribers.[62] The sociability function seems so obviously important today, and yet was ignored or resisted by the industry for almost the first half of its history.

The story of how and why the telephone industry discovered sociability provides a few lessons for understanding the nature of technological diffusion. It suggests that promoters of a technology do not necessarily know or determine its final uses; that they seek problems or "needs" for which their technology is the answer (cf. the home computer business); but that consumers may ultimately determine those uses for the promoters. And the story suggests that, in promoting a technology, vendors are constrained not only by its technical and economic attributes but also by an interpretation of its uses shaped by its and their own histories, a cultural constraint that can be enduring and powerful.

[62]A 1934 survey found that up to 50 percent of women respondents with telephones were "favorable" to shopping by telephone. Presumably, fewer actually did so; see J. M. Shaw, "Buying by Telephone at Department Stores," *Bell Telephone Quarterly* 13 (July 1934): 267–88. This is true despite major emphases on telephone shopping in industry advertising. See also Fischer, "Women" (n. 41 above).

Farmers Deskilled: Hybrid Corn and Farmers' Work

DEBORAH FITZGERALD

In recent years labor historians and, to a lesser extent, historians of technology have begun detailed studies of the ways in which skilled laborers have been supplanted by skilled machines. From the power loom to numerically controlled machine tools, machines have been invented with the express purpose of replicating the mental and manual abilities of craftsmen, a process that often renders such crafts knowledge obsolete. While many workers responded with anger and violence, others seemed to adopt a sort of false consciousness, identifying not with their peers but with managers who argued that the virtues of progress must necessarily, if painfully, include the seemingly inevitable drive to mechanization.[1]

As articulated by labor historians, industrial deskilling usually centers on issues of power, authority, and control. For skilled and semiskilled workers, some measure of power and leverage resides in their possession of specialized knowledge, which in turn allows them to perform work tasks that the untrained are not capable of doing. To deskill such a worker, whether by mechanization or the elaborate subdivison of one big task into many smaller tasks, is thus to

DR. FITZGERALD is Associate Professor in history of technology in the Program in Science, Technology, and Society at the Massachusetts Institute of Technology. She is currently writing a book on engineering and economic models in post–World War I American agriculture. She thanks the *Technology and Culture* referees for their critical analysis and Ruth Cowan who, in commenting on medical technology at a 1989 Sacramento SHOT meeting session, unwittingly gave her the idea for the article.

[1]There is an enormous literature on the general theme of technology as an agent of deskilling, especially in regard to artisans and factory workers. See, e.g., Harry Braverman's seminal work *Labor and Monopoly Capital* (New York, 1974). This genre is also discussed in Philip Scranton, "None-too-Porous Boundaries: Labor History and the History of Technology," *Technology and Culture* 29 (1988):722–43; and in David Noble, *Forces of Production: A Social History of Industrial Automation* (New York, 1984). Jonathan Prude discusses the ambivalent responses to technology by some workers in his *The Coming of Industrial Order: Town and Factory Life in Rural Massachusetts, 1810–1860* (New York, 1983).

disempower him or her. The special knowledge is usurped or discarded, as is the power and leverage that this knowledge provided; workers whose jobs require little or no skill are easily replaced and so are less likely to create problems. For labor historians—who are, after all, primarily interested in laborers—the central point of this is that such deskilling is imposed by a money- and power-hungry management on a threatening and allegedly unstable work force.

While this model has been of enormous value to labor historians in elucidating the relationship between workers and managers, it is problematic for historians of technology interested in the role of technology in work more generally. It is difficult to know how to use this model when the relevant users of technology are not positioned in an adversarial relationship or when the technology appears simply to change the way a job is accomplished, but not the reality of the job itself or even the worker who does it.[2] Yet it seems to me that the concept of deskilling, as a heuristic device for understanding work, is useful even when shorn of its political charge, and can reasonably be extended to work sites off the shop floor.

Several recent works illustrate the way in which technologies change the way tasks are done in a style that might be considered deskilling. Ruth Cowan's work on domestic technology suggests that deskilling may occur without malice from above or protest from below. Bruno Latour's notion of "delegation," in which artifacts of all sorts (e.g., door hinges) are viewed as the nonhuman delegates for human work and activity, likewise emphasizes the point that ordinary, individualistic technologies are deskilling even though their happy-go-lucky human users may not be accustomed to seeing things in that light. While Latour's notion of delegation, at first blush, may seem a trivialization of human work, it facilitates a welcome escape from the shop floor and allows a fresh reassessment of the relations between technology and work of all sorts.[3]

In joining the notions of deskilling and delegation for analyzing work, that work must display several characteristics. First, a human actor or worker must possess manual and mental abilities in regard to

[2]One thinks here of individualistic technologies such as home appliances, personal computers, and the like.

[3]Ruth Schwartz Cowan, *More Work for Mother: The Ironies of Household Technology from the Open Hearth to the Microwave* (New York, 1983); Bruno Latour, "Where Are the Hidden Masses? The Sociology of a Few Mundane Artifacts," in *Shaping Technology/Building Society: Studies in Sociotechnical Change,* ed. Wiebe Bijker and John Law (Cambridge, Mass., 1992). In his *First the Seed: The Political Economy of Plant Biotechnology* (New York, 1988), Jack Kloppenburg, Jr., has suggested similar ways of understanding what plant hybridization represented to farmers and breeders alike in terms of patenting knowledge.

some task—for example, tool and die cutting, cooking, writing a book, or making bricks. Second, these abilities must be at least partly amenable to mechanical replication; for example, hand motions might be replicated by a machine, or mental calculations performed by a computer. Third, other humans (e.g., inventors, managers, engineers) must decide that there are sufficient reasons for creating and marketing the artifact or machine, reasons that may include the deliberate exclusion of the worker from the labor process.

One group of actors whose work has encountered an inordinate number of nonhuman delegates is farmers, although neither farmers themselves nor historians have been inclined to view farmers as laborers. On the face of it, there are several good reasons for this. First, the term "farmer," like the term "worker," implies a social and economic homogeneity that has not only never existed, but that has also suggested a wholly disparate category of activity. Farmers have been seen as a group apart, sharing almost none of the social and economic attributes of artisans, teachers, bankers, entrepreneurs, or even consumers. Ordinary categories of analysis have seemed either inadequate or inappropriate to farmers' special circumstances. And this seeming lack of common currency has made it difficult to consider how technological change might shape the experiences of farmers and laborers in similar ways.

Second, not only is the farmer's position in the social and economic order highly complex—even without considering the great variation among farmers in terms of affluence, education, and geographic location—but many agricultural innovations have been greeted with enthusiasm by farmers. The plain drudgery involved in many farming tasks, the tedious and repetitive aspects of plowing, planting, cultivating, milking, weaning, and harvesting, have led to an unexamined belief that here, at least, mechanization is welcome. Further, the apparent lack of conflict between farmers and agribusiness suggests, if not commiseration, then at least an acquiescence among farmers who are presented with "labor-saving" devices.

Yet I would argue that these factors obscure our understanding of technological change on the farm. If automatic milking machines are greeted with relief and gratitude by farmers, the machines are no less agents of deskilling. Likewise, tractors, combines, and other implements may have saved one kind of labor, but they often required that farmers take on a new set of tasks (such as maintaining the equipment, building storage sheds, and increasing production in an effort to pay for the equipment purchases). Thus the adoption of machinery, even if welcome, may eliminate the farmer's knowledge and abilities with regard to a particular activity and nearly always requires that he or she learn a new set of skills or adapt his or her abilities to the new situation. Furthermore, it has been difficult to assess the deskilling components of agricultural technology because farmers, unlike industrial workers, are not immediately thrown out of work when the technology becomes available. Since farmers are generally their own managers, their choice is initially between adopting or not adopting the implement; industrial workers have no such choice. It is probably true that the decline in the numbers of farmers in America is at least partly due to the rising cost of farming and that these costs have resulted from farmers' increasing use of both machines and chemicals. But it is difficult to demonstrate an immediate causal relationship between the introduction of mechanical and chemical technologies and a farmer's decision to abandon farming.[4]

In this article, I would like to pursue the argument that hybrid corn was an agent by which farmers were effectively deskilled. As a biological artifact, hybrid corn presented farmers with a novel set of opportunities and problems that were not apparent in their previous experience with mechanical innovations. Despite the obvious differences between tractors and seeds, however, the similarities between the two sorts of innovations are striking. If the farmer is viewed as a laborer who possesses a set of manual and mental skills, then hybrids were perhaps more profoundly deskilling than any mechanical implement.

Farmers' Skills before Hybrids

Our notions about what sort of labor farmers did in the early 20th century has been shaped by the "yeoman myth" as much as by reality. The dominant images center on physical tasks such as hand plowing and planting, barn raising, milking cows, and other forms of fairly strenuous and incessant labor ("backbreaking toil," in bucolic jargon). Yet as anyone acquainted with farm work can testify, the mental labor of farming embodied the more important set of skills. The farmer's crucial task was, not so much in performing physical work, but in figuring out what to do, how to do it, and when. Decisions about when to begin spring planting, which make of tractor to buy, whether to diversify the crops, when to market grain and livestock, and so on, could be made by the farmer only with experience and judgment regarding the stakes involved in each misstep. The farmer's knowl-

[4]The relationship between technology and science and the increasingly difficult farm situation has never been fully evaluated, although the causal role of technology in social upheaval is often implied. See, e.g., Dirk Johnson, "Population Decline in Rural America: A Product of Advances in Technology," *New York Times*, September 11, 1990, p. 20A.

edge and judgment regarding proper farm practice rested solidly on experience, observation, and usually an "apprenticeship" with his or her father. Few farmers had formal agricultural education in the 1930s, and indeed most of them were suspicious of "book farming." For both master craftsman and farmer, the most obvious manifestation of their skill was a product, whether bricks or crops, but their real power lay in the knowledge they had acquired through long experience.[5]

The process of growing corn was at once straightforward and infinitely complex. From at least the late 19th century, farmers were attentive to the many factors that could affect their corn crop, many of which were ecological. For example, students of Justus von Liebig persuaded thousands of farmers that the particular composition of soil on a farm would determine the yield a farmer would get and the quality of that yield. Adjacent farms might easily require different soil treatments, such as the application of limestone or phosphorus. Likewise, insect pests were not predictable in their choice of which farms to attack, and a farmer might struggle for years before discovering a means of partially protecting his crops from invasion. The vagaries of weather were fundamentally unmanageable. Farmers in the northern corn belt were routinely faced with a shorter growing season than other farmers, and farmers who had been caught in a drought were usually receptive to adopting new corn strains that promised to be drought-resistant.[6]

The corn varieties available to midwestern farmers by the early 1900s were highly distinctive, exhibiting a wide variety of observable physical characteristics and requiring diverse growing conditions. A list of corn varieties compiled by the University of Illinois Department of Agronomy in 1925 included seven white strains and twelve yellow strains, each of which was considered distinctive enough to warrant the designation "variety." These strains differed from each other as well in the length of time it took for each to mature: a 90-day corn was considered moderately early, a 120-day rather late, and this was an important characteristic, especially for farmers in the northern corn

[5]For a discussion of the yeoman myth, see Richard Hofstadter, *The Age of Reform* (New York, 1955). By far the best account of what farmers actually do is in Richard Rhodes, *Farm: A Year in the Life of an American Farmer* (New York, 1989). Farmers' distaste for agricultural experts is discussed in Roy V. Scott, *The Reluctant Farmer: The Rise of Agricultural Extension to 1914* (Urbana, Ill., 1970). The following discussion draws heavily on Deborah Fitzgerald, *The Business of Breeding: Hybrid Corn in Illinois* (Ithaca, N.Y., 1990).

[6]Margaret Rossiter, *The Emergence of Agricultural Science: Justus Liebig and the Americans, 1840–1880* (New Haven, Conn., 1975).

belt. Finally, different strains often had different purposes. For instance, sweet corn that would be sold to canneries was quite different than feed corn, and feed corn that was grown for the market might well differ from corn grown for the farmer's own livestock.[7]

But it was in the unusually distinctive visual characteristics of corn that a farmer's experience was most notable. While a nonfarmer might have difficulty telling the difference between Reid's Yellow Dent and Western Plowman, an experienced farmer would know at a glance which was which. Further, the difference mattered to farmers, who held strong opinions—about the superiority of, say, roughly indented over smooth corn, or tapered ears over cylindrical ears. Farmers learned how to "read" corn, and how to translate these physical characteristics into meaningful indicators of yield, quality, insect resistance, or simple aesthetic value.

The sensitivity of farmers to ear characteristics was vividly illustrated in the annual corn shows that were often held in conjunction with the state fair in corn-growing states. In Illinois, a group of prominent corn breeders and growers established the Illinois Corn Breeders' Association in 1890, with the purpose of "developing an interest in better seed corn." To this end the association inaugurated corn contests and invented a "scorecard" for judging farmers' entries. The scorecard was an idealized list of what a good ear of corn should look like. Each ear would be judged by its adherence to the ideal on a twelve-point scale; among the points were ear shape (ears should be cylindrical with straight rows), uniformity of kernels (kernels should be of uniform shape and color), kernel shape (kernels should be wedge-shaped with straight edges), and so on. Ears that conformed to this standard received a perfect score, five points for wedge-shaped kernels, ten points for ears in which the tips were filled out with regular-shaped kernels, and so forth.[8]

In addition to the scorecard ideal for corn in general, however, there was an additional "standard of perfection" for the seven most prominent corn varieties. That is, each variety had its own idealized form that made it distinct from the other varieties under consider-

[7]"A Brief History of Certain Varieties of Corn," January 1925, Agronomy department subject file 8/6/2, box 5, file 61, corn history, University of Illinois Archives, Urbana (hereafter 8/6/2, University of Illinois Archives).

[8]See Helen M. Cavenagh, *Seed, Soil and Science: The Story of Eugene D. Funk* (Chicago, 1959), pp. 216–19; for a sample corn scorecard, see "Illinois Corn Growers' Association First Annual Report," Funk Brothers Seed Company files, Bloomington, Ill., 1902; and "Constitution and By-Laws of the Illinois Seed Corn Breeders' Association," May 1911, D. E. Alexander Collection, University of Illinois, Urbana, Department of Maize Genetics.

ation. For example, while the scorecard ideal for each shape was cylindrical, five of the seven varieties were ideally tapered. And while all kernels were ideally wedge-shaped, perfect varietal samples might be long-wedged, broad-wedged, medium-wedged, or very broad-wedged. Similarly, while the scorecard ideal called for the kernel at the butt of the ear to be "swelled out about the shank regularly," this characteristic varied from "deeply rounded compressed" kernels in Reid's Yellow Dent to "shallow rounded depressed" in White Superior. And although the so-called white corns were distinguished in large part by not being yellow, a good corn judge discriminated further: Boone County White should be pearl white, Silver Mine should be cream white, and White Superior should be starch white.[9]

Of course, it was not only corn-show judges who were expected to be able to detect these visual differences between corn varieties, but farmers themselves. Farmers, after all, were the ones entering the contest, and few would enter if ignorant of the standards used by judges. For most farmers, attentiveness to the visual characteristics of different corn strains was important for reasons other than the corn shows. Indeed, knowing which strains yielded well and which poorly was a crucial component in the farmers' economic strategy. This was because farmers obtained most of the corn seed they planted each spring from their own fall harvest. Although a farmer could get enough corn to plant his acreage by simply putting aside the first few bushels he came upon, most farmers did not choose corn so haphazardly, at least not after the state agricultural colleges demonstrated the riskiness of such an approach. Rather, farmers usually selected their field seed either in the field or in the silo according to a set of guidelines that may have been learned from experience, the corn-show scorecard, or the agricultural college recommendations. Whatever the guidelines may have been (and they could vary dramatically), the farmer's selection of one ear but not another reflected his or her understanding that a successful farming operation necessitated both knowledge and judgment.[10]

I do not mean to suggest that the scorecard was a "correct" representation of what good corn should look like, or even that most farmers adhered to such a specification in selecting their own corn for planting. Rather, my point is that open-pollinated corn strains differed from each other both in their observable features and in the way

[9]Ibid.

[10]While there is no way to gauge exactly how many farmers selected their seed, annual reports from both the U.S. Department of Agriculture (USDA) Bureau of Plant Industry and the University of Illinois Crops Extension specialist indicate that most farmers did select.

they behaved in the field and that farmers paid attention to these differences. Show corn—that is, corn that was grown in conformity with the scorecard for the express purpose of entering in the corn contest—was not so different from other state fair-type contests. Those who grew the largest potato, most oddly shaped zucchini, or most variegated rabbit, were all engaged in a sort of "norm-bashing," in which awareness of the standard forms allowed one to push the limits of recognizable types. But such exercises should not be confused with everyday practice, which is informed by a different set of goals. In its early years, the corn contest was similarly an exercise in selecting for particular and highly formalized characteristics, emphasizing aesthetics rather than function; it was the appearance rather than the performance of corn that won prizes.

Farmers who grew corn for a living may or may not have tried to grow show corn, but all knew which kinds of corn grew well on their own farms and which did not. Because the yield of a corn strain depended almost entirely on such things as climate, soil type, time of planting, presence of pests and disease, and a farmer's habits of planting and cultivating, no strain could be identified definitively as better than any other strain. While one farmer might get good results with Western Plowman, his neighbor might not. Further, as the scorecard emphasized, there was considerable variation within a single strain. Hence a farmer could select his own corn only by experience, which required him to correlate the appearance of corn with field results on an individual basis.[11]

By 1906 it was becoming apparent both to breeders and to farmers that while the scorecard was fine as an exercise in breeding, it was less satisfactory in helping farmers select corn for high yield. Corn breeder Eugene Funk, who had helped derive the Illinois scorecard ten years before, found that yields of scorecard corn "went all to pieces" after a few years. Consulting with other corn breeders in the region, he discovered that they too were disenchanted with the results. One breeder reported that his field produced excellent yields although the ear length exceeded the scorecard ideal; another

[11]I have avoided referring to yields in any numerical way for two reasons. First, corn strains such as Western Plowman were types rather than unchanging entities, and no two samples would be quite the same; hence, it is nonsensical to refer to the yield of Western Plowman. Second, the ecological and practical variations among farmers' fields, even within a small area, were quite substantial, and the yield of each farmer tells little about the inherent superiority or inferiority of any given strain. Numerical references to yield characteristics that later appeared in USDA documents referred to average yields for a strain, which was a function of all the reported yields taken together.

suggested that no one interested in yield took the scorecard seriously anyway. Over the next fifteen years, Funk worked to revise the scorecard according to his own tests that correlated visual features with yield, and by 1921 the Illinois scorecard had been rewritten to take into account this new evidence.[12]

H. A. Wallace, too, insisted that corn should be selected not by the corn-show scorecard but by productivity of ears. H. A. shared with his father, H. C. Wallace (who became secretary of agriculture in 1920), a passionate interest in corn improvement. In the pages of the family farm paper, *Wallaces' Farmer,* H. A. barraged farmers with the latest news on the subject. He reported at length on the results of an Ohio experiment that explored the correlation between show corn and high yield and was gleeful when the Ohio agricultural experiment station verified his own dissatisfaction with show standards: "These experiments upset a lot of cherished convictions in the minds of seed corn breeders. Many farmers, however, after reading the account . . . will say: 'I told you so.'" At the USDA's Bureau of Plant Industry, Physiologist in Charge of Corn Investigations C. P. Hartley had also warned his cooperating breeders against selecting for scorecard standards. It must have been a hard lesson to teach, nonetheless; as late as 1917 H. A. Wallace again urged farm boys to grow corn for yield rather than appearance, appealing to their desire to be superior to the other boys: ordinary boys could grow for corn-show appearance, he suggested, but clever boys would grow for yield.[13]

Beginning around 1910, farmers were exhorted by agricultural experiment stations, the USDA, and the farm press to take a more active role in corn growing, and this meant paying more attention to the selection process. For example, Hartley worked tirelessly in encouraging midwestern farmers to improve their yields through judicious selection. He visited an Ohio farmers' group several times a year to supervise their efforts and boasted in 1912 that he had helped these farmers increase their yield by 10 to 20 bushels an acre since

[12]Eugene D. Funk to corn breeders, June 7, 1906; G. O. Sutton to Funk, September 7, 1906; C. G. Hopkins to Funk, June 12, 1906; E. Davenport to Funk, June 15, 1906; A. D. Shamel to Funk, June 19, 1906; and Albert Hume to Funk [n.d. but concurrent with others], all in Funk Brothers files, Bloomington, Ill.; Funk Brothers Seed Company, 1913–14 catalog, p. 7, Bloomington, Ill.; Cavenagh, p. 407.

[13]"Corn Experiments," *Wallaces' Farmer* 40, April 30, 1915, p. 700; "Telling the Yield of Corn by Its Looks," *Wallaces' Farmer* 41, December 22, 1916, p. 1676; C. P. Hartley, "Directions to Cooperative Corn Breeders," *Bureau of Plant Industry 564,* April 23, 1910; Funk's views were reported by Wallace in "Show Corn and Yield," *Wallaces' Farmer* 42, April 13, 1917, pp. 656–57; "The Corn Breeding Plot," *Wallaces' Farmer* 42, January 12, 1917, p. 58.

1900, with several Ohio farms breaking the 100-bushels-per-acre mark in 1912. Secretary of Agriculture "Tama Jim" Wilson was also a booster, claiming that selection alone had increased the yield of some varieties by 25–30 percent.[14]

In the midteens, Funk and others began to notice that yields of corn were declining in the Midwest because of the growing prevalence of corn disease. Funk selected corn on the basis of its general healthy appearance and then tested a few kernels of each good ear on a germinator to ensure that healthy-looking ears would produce healthy corn. Funk considered the germinator the "acid test" in distinguishing good from bad corn seed, and all seed that failed this test was destroyed. Unfortunately, after a few years it appeared that this system had shortcomings, too: seed that passed the germinator test often produced diseased corn in the field. With the assistance of the USDA and G. N. Hoffer at Purdue University, Funk learned that several discrete organisms were responsible for the problem and that these diseases might manifest themselves, not at germination, but later in the plant's growth. Thus, by 1917 it was clear that farmers could not expect high yield if selecting by the scorecard, by selecting healthy-looking ears, or even by using home-built germinators. The problem of correlating visual characteristics with yield was getting more mysterious all the time.[15]

By the early 1920s the disease studies had generated two important programs that empowered farmers in the short run but deskilled them in the long run. The first program grew out of the realization that some visual features of corn could be linked to the presence of corn disease and that, by selecting against such features, farmers could significantly increase their yields. In cooperation with the University of Illinois' College of Agriculture, Funk breeder J. R. Holbert and Crops Extension Specialist J. C. Hackleman devised a new scorecard that codified the correlation between corn disease and corn plant characteristics. While nearly half the percentage points on the new scorecard were reserved for "general appearance," perhaps in deference to those who could not resist the old "fancy" corn that won honors on purely aesthetic grounds, 45 percentage points were concerned with characteristics especially indicative of disease resis-

[14]C. P. Hartley, "How Can the USDA Best Help the Corn Growers of the United States?" address to Butler County, Ohio, Corn Improvement Association, delivered January 10 or 11, 1912, Record Group 54, entry 2, box 319, National Archives, Washington, D.C.; Hartley to Beverly T. Galloway, October 14, 1912, ibid.; James Wilson, "Report of the Secretary of Agriculture," in *Yearbook of Agriculture* (Washington, D.C., 1912), p. 128.

[15]James R. Holbert, "Germination Story," 1950, Funk Brothers files, Bloomington, Ill.

tance. For example, 5 percent of the score was for indentation; judges were instructed to discriminate against roughly indented kernels, which indicated slow maturing because of disease. Five percent went to the chemical composition of the kernel, with discrimination against starchy kernels because they were more susceptible to disease. Disease indicated by ear butt and shank attachments (where the ear hooked onto the stalk) that were pink, brown, or cracked could result in the loss of up to 10 percentage points.[16]

In 1923 Hackleman initiated a project aptly called "Better Seed Corn," through which he trained county farm advisers to identify corn disease in the field. The advisers then passed the lesson on to the farmers themselves. By means of an elaborate series of meetings, classes, and demonstrations, as well as a rural media blitz, Hackleman sought to generate interest in the new methods by instilling enthusiasm in his corps of advisers and in the farmers, who, he hoped, would pester advisers for more information. The main incentive for farmers, naturally, was higher yields. Corn yields after selection tended to average 5–10 bushels an acre more than fields planted with unselected seed, although the amount varied from year to year. In 1927, for example, corn disease was especially prevalent in Illinois and decreased yields by 12–15 bushels an acre. Those farmers who practiced selection, however, suffered less than those who did not. Hackleman rather optimistically concluded that farmers selecting against disease lost on average only 2–5 bushels an acre.[17]

During this period before 1920, then, it is clear that farmers themselves were the linchpin of corn improvement efforts. Their knowledge of the corn plant and its range of variation, and of their own fields and ecological peculiarities, were powerful tools in shaping their own economic stability. While some farmers no doubt put less stock than others in the potential return from these crop-improvement efforts, all shared a form of knowledge and a sense of perspective that many of the experts/breeders, especially a generation later, did not possess. Like their counterparts in the trades, farmers acted out of pride and economic self-interest; an indifferent farmer, like an indifferent carpenter, could not expect to prosper.

[16]James R. Holbert, W. L. Burlison, B. Koehler, C. M. Woodworth, and G. Dungan, "Corn Root, Stalk, and Ear Rot Diseases and Their Control through Selection and Breeding," *Illinois Agricultural Experiment Station Bulletin 255,* 1924.

[17]J. C. Hackleman, "Annual Report of the Crops Extension Specialist for 1922," p. 5; Hackleman, "Annual Report for 1924," pp. 19–22; Hackleman, "Annual Report for 1925," pp. 6–7; Hackleman, "Annual Report for 1927," p. 4; University of Illinois Department of Agronomy, "Historical Data for President Kinley, February 1941," p. 1, all in 8/6/2, University of Illinois Archives.

Hybrid Corn and Useless Knowledge

The second outgrowth of the disease program ran directly counter to the notion that farmers could and should educate themselves to improve their corn. This emphasized the inbreeding and crossbreeding of corn plants in an effort to take advantage of some plants' apparent disease resistance. Corn breeding itself was not new. Since the rediscovery of Mendel's laws of inheritance in 1900, breeders had been experimenting with their ability to control and predict plant characteristics by inbreeding and crossbreeding, and corn was an especially common vehicle for research both by geneticists and by crop specialists. While the goals of these two groups differed, by 1915 or so the net result of their work served to indicate that the large-scale manipulation of corn plants could generate specialized strains, strains with characteristics such as drought or insect resistance. And since the yield of corn was a function of how much good corn was left standing in the field after such natural menaces had abated, the increase in corn yield was the guiding principle of corn breeding.[18]

Although the Mendelian and statistical theories that underlay big inbreeding and crossbreeding programs were not easily grasped, the actual practice was neither scientifically complex nor mysterious. To inbreed the corn plant, a breeder had to prevent random pollination. He or she did so by tying a bag over the tassels when pollen began to appear. The bag of pollen was then tied on to the ear so that the ear silks would be fertilized with its own pollen. This process could be repeated for several generations, or until the breeder thought the plant was sufficiently "reduced" to its most elemental characters. Crossbreeding these inbred plants to each other was no different than it had been before Mendel, although more convenient and efficient methods were developed as time went on.[19]

The simplicity of the method should not, however, obscure several other requirements of breeding programs that effectively locked farmers out of the process. The first was scale. As the early Mendelians had discovered, while inbreeding and crossing were easy to do, the resulting corn might have no redeeming features at all. There seemed to be no way to predict whether two inbreds would "nick" properly by exhibiting the desirable features of their ancestors rather than the undesirable ones. Corn with desirable features might not

[18]For a fuller discussion, see Fitzgerald, *The Business of Breeding* (n. 5 above), pp. 56–69, 154–61; Diane Paul and Barbara Kimmelman, "Mendel in America: Theory and Practice, 1900–1919," in *The American Development of Biology*, ed. Ronald Rainger, Keith R. Benson, and Jane Maienschein (Philadelphia, 1988), pp. 281–310.

[19]Deborah Fitzgerald, "Hybrid Corn: Where's the Science?" (paper presented at the annual meeting of the History of Science Society, Cincinnati, December 1988).

survive inbreeding or might not cross well with other inbreds. One result of this discovery was massive breeding programs aimed at uncovering that one-in-a-million inbred strain. The experiment stations carried on fairly sizable field trials, but the private breeders such as Funk Brothers had far more breeding acres at their disposal.[20]

The second factor was record keeping. Again, while this was not terribly complex or sophisticated, it was essential that each corn ear be identified, labeled, and stored properly and that the kernels planted in the field be matched to the mother ears in storage. Careful records were also needed to keep track of the various generations of each plant. With hundreds of thousands of ears in test each year, simply keeping track of the status of each was both Herculean and tedious, and mistakes could be devastating.

For some of the early breeders, the farmers' skills and knowledge of corn growing were considered advantageous to the breeders' job. H. A. Wallace, for example, felt that individual farmers were the best source of the open-pollinated corn from which inbreds could be developed. Since it was impossible to predict which corn plants would generate good inbreds, Fred Richey of the USDA's Bureau of Plant Industry argued that there was no reason to start with "good" rather than "poor" open-pollinates. But Wallace insisted that farmers who entered the Iowa Yield Contest often had selected strains that featured just those characteristics corn breeders were looking for, such as early maturity and disease resistance. For Wallace, the only way for breeders to discover new strains was to rely on the expertise of the knowledgeable corn farmers themselves.[21]

At the University of Illinois, reliance on farmers was to some extent a necessary evil. In their own breeding work, university researchers were hampered by a lack of fields for trials and literally "farmed out" their inbreds for crossing to farmers who volunteered for the task. This did not sit well with the commercial breeders, who felt that farmers could not be relied on to maintain accurate records or keep the lines pure. At issue was the question of whether ordinary farmers were competent to manage the crossing of corn. Twenty years earlier this was a task that Wallace had claimed "anyone" could do.[22]

[20]Funk Brothers Farms, for example, consisted of 22,000 acres as a result of the pooling efforts of many Funk family members.

[21]Frederick D. Richey to H. A. Wallace, March 17, 1921; Wallace to Carleton R. Ball, January 5, 1922; Ball to Wallace, January 10, 1922, all in Record Group 54, entry 31, box 166-Wallace, National Archives, Washington, D.C.

[22][Henry Wallace], "The Cross Breeding Plot," *Wallaces' Farmer* 43, March 29, 1918, p. 578. Wallace encouraged farmers to experiment with crossing varieties of corn, a

In 1938 some of the more prominent commercial breeders in Illinois met with the University of Illinois agronomy department staff to discuss their concerns about the university giving inbreds to farmers. O. J. Sommer offered the university agronomists an anecdote that underscored the commercial breeders' attitudes toward the farmers:

> Two years ago, a young man came up to me and said, "Do you know I had a letter from the University of Illinois, Crop Improvement Association. They asked me if I want to grow some pure lines and make some hybrid corn." He then said to me, "What is this hybrid corn? Do you think I could do the work—pulling off the tassels, growing it in rows, etc.?" I took the time and explained it to him the best I could. "That sounds interesting," he said, "I think I will try it." My reaction to that was that there was something wrong right then.

Sommer, who lacked scientific training himself, but was in the business of growing and selling corn, agreed with other commercial men that ordinary farmers could not be trusted to understand the intricacies of producing hybrids and must instead adjust to the expertise of the breeders themselves.[23]

But farmers' alleged incompetence was not the only reason private breeders sought to exclude them from the hybrid corn process. Commercial seed producers had bemoaned their inability to control corn varieties well before the advent of hybrids. Many felt that if they went to the trouble of developing and maintaining a particular line, they should be paid accordingly, and as a group seed producers opposed the idea of farmers saving their own seed. After the seed shortage of World War I, for example, they protested when the USDA continued telling farmers to save seed for planting. Claiming that farmers were not really capable of producing good seed themselves, commercial producers tried to think of ways to stop such practices. As one wrote, "There is no reason why the business of legitimate seedsmen should be interfered with and farmers and gardeners urged to produce their own." In essays titled "Beating the Farmer at His Own Game," and "Discourage Home Seed Saving," seed produc-

---

process that involved collecting pollen from the tassel in a bag and tying the bag on the desired ear.

[23]Advisory Committee and Staff, Agronomy Department Meeting, November 1, 1938, p. 15, D. E. Alexander Collection, University of Illinois, Urbana, Department of Maize Genetics.

ers expressed their growing frustration. Nonetheless, while for some breeders the exclusion was deliberate, for most it was simply an unimportant consequence of the new technology.[24]

Like machinists on the shop floor presented with numerically controlled machine tools, farmers were presented with hybrid corn in the mid-1930s. And as other workers had found before and would find afterward, the deskilling properties of the technology seemingly lay, not with the human agent proffering it, but rather within the inert technology itself. First, different strains of hybrid corn were visually indistinguishable from one another due to their genetic similarity. Farmers who had earlier been able to select corn, whether from the field or from the seed dealer, according to visual characteristics, now had no concept of what to look for. Second, the growing properties of a particular line were not apparent in the hybrid seed. With open-pollinates, farmers had been able to correlate visual characteristics such as indentation and color with their particular growing conditions (e.g., short growing season or heavy soil), but with hybrids they could not predict how any line would behave on their own farms. If their experience taught them that, on their farm (but not their neighbor's), Reid's Yellow Dent produced the best yield, this was of little help in selecting a hybrid that combined an inbred of Reid's with three other unknown lines. Third, since hybrids could not be grown successfully for more than one year due to their narrow genetic base, farmers had to purchase and plant seed anew each year. Farmers accustomed to selecting their corn over the years to create a strain suited to their own farm conditions were shocked to discover that saving and planting hybrid seed resulted in distinct decreases in yield the second year. In this sense, a farmer's interest in the long-term quality of his seed was replaced by a short, annual interest; there was no longer anything to build toward. For better or worse, each year's seed was an unknown quantity.

This brute shift in the appearance of corn seed led to a range of misunderstandings between farmers and seedsmen, as well as some outright duplicity. For example, the central problem for farmers and seedsmen alike in the early years of hybrids was the temperamental character of hybrids, most of which were developed for a very small ecological area. That is, no single hybrid would grow well on the many different soil types and climatic regions of the corn belt. Each was developed in a particular region under rather fussy conditions, yet

[24]"Home Seed Saving Again a Factor," *Seed World,* August 15, 1919, pp. 22–23; "Beating the Farmer at His Own Game," *Seed World,* April 18, 1919, p. 512; "Discourage Home Seed Saving," *Seed World,* April 4, 1919, p. 422.

the hoopla surrounding the first hybrids available in the mid-1930s virtually ignored this adaptability problem.[25]

The difficulty of finding a hybrid suitably adapted to a particular farmer's needs was highlighted by the statewide corn-yield tests conducted by the University of Illinois. In some regions, open-pollinates performed distinctly better than hybrids because they were adapted to local conditions, whereas the hybrids frequently seemed temperamental when removed from their original breeding ground. Open-pollinates were also superior as a rule in areas where the soil was poor or unfertilized, whereas hybrids did better on rich, highly fertilized soil. Farmers who tried hybrid corn before the development of adapted lines were often discouraged by their poor showing and were disinclined to risk another crop failure by trying another hybrid.[26]

Some of the large seed companies could turn the adaptation problem to their advantage. Funk Brothers, like other established seed companies, invoked its long experience in corn breeding to reassure farmers who were concerned that a dealer would take advantage of their lack of familiarity with hybrids. Indeed, Funk exacerbated suspicions with warnings such as this: "Prospective purchasers must beware of seed offered as Hybrid and make sure it is good Hybrid of tested performance before they buy. The situation as it exists, large demand and limited supply, is opportune for unethical parties to throw seed corn on the market that is not truly Hybrid Seed." Since farmers could not possibly know how to identify "good" hybrids, Funk helpfully suggested, "You may not know which strain to order. Just order FUNK'S HYBRID CORN. We will supply the hybrid best adapted to your locality."[27]

[25]J. C. Hackleman, "Annual Report for 1926," 8/6/2, University of Illinois Archives, p. 6. Although I have not found any specific instances of fraud, anecdotal evidence suggests that it did exist. It appears that during the hybrid corn mania from 1936 to 1942, some considered selling fake hybrids an ideal get-rich-quick scheme. The farm press devoted much more attention to hybrids than the popular press; nevertheless, the range of popular interest can be seen in such articles as Andrew M. House, "Corn Belt Revolution," *Printer's Ink,* December 15, 1938, pp. 11–14, 88–91; "Miracle Men of the Corn Belt," *Popular Mechanics,* August 1940, pp. 226–29; "Hybrid Corn's Empire Grows," *Business Week,* January 1941, pp. 28–30; and F. Thone, "Hybrid Corn's Conquests," *Science,* April 26, 1941, p. 271. Among farm publications, *Wallaces' Farmer* was the primary forum for hybrid-corn issues; see esp. "The Story of Hybrid Corn," *Wallaces' Farmer,* August 13, 1938, pp. 516–26.

[26]Unidentified farm adviser to Hackleman in J. C. Hackleman, "Annual Report for 1937," 8/6/2, University of Illinois Archives, p. 22; see also p. 21, and "Annual Report for 1936," p. 56.

[27]Funk Brothers Seed Company, 1936 catalog, p. 1; 1938 catalog; 1935 catalog, p. 1, Bloomington, Ill.

And indeed, farmers faced an alarming set of indistinct options in selecting among hybrid lines. By 1937 the University of Illinois had developed one hundred hybrid combinations, all of which were available to farmers who asked to try them. The Funk Brothers catalog for 1938 listed thirty-six of the company's own hybrids plus twenty-nine public, University of Illinois lines. Other seed companies offered similar listings. While these many different hybrids did not represent dramatically different field characteristics for the most part, they nonetheless served to convince farmers that their commonsense experience had no applicability in the newly mysterious area of seed selection.[28]

Even the seed salesmen out in the counties were perplexed by the invisibility of crucial properties of hybrids. One of Funk's associate producers, who was not told the pedigree of the corn he was selling, complained:

> I believe the research department has been too close with their information about the makeup of the actual hybrids we are now producing. It is not material to me, or to those in our company, what the actual inbreds are that go into the makeup of a given hybrid. I don't particularly care to know what [*sic*] R4 and Hy make up one of the single crosses in a particular number. But I do feel that we are entirely too ignorant as to the relation between certain hybrids; the reason why they may be drought resistant, grasshopper resistant, or chinch bug resistant. I feel that I should be able to talk to a customer and tell him that Funk's inbred 21, or single-cross 112, has certain characteristics, and that this single cross is contained in such and such hybrids.[29]

Despite the confusion and the problems, however, hybrid corn replaced open-pollinated within about ten years of its introduction in the mid-1930s. Whereas hybrids accounted for only 0.4 percent of corn acreage in the United States in 1933, by 1945 they made up 90 percent. Aside from the fact that most seed companies stopped selling open-pollinates around 1940, there were several incentives to switch from open-pollinates. First, some hybrids did seem to offer advantages over open-pollinates, particularly in the increased yield that

[28]University of Illinois Agricultural Experiment Station, "Illinois Cooperative Corn Improvement Program — 1937 Policy for the Distribution of Available Supplies of Foundation Single Crosses and Inbred Lines of Corn," n.d., winter 1937, 8/6/2, University of Illinois Archives; Funk Brothers Seed Company, 1938 catalog, Bloomington, Ill.

[29]R. D. Herrington (J. C. Robinson Seed Company, Waterloo, Nebr.) to E. D. Funk, December 13, 1941, Funk Brothers files, Bloomington, Ill.

came from resistance to common pests, diseases, and droughts.[30] Second, a series of droughts in the mid-1930s, and especially a severe drought in 1936, made seed corn of any kind difficult to locate; hence farmers were happy to buy whatever was available.[31] And third, the Agricultural Adjustment Administration's program in acreage reduction, which paid farmers to take land out of corn production, tempted clever farmers to try hybrids, which yielded more per acre. They could take a few acres out of production, receive a government payment, *and* grow as much corn as ever—obeying the letter, if not the spirit, of the law.[32] Thus, to say that hybrids were adopted simply because they were better than open-pollinates is to ignore why farmers might make such a choice, using reasons that were perhaps more compelling.

Conclusion

It is notoriously difficult to pin down just what human characteristics one is referring to when designating a "skill." This article makes no pretense of clarifying that matter. Rather, it suggests that, even though historians' studies of the relation between technology and work have broken new ground in evaluating the transformation of work, they have created an unnecessarily narrow understanding of what kinds of work might be so transformed *and* what sorts of

[30]While it may not have been wise in the long run for farmers to seek big yield increases in terms of market capacity to handle so much corn, individual farmers were accustomed to a simpler calculus that equated increased yield on their farm with increased money in their pockets.

[31]The census estimated that corn production dropped from 2,398 million bushels in 1933 to 1,449 million bushels in 1934. In 1935, which was a good year in terms of weather, the production figure rose to 2,299 million bushels, then fell again in 1936 to 1,506 million bushels. The figures for 1934 and 1936 were the lowest for the century. See *Historical Statistics of the United States* (Washington, D.C., 1976), p. 511. Although Shaw and Durost maintain that the weather had a negligible effect on overall yields from 1929 to 1962, their records indicate that in 1934 and 1936 weather distinctly reduced yields. See Lawrence H. Shaw and Donald D. Durost, *The Effect of Weather and Technology on Corn Yields in the Corn Belt, 1929–1962* (Washington, D.C., 1965). See also J. C. Hackleman, "Annual Report for 1935," 8/6/2, University of Illinois Archives, p. 15. In his 1936 report, Hackleman noted that the scarcity of seed had created a "mild panic" for hybrids (pp. 49–50).

[32]For the effect of the Agricultural Adjustment Administration on corn growing, see, e.g., Theodore Saloutos, *The American Farmer and the New Deal* (Ames, Iowa, 1982), esp. pp. 72–75; Gilbert Fite, *American Farmers: The New Minority* (Bloomington, Ind., 1981), pp. 71–72. In 1932 corn acreage harvested reached a peak of 110,577,000 acres; by 1940 it was down to 86,429,000 and by 1950, 81,818,000. See also J. C. Hackleman, "Annual Report for 1940," 8/6/2, University of Illinois Archives, p. 12; and H. E. Klinefelter, "The Coming Revolution in Corn Production," *Missouri Farmer,* October 15, 1938, p. 5.

technology might be operative. While it is unenlightening simply to encompass in our definition of skilled work all activities that can be reproduced by Latour's "nonhuman actors," this notion can be useful, especially in identifying the nonindustrial sectors of the skilled work force. It is useful as well in locating the technologies that lie outside the usual nuts-and-bolts arena of industrial labor.

Farmers in the 1920s and 1930s performed hard manual labor; they owned their own tools and other means of production; they worked without supervision, making all decisions according to their own particular standards and desires. They were in many ways more like skilled artisans than industrial workers. Unlike many artisans and most industrial workers, however, farmers generally have chosen not to organize themselves into a united front. Faced again and again with adverse conditions, whether economic, legislative, social, or political in nature, farmers have clung stubbornly to their status as independent producers.

The nearly invisible status of farmers as producers and workers is furthered by their amiable adaptation to technological change. As I discussed earlier, much of the manual work farmers did was unpleasant, strenuous, and tedious, the sort of labor most of them were only too happy to eliminate. In addition, most of the technological innovations introduced before hybrids were small, piecemeal changes that combined modernity with a tradition of self-sufficiency. An example of this was the small electric farm plant, which provided a single farm with power and could be upgraded as needs and finances dictated. Another would be the myriad of commercial feeds, innoculants, and fertilizers advertised in the farm press. Even the tractor, the most dramatic innovation before hybrids, was adopted only gradually in the corn belt, both because of its expense and because comparisons of its efficiency against horses were inconclusive.[33]

The amiability of farmers notwithstanding, however (and there is little evidence that farmers were fundamentally hostile to hybrids), hybrid corn represented a different order of innovation. There was nothing gradual about using hybrids, and no way to combine the skills needed for open-pollinates with those needed for hybrids—it was an all-or-nothing proposition. Most important, hybrids effectively locked farmers out from an understanding of their own operations without the aid of experts. To borrow a metaphor from Sherry Turkle, where open-pollinates were transparent, hybrids were opaque. It matters

[33]See, e.g., I. W. Dickerson, "Electricity in the Farm Home," *Wallaces' Farmer* 45, January 16, 1920, pp. 164, 171, which discusses how the farm and farm home could be electrified in a piecemeal way.

little that farmers acquiesced in the shift. What did matter was that their authority and knowledge were thereby delegated to geneticists and seed dealers.[34]

My intent in drawing this out is neither to ennoble nor to denigrate the work farmers did but, rather, to point out the obvious but unexamined fact that hybrid corn replaced an activity that farmers had previously done for themselves. The heroic literature that surrounds this innovation neglects to report that hybrids were a replacement—a delegation—rather than an entirely new phenomenon. The knowledge farmers had created over the years—regarding the differences between various lines of corn, their growing range, maturing rates, value as livestock feed or on the open market, susceptibility to bugs or disease, and in particular the visible features that good corn should have for their particular farms—this sort of knowledge was made obsolete for farmers (although, ironically, breeders needed it more than ever in combining inbred lines). Indeed, it was of negative value if farmers attempted to select seed from their hybrid fields, a move that courted crop failure the following year.

Finally, while there is little to gain at this stage in debating whether or not a technology like hybrid corn was a "good" or "bad" technology, there is merit in understanding how such technologies usurp the skill and knowledge of existing workers. Even the most benign technologies generally replicate the activity of some worker, and the evaluation of a technology needs to consider not simply the economic advantages and disadvantages of deskilling but also the less tangible features comprised in the knowledge workers possess. Only then can the character of technological change be viewed, not as a bureaucratic triumph or political conspiracy, but as an ambiguous stand-in for the work people do.

[34]Sherry Turkle, *The Second Self* (New York, 1984), esp. pp. 174–82.

The "Industrial Revolution" in the Home: Household Technology and Social Change in the 20th Century

RUTH SCHWARTZ COWAN

When we think about the interaction between technology and society, we tend to think in fairly grandiose terms: massive computers invading the workplace, railroad tracks cutting through vast wildernesses, armies of woman and children toiling in the mills. These grand visions have blinded us to an important and rather peculiar technological revolution which has been going on right under our noses: the technological revolution in the home. This revolution has transformed the conduct of our daily lives, but in somewhat unexpected ways. The industrialization of the home was a process very different from the industrialization of other means of production, and the impact of that process was neither what we have been led to believe it was nor what students of the other industrial revolutions would have been led to predict.

* * *

Some years ago sociologists of the functionalist school formulated an explanation of the impact of industrial technology on the modern family. Although that explanation was not empirically verified, it has become almost universally accepted.[1] Despite some differences in emphasis, the basic tenets of the traditional interpretation can be roughly summarized as follows:

Before industrialization the family was the basic social unit. Most families were rural, large, and self-sustaining; they produced and processed almost everything that was needed for their own support and for trading in the marketplace, while at the same time perform-

DR. COWAN, professor of history at the State University of New York at Stony Brook, is the author of *A Social History of American Technology* (1997) and *More Work for Mother: The Ironies of Household Technology from the Open Hearth to the Microwave* (1983).

[1]For some classic statements of the standard view, see W. F. Ogburn and M. F. Nimkoff, *Technology and the Changing Family* (Cambridge, Mass., 1955); Robert F. Winch, *The Modern Family* (New York, 1952); and William J. Goode, *The Family* (Englewood Cliffs, N.J., 1964).

ing a host of other functions ranging from mutual protection to entertainment. In these preindustrial families women (adult women, that is) had a lot to do, and their time was almost entirely absorbed by household tasks. Under industrialization the family is much less important. The household is no longer the focus of production; production for the marketplace and production for sustenance have been removed to other locations. Families are smaller and they are urban rather than rural. The number of social functions they perform is much reduced, until almost all that remains is consumption, socialization of small children, and tension management. As their functions diminished, families became atomized; the social bonds that had held them together were loosened. In these postindustrial families women have very little to do, and the tasks with which they fill their time have lost the social utility that they once possessed. Modern women are in trouble, the analysis goes, because modern families are in trouble; and modern families are in trouble because industrial technology has either eliminated or eased almost all their former functions, but modern ideologies have not kept pace with the change. The results of this time lag are several: some women suffer from role anxiety, others land in the divorce courts, some enter the labor market, and others take to burning their brassieres and demanding liberation.

This sociological analysis is a cultural artifact of vast importance. Many Americans believe that it is true and act upon that belief in various ways: some hope to reestablish family solidarity by relearning lost productive crafts—baking bread, tending a vegetable garden—others dismiss the women's liberation movement as "simply a bunch of affluent housewives who have nothing better to do with their time." As disparate as they may seem, these reactions have a common ideological source—the standard sociological analysis of the impact of technological change on family life.

As a theory this functionalist approach has much to recommend it, but at present we have very little evidence to back it up. Family history is an infant discipline, and what evidence it has produced in recent years does not lend credence to the standard view.[2] Phillippe Ariès has shown, for example, that in France the ideal of the small nuclear family predates industrialization by more than a century.[3] Historical demographers working on data from English and French families have been surprised to find that most families were quite small and

[2]This point is made by Peter Laslett in "The Comparative History of Household and Family," in *The American Family in Social Historical Perspective*, ed. Michael Gordon (New York, 1973), pp. 28–29.

[3]Phillippe Ariès, *Centuries of Childhood: A Social History of Family Life* (New York, 1960).

that several generations did not ordinarily reside together; the extended family, which is supposed to have been the rule in preindustrial societies, did not occur in colonial New England either.[4] Rural English families routinely employed domestic servants, and even very small English villages had their butchers and bakers and candlestick makers; all these persons must have eased some of the chores that would otherwise have been the housewife's burden.[5] Preindustrial housewives no doubt had much with which to occupy their time, but we may have reason to wonder whether there was quite as much pressure on them as sociological orthodoxy has led us to suppose. The large rural family that was sufficient unto itself back there on the prairies may have been limited to the prairies—or it may never have existed at all (except, that is, in the reveries of sociologists).

Even if all the empirical evidence were to mesh with the functionalist theory, the theory would still have problems, because its logical structure is rather weak. Comparing the average farm family in 1750 (assuming that you knew what that family was like) with the average urban family in 1950 in order to discover the significant social changes that had occurred is an exercise rather like comparing apples with oranges; the differences between the fruits may have nothing to do with the differences in their evolution. Transferring the analogy to the case at hand, what we really need to know is the difference, say, between an urban laboring family of 1750 and an urban laboring family 100 and then 200 years later, or the difference between the rural nonfarm middle classes in all three centuries, or the difference between the urban rich yesterday and today. Surely in each of these cases the analyses will look very different from what we have been led to expect. As a guess we might find that for the urban laboring families the changes have been precisely the opposite of what the model predicted; that is, that their family structure is much firmer today than it was in centuries past. Similarly, for the rural nonfarm middle class the results might be equally surprising; we might find that married women of that class rarely did any housework at all in 1890 because they had farm girls as servants, whereas in 1950 they bore the full brunt of the work themselves. I could go on, but the point is, I hope, clear: in order to verify or falsify the functionalist theory, it will be necessary to know more than we presently do about the impact of industrialization on families of similar classes and geographical locations.

* * *

[4]See Laslett, pp. 20–24; and Philip J. Greven, "Family Structure in Seventeenth Century Andover, Massachusetts," *William and Mary Quarterly* 23 (1966): 234–56.

[5]Peter Laslett, *The World We Have Lost* (New York, 1965), passim.

With this problem in mind I have, for the purposes of this initial study, deliberately limited myself to one kind of technological change affecting one aspect of family life in only one of the many social classes of families that might have been considered. What happened, I asked, to middle-class American women when the implements with which they did their everyday household work changed? Did the technological change in household appliances have any effect upon the structure of American households, or upon the ideologies that governed the behavior of American women, or upon the functions that families needed to perform? Middle-class American women were defined as actual or potential readers of the better-quality women's magazines, such as the *Ladies' Home Journal, American Home, Parents' Magazine, Good Housekeeping,* and *McCall's*.[6] Nonfictional material (articles and advertisements) in those magazines was used as a partial indicator of some of the technological and social changes that were occurring.

The *Ladies' Home Journal* has been in continuous publication since 1886. A casual survey of the nonfiction in the *Journal* yields the immediate impression that that decade between the end of World War I and the beginning of the depression witnessed the most drastic changes in patterns of household work. Statistical data bear out this impression. Before 1918, for example, illustrations of homes lit by gaslight could still be found in the *Journal;* by 1928 gaslight had disappeared. In 1917 only one-quarter (24.3 percent) of the dwellings in the United States had been electrified, but by 1920 this figure had doubled (47.4 percent—for rural nonfarm and urban dwellings), and by 1930 it had risen to four-fifths percent).[7] If electrification had meant simply the change from gas or oil lamps to electric lights, the changes in the housewife's routines might not have been very great (except for eliminating the chore of cleaning and filling oil lamps);

[6]For purposes of historical inquiry, this definition of middle-class status corresponds to a sociological reality, although it is not, admittedly, very rigorous. Our contemporary experience confirms that there are class differences reflected in magazines, and this situation seems to have existed in the past as well. On this issue see Robert S. Lynd and Helen M. Lynd, *Middletown: A Study in Contemporary American Culture* (New York, 1929), pp. 240–44, where the marked difference in magazines subscribed to by the business-class wives as opposed to the working-class wives is discussed; Salme Steinberg, "Reformer in the Marketplace: E. W. Bok and *The Ladies Home Journal*" (Ph.D. diss., Johns Hopkins University, 1973), where the conscious attempt of the publisher to attract a middle-class audience is discussed; and Lee Rainwater et al., *Workingman's Wife* (New York, 1959), which was commissioned by the publisher of working-class women's magazines in an attempt to understand the attitudinal differences betweeen working-class and middle-class women.

[7]*Historical Statistics of the United States, Colonial Times to 1957* (Washington, D.C., 1960), p. 510.

but changes in lighting were the least of the changes that electrification implied. Small electric appliances followed quickly on the heels of the electric light, and some of those augured much more profound changes in the housewife's routine.

Ironing, for example, had traditionally been one of the most dreadful household chores, especially in warm weather when the kitchen stove had to be kept hot for the better part of the day; irons were heavy and they had to be returned to the stove frequently to be reheated. Electric irons eased a good part of this burden.[8] They were relatively inexpensive and very quickly replaced their predecessors; advertisements for electric irons first began to appear in the ladies' magazines after the war, and by the end of the decade the old flatiron had disappeared; by 1929 a survey of 100 Ford employees revealed that ninety-eight of them had the new electric irons in their homes.[9]

Data on the diffusion of electric washing machines are somewhat harder to come by; but it is clear from the advertisements in the magazines, particularly advertisements for laundry soap, that by the middle of the 1920s those machines could be found in a significant number of homes. The washing machine is depicted just about as frequently as the laundry tub by the middle of the 1920s; in 1929, forty-nine out of those 100 Ford workers had the machines in their homes. The washing machines did not drastically reduce the time that had to be spent on household laundry, as they did not go through their cycles automatically and did not spin dry; the housewife had to stand guard, stopping and starting the machine at appropriate times, adding soap, sometimes attaching the drain pipes, and putting the clothes through the wringer manually. The machines did, however, reduce a good part of the drudgery that once had been associated with washday, and this was a matter of no small consequence.[10] Soap powders appeared on the market in the early 1920s, thus eliminating the need to scrape and boil bars of laundry soap.[11] By the end of the

[8]The gas iron, which was available to women whose homes were supplied with natural gas, was an earlier improvement on the old-fashioned flatiron, but this kind of iron is so rarely mentioned in the sources that I used for this survey that I am unable to determine the extent of its diffusion.

[9]Hazel Kyrk, *Economic Problems of the Family* (New York, 1933), p. 368, reporting a study in *Monthly Labor Review* 30 (1930): 1209–52.

[10]Although this point seems intuitively obvious, there is some evidence that it may not be true. Studies of energy expenditure during housework have indicated that by far the greatest effort is expended in hauling and lifting the wet wash, tasks which were not eliminated by the introduction of washing machines. In addition, if the introduction of the machines served to increase the total amount of wash that was done by the housewife, this would tend to cancel the energy-saving effects of the machines themselves.

[11]Rinso was the first granulated soap; it came on the market in 1918. Lux Flakes had been available since 1906; however it was not intended to be a general laundry product

1920s Blue Monday must have been considerably less blue for some housewives—and probably considerably less "Monday," for with an electric iron, a washing machine, and a hot water heater, there was no reason to limit the washing to just one day of the week.

Like the routines of washing the laundry, the routines of personal hygiene must have been transformed for many households during the 1920s—the years of the bathroom mania.[12] More and more bathrooms were built in older homes, and new homes began to include them as a matter of course. Before the war most bathroom fixtures (tubs, sinks, and toilets) were made out of porcelain by hand; each bathroom was custom-made for the house in which it was installed. After the war industrialization descended upon the bathroom industry; cast iron enamelware went into mass production and fittings were standardized. In 1921 the dollar value of the production of enameled sanitary fixtures was $2.4 million, the same as it had been in 1915. By 1923, just two years later, that figure had doubled to $4.8 million; it rose again, to $5.1 million, in 1925.[13] The first recessed, double-shell cast iron enameled bathtub was put on the market in the early 1920s. A decade later the standard American bathroom had achieved its standard American form: the recessed tub, plus tiled floors and walls, brass plumbing, a single-unit toilet, an enameled sink, and a medicine chest, all set into a small room which was very often 5 feet square.[14] The bathroom evolved more quickly than any other room of the house; its standardized form was accomplished in just over a decade.

Along with bathrooms came modernized systems for heating hot water: 61 percent of the homes in Zanesville, Ohio, had indoor plumbing with centrally heated water by 1926, and 83 percent of the homes valued over $2,000 in Muncie, Indiana, had hot and cold running

---

but rather one for laundering delicate fabrics. "Lever Brothers," *Fortune* 26 (November 1940): 95.

[12] I take this account, and the term, from Lynd and Lynd, p. 97. Obviously, there were many American homes that had bathrooms before the 1920s, particularly urban row houses, and I have found no way of determining whether the increases of the 1920s were more marked than in previous decades. The rural situation was quite different from the urban; the President's Conference on Home Building and Home Ownership reported that in the late 1920s, 71 percent of the urban families surveyed had bathrooms, but only 33 percent of the rural families did (John M. Gries and James Ford, eds., *Homemaking, Home Furnishing and Information Services,* President's Conference on Home Building and Home Ownership, vol. 10 [Washington, D.C., 1932], p. 13).

[13] The data above come from Siegfried Giedion, *Mechanization Takes Command* (New York, 1948), pp. 685–703.

[14] For a description of the standard bathroom see Helen Sprackling, "The Modern Bathroom," *Parents' Magazine* 8 (February 1933): 25.

water by 1935.[15] These figures may not be typical of small American cities (or even large American cities) at those times, but they do jibe with the impression that one gets from the magazines: after 1918 references to hot water heated on the kitchen range, either for laundering or for bathing, become increasingly difficult to find.

Similarly, during the 1920s many homes were outfitted with central heating; in Muncie most of the homes of the business class had basement heating in 1924; by 1935 Federal Emergency Relief Administration data for the city indicated that only 22.4 percent of the dwellings valued over $2,000 were still heated by a kitchen stove.[16] What all these changes meant in terms of new habits for the average housewife is somewhat hard to calculate; changes there must have been, but it is difficult to know whether those changes produced an overall saving of labor and/or time. Some chores were eliminated—hauling water, heating water on the stove, maintaining the kitchen fire—but other chores were added—most notably the chore of keeping yet another room scrupulously clean.

It is not, however, difficult to be certain about the changing habits that were associated with the new American kitchen—a kitchen from which the coal stove had disappeared. In Muncie in 1924, cooking with gas was done in two out of three homes; in 1935 only 5 percent of the homes valued over $2,000 still had coal or wood stoves for cooking.[17] After 1918 advertisements for coal and wood stoves disappeared from the *Ladies' Home Journal;* stove manufacturers purveyed only their gas, oil, or electric models. Articles giving advice to homemakers on how to deal with the trials and tribulations of starting, stoking, and maintaining a coal or a wood fire also disappeared. Thus it seems a safe assumption that most middle-class homes had switched to the new method of cooking by the time the depression began. The change in routine that was predicated on the change from coal or wood to gas or oil was profound; aside from the elimination of such chores as loading the fuel and removing the ashes, the new stoves were much easier to light, maintain, and regulate (even when they did not have thermostats, as the earliest models did not).[18] Kitchens were, in addition, much easier to clean when they did not have coal dust regularly tracked through them; one writer in the *Ladies'*

[15]*Zanesville, Ohio and Thirty-six Other American Cities* (New York, 1927), p. 65. Also see Robert S. Lynd and Helen M. Lynd, *Middletown in Transition* (New York, 1936), p. 537. Middletown is Muncie, Indiana.

[16]Lynd and Lynd, *Middletown,* p. 96, and *Middletown in Transition,* p. 539.

[17]Lynd and Lynd, *Middletown,* p. 98, and *Middletown in Transition,* p. 562.

[18]On the advantages of the new stoves, see *Boston Cooking School Cookbook* (Boston, 1916), pp. 15–20; and Russell Lynes, *The Domesticated Americans* (New York, 1957), pp. 119–20.

Home Journal estimated that kitchen cleaning was reduced by one-half when coal stoves were eliminated.[19]

Along with new stoves came new foodstuffs and new dietary habits. Canned foods had been on the market since the middle of the 19th century, but they did not become an appreciable part of the standard middle-class diet until the 1920s—if the recipes given in cookbooks and in women's magazines are a reliable guide. By 1918 the variety of foods available in cans had been considerably expanded from the peas, corn, and succotash of the 19th century; an American housewife with sufficient means could have purchased almost any fruit or vegetable and quite a surprising array of ready-made meals in a can —from Heinz's spaghetti in meat sauce to Purity Cross's lobster à la Newburg. By the middle of the 1920s home canning was becoming a lost art. Canning recipes were relegated to the back pages of the women's magazines; the business-class wives of Muncie reported that, while their mothers had once spent the better part of the summer and fall canning, they themselves rarely put up anything, except an occasional jelly or batch of tomatoes.[20] In part this was also due to changes in the technology of marketing food; increased use of refrigerated railroad cars during this period meant that fresh fruits and vegetables were in the markets all year round at reasonable prices.[21] By the early 1920s convenience foods were also appearing on American tables: cold breakfast cereals, pancake mixes, bouillon cubes, and packaged desserts could be found. Wartime shortages accustomed Americans to eating much lighter meals than they had previously been wont to do; and as fewer family members were taking all their meals at home (businessmen started to eat lunch in restaurants downtown, and factories and schools began installing cafeterias), there was simply less cooking to be done, and what there was of it was easier to do.[22]

* * *

Many of the changes just described—from hand power to electric power, from coal and wood to gas and oil as fuels for cooking, from one-room heating to central heating, from pumping water to running water—are enormous technological changes. Changes of a similar dimension, either in the fundamental technology of an industry, in the diffusion of that technology, or in the routines of workers, would have long since been labeled an "industrial revolution." The change from the laundry tub to the washing machine is no less profound than

[19]"How to Save Coal While Cooking," *Ladies' Home Journal* 25 (January 1908): 44.
[20]Lynd and Lynd, *Middletown,* p. 156.
[21]Ibid.; see also "Safeway Stores," *Fortune* 26 (October 1940): 60.
[22]Lynd and Lynd, *Middletown,* pp. 134–35 and 153–54.

the change from the hand loom to the power loom; the change from pumping water to turning on a water faucet is no less destructive of traditional habits than the change from manual to electric calculating. It seems odd to speak of an "industrial revolution" connected with housework, odd because we are talking about the technology of such homely things, and odd because we are not accustomed to thinking of housewives as a labor force or of housework as an economic commodity—but despite this oddity, I think the term is altogether appropriate.

In this case other questions come immediately to mind, questions that we do not hesitate to ask, say, about textile workers in Britain in the early 19th century, but we have never thought to ask about housewives in America in the 20th century. What happened to this particular work force when the technology of its work was revolutionized? Did structural changes occur? Were new jobs created for which new skills were required? Can we discern new ideologies that influenced the behavior of the workers?

The answer to all of these questions, surprisingly enough, seems to be yes. There were marked structural changes in the work force, changes that increased the work load and the job description of the workers that remained. New jobs were created for which new skills were required; these jobs were not physically burdensome, but they may have taken up as much time as the jobs they had replaced. New ideologies were also created, ideologies which reinforced new behavioral patterns, patterns that we might not have been led to expect if we had followed the sociologists' model to the letter. Middle-class housewives, the women who must have first felt the impact of the new household technology, were not flocking into the divorce courts or the labor market or the forums of political protest in the years immediately after the revolution in their work. What they were doing was sterilizing baby bottles, shepherding their children to dancing classes and music lessons, planning nutritious meals, shopping for new clothes, studying child psychology, and hand stitching color-coordinated curtains—all of which chores (and others like them) the standard sociological model has apparently not provided for.

The significant change in the structure of the household labor force was the disappearance of paid and unpaid servants (unmarried daughters, maiden aunts, and grandparents fall in the latter category) as household workers—and the imposition of the entire job on the housewife herself. Leaving aside for a moment the question of which was cause and which effect (did the disappearance of the servant create a demand for the new technology, or did the new technology make the servant obsolete?), the phenomenon itself is relatively easy to document. Before World War I, when illustrators in the women's magazines depicted women doing housework, the women were very often servants. When the lady of the house was drawn, she was often the person being served, or she was supervising the serving, or she was adding an elegant finishing touch to the work. Nursemaids diapered babies, seamstresses pinned up hems, waitresses served meals, laundresses did the wash, and cooks did the cooking. By the end of the 1920s the servants had disappeared from those illustrations; all those jobs were being done by housewives—elegantly manicured and coiffed, to be sure, but housewives nonetheless (compare figs. 1 and 2).

If we are tempted to suppose that illustrations in advertisements are not a reliable indicator of structural changes of this sort, we can corroborate the changes in other ways. Apparently, the illustrators really did know whereof they drew. Statistically the number of persons throughout the country employed in household service dropped from 1,851,000 in 1910 to 1,411,000 in 1920, while the number of households enumerated in the census rose from 20.3 million to 24.4 million.[23] In Indiana the ratio of households to servants increased from 13.5/1 in 1890 to 30.5/1 in 1920, and in the country as a whole the number of paid domestic servants per 1,000 population dropped from 98.9 in 1900 to 58.0 in 1920.[24] The business-class housewives of Muncie reported that they employed approximately one-half as many woman-hours of domestic service as their mothers had done.[25]

In case we are tempted to doubt these statistics (and indeed statistics about household labor are particularly unreliable, as the labor is often transient, part-time, or simply unreported), we can turn to articles on the servant problem, the disappearance of unpaid family workers, the design of kitchens, or to architectural drawings for houses. All of this evidence reiterates the same point: qualified servants were difficult to find; their wages had risen and their numbers fallen; houses were being designed without maid's rooms; daughters and unmarried aunts were finding jobs downtown; kitchens were being designed for housewives, not for servants.[26] The first home with a

[23]*Historical Statistics,* pp. 16 and 77.

[24]For Indiana data, see Lynd and Lynd, *Middletown,* p. 169. For national data, see D. L. Kaplan and M. Claire Casey, *Occupational Trends in the United States, 1900–1950,* U.S. Bureau of the Census Working Paper no. 5 (Washington, D.C., 1958), table 6. The extreme drop in numbers of servants between 1910 and 1920 also lends credence to the notion that this demographic factor stimulated the industrial revolution in housework.

[25]Lynd and Lynd, *Middletown,* p. 169.

[26]On the disappearance of maiden aunts, unmarried daughters, and grandparents, see Lynd and Lynd, *Middletown,* pp. 25, 99, and 110; Edward Bok, "Editorial," *American Home* 1 (October 1928): 15; "How to Buy Life Insurance," *Ladies' Home Journal* 45

The Ladies' Home Journal for April, 1918 107

The things you'd never put in the family laundry

YES, it's beginning to look dusky around the edges of the cuff and along the roll of the collar. Your precious new Georgette—you'd never dream of putting it in with the general laundry.

Your silk underwear, silk stockings, white satin collars—how they discolor, or yellow—how the threads break and grow weak when they are washed in the family laundry.

If only you *knew* the heartless laundress would not rub the life and the newness out of your dainty things!

You cannot afford to have your nicest things go so fast.

You, yourself, with a fraction of the energy you once spent hating the laundress, can now gently rinse the dirt out of your filmiest things—take them from the pure Lux suds soft and gleaming and *new!*

The secret? No ruinous rubbing of a cake of soap on fine fabrics! No rubbing again to get the soap and the dirt out. Just the gentle, tender cleansing with pure Lux suds that frail things must have to keep them unhurt.

No ruinous rubbing of fine fabrics

Lux is the most modern form of soap. There is nothing else like it. Lux comes in wonderful, delicate white flakes—pure and transparent. They dissolve instantly in hot water. You whisk them into the richest, sudsiest lather that loosens all the dirt—leaves the finest silk fabric clean and new—not a fiber roughened or torn or weakened in any way.

Write for free booklet and simple Lux directions for laundering. Learn how easy it is to launder perfectly the most delicate fabrics.

Be sure to get your package of Lux today. Your grocer, druggist or department store has it—Lever Bros. Co., Dept. A1, Cambridge, Mass.

LUX

These things need never be spoiled by washing. Try washing them the Lux way.

Lux is so pure that it will not harm anything that pure water alone will not injure.

To wash silk blouses

FIG. 1.—The housewife as manager. (*Ladies' Home Journal,* April 1918. Courtesy of Lever Brothers Co.)

FIG. 2.—The housewife as laundress. (*Ladies' Home Journal*, August 1928. Courtesy of Colgate-Palmolive-Peet.)

kitchen that was not an entirely separate room was designed by Frank Lloyd Wright in 1934.[27] In 1937 Emily Post invented a new character for her etiquette books: Mrs. Three-in-One, the woman who is her own cook, waitress, and hostess.[28] There must have been many new Mrs. Three-in-Ones abroad in the land during the 1920s.

As the number of household assistants declined, the number of household tasks increased. The middle-class housewife was expected to demonstrate competence at several tasks that previously had not been in her purview or had not existed at all. Child care is the most obvious example. The average housewife had fewer children than her mother had had, but she was expected to do things for her children that her mother would never have dreamed of doing: to prepare their special infant formulas, sterilize their bottles, weigh them every day, see to it that they ate nutritionally balanced meals, keep them isolated and confined when they had even the slightest illness, consult with their teachers frequently, and chauffeur them to dancing lessons, music lessons, and evening parties.[29] There was very little Freudianism in this new attitude toward child care: mothers were not spending more time and effort on their children because they feared the psychological trauma of separation, but because competent nursemaids could not be found, and the new theories of child care required constant attention from well-informed persons—persons who were willing and able to read about the latest discoveries in nutrition, in the control of contagious diseases, or in the techniques of behavioral psychology. These persons simply had to be their mothers.

Consumption of economic goods provides another example of the housewife's expanded job description; like child care, the new tasks associated with consumption were not necessarily physically burdensome, but they were time consuming, and they required the acquisi-

---

(March 1928): 35. The house plans appeared every month in *American Home,* which began publication in 1928. On kitchen design, see Giedion, pp. 603–21; "Editorial," *Ladies' Home Journal* 45 (April 1928): 36; advertisement for Hoosier kitchen cabinets, *Ladies' Home Journal* 45 (April 1928): 117. Articles on servant problems include "The Vanishing Servant Girl," *Ladies Home Journal* 35 (May 1918): 48; "Housework, Then and Now," *American Home* 8 (June 1932): 128; "The Servant Problem," *Fortune* 24 (March 1938): 80–84; and *Report of the YWCA Commission on Domestic Service* (Los Angeles, 1915).

[27]Giedion, p. 619. Wright's new kitchen was installed in the Malcolm Willey House, Minneapolis.

[28]Emily Post, *Etiquette: The Blue Book of Social Usage,* 5th ed. rev. (New York, 1937), p. 823.

[29]This analysis is based upon various child-care articles that appeared during the period in the *Ladies' Home Journal, American Home,* and *Parents' Magazine.* See also Lynd and Lynd, *Middletown,* chap. 11.

tion of new skills.[30] Home economists and the editors of women's magazines tried to teach housewives to spend their money wisely. The present generation of housewives, it was argued, had been reared by mothers who did not ordinarily shop for things like clothing, bed linens, or towels; consequently modern housewives did not know how to shop and would have to be taught. Furthermore, their mothers had not been accustomed to the wide variety of goods that were now available in the modern marketplace; the new housewives had to be taught not just to be consumers, but to be informed consumers.[31] Several contemporary observers believed that shopping and shopping wisely were occupying increasing amounts of housewives' time.[32]

Several of these contemporary observers also believed that standards of household care changed during the decade of the 1920s.[33] The discovery of the "household germ" led to almost fetishistic concern about the cleanliness of the home. The amount and frequency of laundering probably increased, as bed linen and underwear were changed more often, children's clothes were made increasingly out of washable fabrics, and men's shirts no longer had replaceable collars and cuffs.[34] Unfortunately all these changes in standards are difficult to document, being changes in the things that people regard as so insignificant as to be unworthy of comment; the improvement in standards seems a likely possibility, but not something that can be proved.

In any event we do have various time studies which demonstrate somewhat surprisingly that housewives with conveniences were spending just as much time on household duties as were housewives without them—or, to put it another way, housework, like so many

[30]John Kenneth Galbraith has remarked upon the advent of woman as consumer in *Economics and the Public Purpose* (Boston, 1973), pp. 29–37.

[31]There was a sharp reduction in the number of patterns for home sewing offered by the women's magazines during the 1920s; the patterns were replaced by articles on "what is available in the shops this season." On consumer education see, for example, "How to Buy Towels," *Ladies' Home Journal* 45 (February 1928): 134; "Buying Table Linen," *Ladies' Home Journal* 45 (March 1928): 43; and "When the Bride Goes Shopping," *American Home* 1 (January 1928): 370.

[32]See, for example, Lynd and Lynd, *Middletown,* pp. 176 and 196; and Margaret G. Reid, *Economics of Household Production* (New York, 1934), chap. 13.

[33]See Reid, pp. 64–68; and Kyrk, p. 98.

[34]See advertisement for Cleanliness Institute—"Self-respect thrives on soap and water," *Ladies' Home Journal* 45 (February 1928): 107. On changing bed linen, see "When the Bride Goes Shopping," *American Home* 1 (January 1928): 370. On laundering children's clothes, see, "Making a Layette," *Ladies' Home Journal* 45 (January 1928): 20; and Josephine Baker, "The Youngest Generation," *Ladies' Home Journal* 45 (March 1928): 185.

other types of work, expands to fill the time available.[35] A study comparing the time spent per week in housework by 288 farm families and 154 town families in Oregon in 1928 revealed 61 hours spent by farm wives and 63.4 hours by town wives; in 1929 a U.S. Department of Agriculture study of families in various states produced almost identical results.[36] Surely if the standard sociological model were valid, housewives in towns, where presumably the benefits of specialization and electrification were most likely to be available, should have been spending far less time at their work than their rural sisters. However, just after World War II economists at Bryn Mawr College reported the same phenomenon: 60.55 hours spent by farm housewives, 78.35 hours by women in small cities, 80.57 hours by women in large ones—precisely the reverse of the results that were expected.[37] A recent survey of time studies conducted between 1920 and 1970 concludes that the time spent on housework by nonemployed housewives has remained remarkably constant throughout the period.[38] All these results point in the same direction: mechanization of the household meant that time expended on some jobs decreased, but also that new jobs were substituted, and in some cases—notably laundering—time expenditures for old jobs increased because of higher standards. The advantages of mechanization may be somewhat more dubious than they seem at first glance.

* * *

As the job of the housewife changed, the connected ideologies also changed; there was a clearly perceptible difference in the attitudes that women brought to housework before and after World War I.[39]

[35]This point is also discussed at length in my paper "What Did Labor-saving Devices Really Save?" (unpublished).

[36]As reported in Lyrk, p. 51.

[37]Bryn Mawr College Department of Social Economy, *Women During the War and After* (Philadelphia, 1945); and Ethel Goldwater, "Woman's Place," *Commentary* 4 (December 1947): 578–85.

[38]JoAnn Vanek, "Keeping Busy: Time Spent in Housework, United States, 1920–1970" (Ph.D. diss., University of Michigan, 1973). Vanek reports an average of 53 hours per week over the whole period. This figure is significantly lower than the figures reported above, because each time study of housework has been done on a different basis, including different activities under the aegis of housework, and using different methods of reporting time expenditures; the Bryn Mawr and Oregon studies are useful for the comparative figures that they report internally, but they cannot easily be compared with each other.

[39]This analysis is based upon my reading of the middle-class women's magazines between 1918 and 1930. For detailed documentation see my paper "Two Washes in the Morning and a Bridge Party at Night: The American Housewife between the Wars," *Women's Studies* (in press). It is quite possible that the appearance of guilt as a strong

Before the war the trials of doing housework in a servantless home were discussed and they were regarded as just that—trials, necessary chores that had to be got through until a qualified servant could be found. After the war, housework changed: it was no longer a trial and a chore, but something quite different—an emotional "trip." Laundering was not just laundering, but an expression of love; the housewife who truly loved her family would protect them from the embarrassment of tattletale gray. Feeding the family was not just feeding the family, but a way to express the housewife's artistic inclinations and a way to encourage feelings of family loyalty and affection. Diapering the baby was not just diapering, but a time to build the baby's sense of security and love for the mother. Cleaning the bathroom sink was not just cleaning, but an exercise of protective maternal instincts, providing a way for the housewife to keep her family safe from disease. Tasks of this emotional magnitude could not possibly be delegated to servants, even assuming that qualified servants could be found.

Women who failed at these new household tasks were bound to feel guilt about their failure. If I had to choose one word to characterize the temper of the women's magazines during the 1920s, it would be "guilt." Readers of the better-quality women's magazines are portrayed as feeling guilty a good lot of the time, and when they are not guilty they are embarrassed: guilty if their infants have not gained enough weight, embarrassed if their drains are clogged, guilty if their children go to school in soiled clothes, guilty if all the germs behind the bathroom sink are not eradicated, guilty if they fail to notice the first signs of an oncoming cold, embarrassed if accused of having body odor, guilty if their sons go to school without good breakfasts, guilty if their daughters are unpopular because of old-fashioned, or unironed, or—heaven forbid—dirty dresses (see figs. 3 and 4). In earlier times women were made to feel guilty if they abandoned their children or were too free with their affections. In the years after World War I, American women were made to feel guilty about sending their children to school in scuffed shoes. Between the two kinds of guilt there is a world of difference.

* * *

Let us return for a moment to the sociological model with which this essay began. The model predicts that changing patterns of

element in advertising is more the result of new techniques developed by the advertising industry than the result of attitudinal changes in the audience—a possibility that I had not considered when doing the initial research for this paper. See A. Michael McMahon, "An American Courtship: Psychologists and Advertising Theory in the Progressive Era," *American Studies* 13 (1972): 5–18.

August, 1928 LADIES' HOME JOURNAL 9

on't fool yourself

Since halitosis never announces itself, to the victim, you simply cannot know when you have it.

Are you unpopular with your own children?

Make sure that you don't have halitosis. It is inexcusable. And unnecessary.

Some mothers blame everything but halitosis (unpleasant breath) when children show resentment or lack of affection.

More often than you would imagine, however, halitosis is at fault. Children are quick to resent it.

Intelligent people realize that as a result of modern habits halitosis is very common and that anyone may have it, *and yet be ignorant of the fact.*

Realizing this, they eliminate any risk of offending, by the systematic use of Listerine in the mouth. Every morning. Every night. And between times when necessary, especially before meeting people. Keep a bottle handy in home and office for this purpose.

Listerine checks halitosis instantly. Being antiseptic, it strikes at its commonest cause—fermentation in the oral cavity. Then, being a powerful deodorant, it destroys the odors themselves.

If you have any doubt of Listerine's power to deodorize, make this test: Rub a slice of onion on your hand. Then apply Listerine clear. Immediately, every trace of onion odor is gone. Lambert Pharmacal Company, St. Louis, Mo.

INTENDED FOR MEN
but try and keep the women away

Have you tried the new LISTERINE SHAVING CREAM? Keeps your face marvelously cool long after shaving. Incidentally, women think rather highly of it as a shampoo.

LISTERINE
—*the safe antiseptic*

FIG. 3.—Sources of housewifely guilt: the good mother smells sweet. (*Ladies' Home Journal*, August 1928. Courtesy of Warner-Lambert, Inc.)

FIG. 4.—Sources of housewifely guilt: the good mother must be beautiful. (*Ladies' Home Journal*, July 1928. Courtesy of Colgate-Palmolive-Peet.)

household work will be correlated with at least two striking indicators of social change: the divorce rate and the rate of married women's labor force participation. That correlation may indeed exist, but it certainly is not reflected in the women's magazines of the 1920s and 1930s: divorce and full-time paid employment were not part of the life-style or the life pattern of the middle-class housewife as she was idealized in her magazines.

There were social changes attendant upon the introduction of modern technology into the home, but they were not the changes that the traditional functionalist model predicts; on this point a close analysis of the statistical data corroborates the impression conveyed in the magazines. The divorce rate was indeed rising during the years between the wars, but it was not rising nearly so fast for the middle and upper classes (who had, presumably, easier access to the new technology) as it was for the lower classes. By almost every gauge of socioeconomic status—income, prestige of husband's work, education—the divorce rate is higher for persons lower on the socioeconomic scale—and this is a phenomenon that has been constant over time.[40]

The supposed connection between improved household technology and married women's labor force participation seems just as dubious, and on the same grounds. The single socioeconomic factor which correlates most strongly (in cross-sectional studies) with married women's employment is husband's income, and the correlation is strongly negative; the higher his income, the less likely it will be that she is working.[41] Women's labor force participation increased during the 1920s but this increase was due to the influx of single women into the force. Married women's participation increased slightly during those years, but that increase was largely in factory labor —precisely the kind of work that middle-class women (who were, again, much more likely to have labor-saving devices at home) were least likely to do.[42] If there were a necessary connection between the improvement of household technology and either of these two social indicators, we would expect the data to be precisely the reverse of what in fact has occurred: women in the higher social classes should have fewer func-

[40]For a summary of the literature on differential divorce rates, see Winch, p. 706; and William J. Goode, *After Divorce* (New York, 1956) p. 44. The earliest papers demonstrating this differential rate appeared in 1927, 1935, and 1939.

[41]For a summary of the literature on married women's labor force participation, see Juanita Kreps, *Sex in the Marketplace: American Women at Work* (Baltimore, 1971), pp. 19–24.

[42]Valerie Kincaid Oppenheimer, *The Female Labor Force in the United States,* Population Monograph Series, no. 5 (Berkeley, 1970), pp. 1–15; and Lynd and Lynd, *Middletown,* pp. 124–27.

tions at home and should therefore be more (rather than less) likely to seek paid employment or divorce.

Thus for middle-class American housewives between the wars, the social changes that we can document are not the social changes that the functionalist model predicts; rather than changes in divorce or patterns of paid employment, we find changes in the structure of the work force, in its skills, and in its ideology. These social changes were concomitant with a series of technological changes in the equipment that was used to do the work. What is the relationship between these two series of phenomena? Is it possible to demonstrate causality or the direction of that causality? Was the decline in the number of households employing servants a cause or an effect of the mechanization of those households? Both are, after all, equally possible. The declining supply of household servants, as well as their rising wages, may have stimulated a demand for new appliances at the same time that the acquisition of new appliances may have made householders less inclined to employ the laborers who were on the market. Are there any techniques available to the historian to help us answer these questions?

* * *

In order to establish causality, we need to find a connecting link between the two sets of phenomena, a mechanism that, in real life, could have made the causality work. In this case a connecting link, an intervening agent between the social and the technological changes, comes immediately to mind: the advertiser—by which term I mean a combination of the manufacturer of the new goods, the advertising agent who promoted the goods, and the periodical that published the promotion. All the new devices and new foodstuffs that were being offered to American households were being manufactured and marketed by large companies which had considerable amounts of capital invested in their production: General Electric, Procter & Gamble, General Foods, Lever Brothers, Frigidaire, Campbell's, Del Monte, American Can, Atlantic & Pacific Tea—these were all well-established firms by the time the household revolution began, and they were all in a position to pay for national advertising campaigns to promote their new products and services. And pay they did; one reason for the expanding size and number of women's magazines in the 1920s was, no doubt, the expansion in revenues from available advertisers.[43]

Those national advertising campaigns were likely to have been powerful stimulators of the social changes that occurred in the

[43]On the expanding size, number, and influence of women's magazines during the 1920s, see Lynd and Lynd, *Middletown,* pp. 150 and 240–44.

household labor force; the advertisers probably did not initiate the changes, but they certainly encouraged them. Most of the advertising campaigns manifestly worked, so they must have touched upon areas of real concern for American housewives. Appliance ads specifically suggested that the acquisition of one gadget or another would make it possible to fire the maid, spend more time with the children, or have the afternoon free for shopping.[44] Similarly, many advertisements played upon the embarrassment and guilt which were now associated with household work. Ralston, Cream of Wheat, and Ovaltine were not themselves responsible for the compulsive practice of weighing infants and children repeatedly (after every meal for newborns, every day in infancy, every week later on), but the manufacturers certainly did not stint on capitalizing upon the guilt that women apparently felt if their offspring did not gain the required amounts of weight.[45] And yet again, many of the earliest attempts to spread "wise" consumer practices were undertaken by large corporations and the magazines that desired their advertising: mail-order shopping guides, "product-testing" services, pseudoinformative pamphlets, and other such promotional devices were all techniques for urging the housewife to buy new things under the guise of training her in her role as skilled consumer.[46]

Thus the advertisers could well be called the "ideologues" of the 1920s, encouraging certain very specific social changes—as ideologues are wont to do. Not surprisingly, the changes that occurred were precisely the ones that would gladden the hearts and fatten the purses of the advertisers; fewer household servants meant a greater demand for labor and timesaving devices; more household tasks for women meant more and more specialized products that they would need to buy; more guilt and embarrassment about their failure to succeed at their work meant a greater likelihood that they would buy the products that were intended to minimize that failure. Happy,

[44]See, for example, the advertising campaigns of General Electric and Hotpoint from 1918 through the rest of the decade of the 1920s; both campaigns stressed the likelihood that electric appliances would become a thrifty replacement for domestic servants.

[45]The practice of carefully observing children's weight was initiated by medical authorities, national and local governments, and social welfare agencies, as part of the campaign to improve child health which began about the time of World War I.

[46]These practices were ubiquitous, *American Home,* for example, which was published by Doubleday, assisted its advertisers by publishing a list of informative pamphlets that readers could obtain; devoting half a page to an index of its advertisers; specifically naming manufacturer's and list prices in articles about products and services; allotting almost one-quarter of the magazine to a mail-order shopping guide which was not (at least ostensibly) paid advertisement; and as part of its editorial policy, urging its readers to buy new goods.

full-time housewives in intact families spend a lot of money to maintain their households; divorced women and working women do not. The advertisers may not have created the image of the ideal American housewife that dominated the 1920s—the woman who cheerfully and skillfully set about making everyone in her family perfectly happy and perfectly healthy—but they certainly helped to perpetuate it.

The role of the advertiser as connecting link between social change and technological change is at this juncture simply a hypothesis, with nothing much more to recommend it than an argument from plausibility. Further research may serve to test the hypothesis, but testing it may not settle the question of which was cause and which effect—if that question can ever be settled definitively in historical work. What seems most likely in this case, as in so many others, is that cause and effect are not separable, that there is a dynamic interaction between the social changes that married women were experiencing and the technological changes that were occurring in their homes. Viewed this way, the disappearance of competent servants becomes one of the factors that stimulated the mechanization of homes, and this mechanization of homes becomes a factor (though by no means the only one) in the disappearance of servants. Similarly, the emotionalization of housework becomes both cause and effect of the mechanization of that work; and the expansion of time spent on new tasks becomes both cause and effect of the introduction of time-saving devices. For example the social pressure to spend more time in child care may have led to a decision to purchase the devices; once purchased, the devices could indeed have been used to save time— although often they were not.

* * *

If one holds the question of causality in abeyance, the example of household work still has some useful lessons to teach about the general problem of technology and social change. The standard sociological model for the impact of modern technology on family life clearly needs some revision: at least for middle-class nonrural American families in the 20th century, the social changes were not the ones that the standard model predicts. In these families the functions of at least one member, the housewife, have increased rather than decreased; and the dissolution of family life has not in fact occurred.

Our standard notions about what happens to a work force under the pressure of technological change may also need revision. When industries become mechanized and rationalized, we expect certain general changes in the work force to occur: its structure becomes more highly differentiated, individual workers become more specialized, managerial functions increase, and the emotional context of the work disappears. On all four counts our expectations are reversed with regard to household work. The work force became less rather than more differentiated as domestic servants, unmarried daughters, maiden aunts, and grandparents left the household and as chores which had once been performed by commercial agencies (laundries, delivery services, milkmen) were delegated to the housewife. The individual workers also became less specialized; the new housewife was now responsible for every aspect of life in her household, from scrubbing the bathroom floor to keeping abreast of the latest literature in child psychology.

The housewife is just about the only unspecialized worker left in America—a veritable jane-of-all-trades at a time when the jacks-of-all-trades have disappeared. As her work became generalized the housewife was also proletarianized: formerly she was ideally the manager of several other subordinate workers; now she was idealized as the manager and the worker combined. Her managerial functions have not entirely disappeared, but they have certainly diminished and have been replaced by simple manual labor; the middle-class, fairly well educated housewife ceased to be a personnel manager and became, instead, a chauffeur, charwoman, and short-order cook. The implications of this phenomenon, the proletarianization of a work force that had previously seen itself as predominantly managerial, deserve to be explored at greater length than is possible here, because I suspect that they will explain certain aspects of the women's liberation movement of the 1960s and 1970s which have previously eluded explanation: why, for example, the movement's greatest strength lies in social and economic groups who seem, on the surface at least, to need it least—women who are white, well-educated, and middle-class.

Finally, instead of desensitizing the emotions that were connected with household work, the industrial revolution in the home seems to have heightened the emotional context of the work, until a woman's sense of self-worth became a function of her success at arranging bits of fruit to form a clown's face in a gelatin salad. That pervasive social illness, which Betty Friedan characterized as "the problem that has no name," arose not among workers who found that their labor brought no emotional satisfaction, but among workers who found that their work was invested with emotional weight far out of proportion to its own inherent value: "How long," a friend of mine is fond of asking, "can we continue to believe that we will have orgasms while waxing the kitchen floor?"

Research, Engineering, and Science in American Engineering Colleges: 1900–1960

BRUCE SEELY

In 1872 the catalog for the University of Illinois's engineering college told students, "This school is designed to make good practical engineers."[1] At the turn of the century, this goal still guided most American engineering educators, including Embury A. Hitchcock, Professor of Experimental Engineering at Ohio State University. After introducing his mechanical engineering students to machinery, design, and practical problem solving, he assigned senior projects such as calculating heat balances for moving locomotives on the Hocking Valley Railroad. Similarly, students at Cornell in 1899 tested street railway motors and generators in Buffalo; textile engineering students at Georgia Tech ran a factory.[2] College may have replaced the apprenticeships as the locus of engineering education, but the practical focus of the "shop culture" had not disappeared.

DR. SEELY is in the Department of Social Sciences at Michigan Technological University. He wishes to thank a number of people for advice, comments, suggestions, and assistance that significantly improved the article: Terry Reynolds, Walter Vincenti, Edwin Layton, the *Technology and Culture* referees, and conference and seminar participants who heard earlier versions. He also acknowledges support from a number of agencies over a period of more than six years, including the Texas Engineering Experiment Station, the National Science Foundation Program in the History and Philosophy of Science (awards SES-8711164 and SES-8921936), and the National Endowment for the Humanities Travel to Collections program.

[1]University of Illinois, *Report of the Board of Trustees* (1870–71), p. 41, quoted in Ira O. Baker and Everett E. King, *History of the College of Engineering of the University of Illinois, 1868–1945* (Urbana, Ill., ca. 1946), p. 242.

[2]Heat balance calculations required students to measure fuel and boiler water consumed and gallons of water pumped for twenty-four hours. See Embury A. Hitchcock, *My Fifty Years in Engineering: The Autobiography of a Human Engineer* (Caldwell, Idaho, 1939), pp. 75–78, 91–112; Robert C. McMath, Jr., Ronald H. Baylor, James E. Brittain, Lawrence Foster, August W. Giebelhaus, and Germaine M. Reed, *Engineering the New South: Georgia Tech, 1885–1985* (Athens, Ga., 1985), pp. 81–88; and Cornell University, *Annual Report of the President* (Ithaca, N.Y., 1899–1900), p. 53. On school versus shop culture, see Monte Calvert, *The Mechanical Engineer in America, 1830–1910: Professional Cultures in Conflict* (Baltimore, 1967).

By 1960, however, engineering education looked very different. The emphasis on rules of thumb learned through practical experience had given way to an education stressing scientifically derived theory expressed in the language of mathematics. This ideal was not new; it appeared before 1900 in chemical and electrical engineering, and during the early 20th century scientific methodologies appeared in other branches of engineering.[3] A 1955 study for the American Society for Engineering Education recommended that 50 percent of undergraduate instruction be in science and engineering science, with the other half divided equally between engineering technology and general education electives.[4]

Academic engineering research changed in similar ways during this period. In 1900, little research was conducted in engineering schools because of the heavy teaching demands on faculty. Where research was conducted, it usually grew out of consulting projects and focused on specific, practical problems. After 1900 a few more professors were likely to conduct research, but those studies remained tied to real-world problems such as sewage treatment, road building, concrete culverts, electrical transmission, and uses of local mineral resources. This pattern prevailed in American engineering colleges through World War II. After 1945, however, academic researchers increasingly attacked, not practical problems, but theoretical questions related to materials or engineering principles. And as theory attracted more attention, many engineering research projects became difficult to distinguish from "pure" scientific studies. Titles do not tell the whole story, but research is obviously changing when engineers at the University of Illinois are working on "Analysis of Stresses in Rotationally Symmetric Pressure Vessels" (1957–58) and "Design of Accurate, Consistent, and Stable Finite-Difference Algorithms for the Integration of Time-dependent Systems of Conservation Laws" (1989).[5] Of central importance to this development was an enormous increase in federal funding for academic research.

[3]See, e.g., Edwin T. Layton, Jr., "Mirror-Image Twins: The Communities of Science and Technology in 19th-Century America," *Technology and Culture* 12 (October 1971): 562–80, "American Ideologies of Science and Engineering," *Technology and Culture* 17 (October 1976): 688–700, "Scientific Technology, 1845–1900: The Hydraulic Turbine and the Origin of American Industrial Research," *Technology and Culture* 20 (January 1979): 64–89; David F. Channell, "The Harmony of Theory and Practice: The Engineering Science of W. J. M. Rankine," *Technology and Culture* 23 (January 1982): 39–52; and Edward W. Constant, "Scientific Theory and Technological Testability: Science, Dynamometers, and Water Turbines in the 19th Century," *Technology and Culture* 24 (April 1983): 183–98.

[4]L. E. Grinter, "Report on the Evaluation of Engineering Education," *Journal of Engineering Education* 46 (April 1956): 25–63; and Eric A. Walker, "The Goals Study," *Engineering Education* 57 (September 1966): 16.

[5]Contrast the statement of Ira O. Baker, "Engineering Education in the United States at the End of the Century," *Proceedings of the Society for the Promotion of Engineering*

This article deals with the parallel transformations that occurred in both engineering education and academic engineering research, but the primary focus is on the latter. The central argument is that changes in academic research were crucially important to the transformation of engineering education during the 20th century. More specifically, the article argues that the emergence of new goals and a new style of academic engineering research encouraged the widespread adoption of an approach to engineering education that derived not only methods, but also its orientation and values, from science.

Practical Engineering Research: Engineering Experiment Stations, 1900–40

In 1900, few American college teachers were serious researchers. The German research university, which emphasized the pursuit of knowledge for its own sake, had emerged as a model, but only a few American schools embraced it. Research by academic engineers was very unusual, especially at the land-grant schools from which the majority of engineering students graduated at the turn of the century.[6] The reason was evident—most professors had enormous teaching loads. Even so, the dean of engineering at the Iowa State College was dissatisfied that research was "carried on spasmodically, with no systematic, prearranged plan." Anson Marston proceeded to make "a plea for systematic original investigation work in technical schools," arguing that research was a social responsibility of engineering educators. He grounded his argument in a land-grant ethos of service that had developed during the last quarter of the 19th century at schools like Wisconsin. The ideal was extended from the classroom and short courses to research by the formation of agriculture experiment stations after 1870. By finding useful solutions to problems confronting farmers, station scientists estab-

Education (hereafter *SPEE Proceedings*) 8 (1900): 23, with the University of Illinois, *Summary of Engineering Research* (Urbana, 1957–58, 1989).

[6]Roger L. Geiger, *To Advance Knowledge: The Growth of American Research Universities, 1900–1940* (New York, 1986), p. 86. In 1900, the ten largest engineering schools were land-grant colleges; 3,398 of the 4,459 bachelor of engineering degrees awarded that year came from land-grant schools. See Donald D. Glower, "Mechanical Engineering in the First Hundred Years of the Ohio State University, 1871–1971," in Ohio State University (OSU), *Centennial History of the College of Engineering,* pt. 2 (1969), pp. 33–37, in Ohio State University Archives, Columbus; and W. E. Dalby, "The Training of Engineers in the United States," *Royal Institute of Naval Architects* 45 (1903): 39.

lished a pattern of practical research activities on land-grant campuses.[7]

Engineers with heavy teaching loads only slowly emulated the research activities of their agricultural colleagues. The first step came when exceptional individuals built ties to industry as consultants and spent summers working in industry. Their efforts reflected the general expectation that engineering teachers needed practical experience, and consulting permitted faculty to remain abreast of their specialty. Dugald Jackson of Wisconsin and later MIT encouraged consulting for another reason, hoping that faculty consulting would attract industrial funds to the colleges. And occasionally it led to small research projects on specific problems for industrial and governmental clients.[8]

Only extraordinary engineering professors were consultants before 1900. One was a professor of mechanical engineering at Purdue, W. F. M. Goss, who purchased a steam locomotive in 1891 and created the first locomotive performance testing plant in the country. At Ohio State, Edward Orton, Jr., launched a ceramic engineering program in the 1890s with close links to Ohio's ceramic industry. Civil engineers Anson Marston at Iowa State and A. N. Talbot and Ira O. Baker at the University of Illinois studied brick, cement, pipe, road-building materials, and sewage treatment in the 1890s. These were exceptional individuals, all of them pioneers in their fields. Indicative of the obstacles research-oriented faculty faced were the problems encountered by Cornell's Robert Thurston, the teacher who inspired Marston and his generation to conduct research. Although Cornell established a hydraulics laboratory of international repute, Thurston

[7]Quotation from Anson Marston, "Original Investigations by Engineering Schools a Duty to the Public and to the Profession," *SPEE Proceedings* 8 (1900): 236, 238. On the land-grant ideal, see Edward D. Eddy, Jr., *Colleges for Our Land and Times: The Land-Grant Idea in American Education* (New York, 1956), pp. 46–112; Earle D. Ross, *Democracy's College: The Land-Grant Movement in the Formative Stage* (Ames, Iowa, 1942); and Michael Bezilla, *Engineering Education at Penn State: A Century in the Land-Grant Tradition* (University Park, Pa., 1981), pp. 13–20. On the agriculture stations, see A. Hunter Dupree, *Science in the Federal Government: A History of Policies and Activities to 1940* (Cambridge, Mass., 1957), pp. 146–83; and Charles E. Rosenberg's essays in his *No Other Gods: On Science and American Social Thought* (Baltimore, 1976), pp. 153–95.

[8]The pattern of faculty as consultants becomes clear in the histories of engineering schools and in autobiographies, including J. Merrill Weed, "The Second Quarter Century of the College of Engineering at the Ohio State University," in OSU, *Centennial History*, pt. 1, p. 7; and Hitchcock. On Jackson, see W. Bernard Carlson, "Academic Entrepreneurship and Engineering Education: Dugald C. Jackson and the MIT-GE Cooperative Engineering Course, 1907–1932," *Technology and Culture* 29 (July 1988): 543–44.

failed to convince the trustees to authorize an industrial and engineering testing laboratory.[9]

Another reason the volume of engineering research increased after 1900 was the appearance of a new institution—engineering experiment stations. The University of Illinois formed the first one in December 1903 "to carry out investigations along various lines of engineering and to study problems of importance to professional engineers and to manufacturing, railway, constructional, and industrial interests of the state. . . . It is believed that this experimental work will result in contributions of value to engineering science and that the presence of such investigations will give inspiration to students and add efficiency to the College of Engineering." L. P. Breckinridge, head of mechanical engineering, won the university trustees' support for this concept and then secured $150,000 in state funds for the station. In 1904, the station released its first research bulletins.[10] By then, Iowa State College had created the second station. It had an identical mission, with one addition in the charter—a requirement "to assist the urban population of the state in solving the technical problems of urban life." This provision, which opened the door for cooperation with municipal government agencies as well as with industrial interests, reflected Dean Marston's view of the station as a public service research center for solving problems that industry ignored.[11] The practical service orientation at both stations showed the influence of the agriculture station model.

[9]On early research, see Baker and King, pp. 118, 263, 390–91; H. B. Knoll, *The Story of Purdue Engineering* (West Lafayette, Ind., 1963), pp. 191–93; "History of the Department of Ceramic Engineering," OSU, *Centennial History*, pt. 4, pp. 1–7; and Arthur S. Watts, "Ceramic Engineering," in U.S. Department of the Interior, Bureau of Education, *Land-Grant College Education, 1910–1920*, pt. 4, *Engineering and the Mechanic Arts*, Bulletin no. 5 (1925), pp. 45–49; Herbert J. Gilkey, *Anson Marston: Iowa State University's First Dean of Engineering* (Ames, Iowa, 1968), pp. 13–15, 21–25; "Arthur Newell Talbot Laboratory," *University of Illinois Bulletin* 35 (April 1, 1938): 37–38. On Thurston, see Calvert, p. 102. Cornell's hydraulics lab is mentioned in Marston, "Original Investigations," p. 237. See also Cornell University, *Annual Report of the President* (1896–97), pp. xl–xli; (1897–98), pp. 42–43.

[10]Information on the station from University of Illinois, *Proceedings of the Board of Trustees*, twenty-second annual report (1904), pp. 238–39; Baker and King, pp. 767–70; L. P. Breckinridge, *The Engineering Experiment Station of the University of Illinois*, University of Illinois Engineering Experiment Station Bulletin no. 3 (March 1906); Ellery B. Paine, "The Engineering Experiment Station of the University of Illinois," *Proceedings of the American Institute of Electrical Engineers* 34 (October 1915): 2421–27. Quotation from the first bulletin issued by the station: A. N. Talbot, *Tests of Reinforced Concrete Beams*, University of Illinois Engineering Experiment Station Bulletin no. 1 (1904), cover.

[11]On the station's founding and purposes, see Anson Marston, "The Engineering Experiment Station at Iowa State College," 1908, Anson Marston Papers, no. 81, Iowa State University Archives, Ames (hereafter cited as Marston Papers); and Gilkey.

The Illinois station embarked on an ambitious program using faculty researchers, new laboratories (built in 1903), and $30,000 to $40,000 per year from the legislature. Although hampered by much smaller budgets ($3,000 a year), engineers at Iowa State also started several projects. At both stations, engineers focused on practical problems facing state and local government—the design of culverts, drainage systems, and reinforced concrete beams and the performance of brick, chain links, and other products and materials. Much of this work involved testing. Investigations of sewage treatment facilities by A. N. Talbot at Illinois and Marston at Iowa State continued earlier projects, but station bulletins carried the results, usually in tabular form, to a wider audience.[12] This became the pattern of station research, and it reflected the Progressive Era's faith in apolitical experts as problem solvers.[13]

"Public service" research dominated the agenda of faculty researchers through 1920 and beyond. But experiment stations also were expected to assist local industries. Engineers at Illinois, for example, built a locomotive testing plant after Goss was hired from Purdue to become dean of engineering in 1907. Iowa State engineers helped the state's cement industry, while faculty researchers at both schools worked on problems hindering the commercial use of local coal, clay, and other mineral resources. Materials testing for industry also became an important activity. But the engineering deans at both schools placed limits on aid to private companies. The director of the Illinois station explained in 1915, "No researches are undertaken with the object of obtaining information of chief value to some individual or company." Guided by a Progressive moral vision, the main industrial client of the stations was to be small industries that could not afford laboratories.[14]

Pursuing both lines of research, the Illinois station grew rapidly, releasing eighty-three technical bulletins by 1913. The historians of

---

Quotation from Iowa State College, Engineering Experiment Station, *The Iowa State College Sewage Disposal Plant and Investigations,* Bulletin no. 1 (1904), back cover.

[12]Station activities can be glimpsed through their published bulletins. Also see Breckinridge; Baker and King (n. 1 above); and Marston Papers, no. 81.

[13]See John C. Burnham, "Essay," in *Progressivism,* ed. John D. Buenker, John C. Burnham, and Robert M. Crunden (Cambridge, Mass., 1977), pp. 3–29; Samuel P. Hays, *Conservation and the Gospel of Efficiency: The Progressive Conservation Movement, 1840–1920* (Cambridge, Mass., 1959); Robert Wiebe, *The Search for Order, 1877–1920* (New York, 1967); and Donald T. Critchlow, *The Brookings Institution, 1916–1952: Expertise and the Public Interest in a Democratic Society* (De Kalb, Ill., 1985), pp. 17–40.

[14]Paine, p. 2421; Anson Marston, "Value to Industries of Engineering Research at Iowa State College," paper read at the thirty-ninth annual meeting of the Association of Land-Grant Colleges, November 17, 1925, Marston Papers, no. 172. See also station publications.

the college also observed that "the unprejudiced character of the publications has gone far in establishing the prestige which the University maintains among the educational institutions of the United States and the world at large."[15] Land-grant engineers hoped that further growth would follow receipt of federal funds like those for agriculture stations, and they campaigned for equal support. Although Congress was unmoved, the prospect of federal assistance induced several land-grant schools to form stations, just in case. Penn State and Kansas State acted in 1912, Wisconsin and Texas A&M in 1914, Maine in 1915, Colorado and Purdue in 1917. These early stations imitated the programs at Illinois and Iowa State, with routine testing of road-building materials emerging as an important topic.[16]

The critical problem facing newer stations was lack of funds. The station at Texas A&M, for example, received no state money until 1921. The station director reported in 1917 that no research had been accomplished "owing to the large amount of [class-room] work carried on by various members of the engineering teaching staff."[17] Even so, by 1931 forty stations were in existence, if often poorly supported. The station at the University of Illinois was preeminent, publishing 237 bulletins by 1930, when its budget reached $246,000. Stations at Ohio State, Washington State, and Texas A&M were smaller and more typical; $10,000 was an average budget for these stations by the end of the decade. (See table 1.) Such limited resources reinforced the tendency to conduct practical, service-oriented research, including routine tests of materials or studies of highway problems, water supply, or sewage treatment.

Yet not all stations were hampered by small state appropriations or limited university funds. A few stations grew rapidly during the 1920s

[15]Baker and King, p. 832.

[16]Efforts to win federal funding for engineering experiment stations peaked during World War I, with Marston playing a leading role. See the Marston Papers; also see Willis R. Whitney, "Engineering Experiment Stations in State Colleges," *Science* 43 (June 23, 1916): 890–91, 895–96; Henry H. Armsby, "A Review of Proposals for Federal Support of Engineering Research in the Colleges," *Engineering Experiment Station Record* 27 (October 1947): 39; Daniel J. Kevles, "Federal Legislation for Engineering Experiment Stations," *Technology and Culture* 12 (April 1971): 112–19; and comment by C. R. Mann in *Proceedings of the Thirty-second Annual Convention of the Association of American Agricultural Colleges and Experiment Stations,* January 8–10, 1919, p. 158.

[17]Henry C. Dethloff, *A Centennial History of Texas A&M University, 1876–1976,* 2 vols. (College Station, Texas, 1975), 1:248–49. See also various issues of the *Engineering Experiment Station Record* (1921–30). Unless otherwise noted, statistical information on station budgets is drawn from the annual summaries that appeared in this quarterly newsletter.

TABLE 1
RESEARCH EXPENDITURES AT SELECTED ENGINEERING EXPERIMENT STATIONS, 1920–40
($ IN THOUSANDS)

Station	Legislative Appropriations	University Funds	Other	Total
Illinois:				
1920	30	85	. . .	115
1926	N.A.	92	72	164
1931	N.A.	101	183	284
1934	N.A.	77	73	150
1940	N.A.	122	120	242
Purdue:				
1920	N.A.	15	. . .	15
1925	N.A.	25	32	57
1926	N.A.	26	127	153
1931	N.A.	23	240	263
1934	N.A.	13	28	41
1940	N.A.	21	131	152
Ohio State:				
1920	10	. . .	. . .	10
1924	10	. . .	3	13
1930	38	37	30	105
1934	N.A.	37	8	45
1940	N.A.	51	18	69
Texas A&M:				
1920	3	. . .	. . .	3
1926	2	2	4	8
1931	2	10	1	13
1934	. . .	.5	10	10.5
1940	. . .	40	. . .	40
Washington State:				
1920	. . .	5	. . .	5
1926	. . .	7.5	. . .	7.5
1931	. . .	6	. . .	6
1933	. . .	3	. . .	3
1940	. . .	10	2.5	12.5

SOURCE.—*Engineering Experiment Station Record,* vols. 1–20 (1920–40).
NOTE.—N.A. = not available; ellipses indicate that there are no funds from that source.

by finding industrial patrons. This was an almost natural extension of ties that had been growing between industry and engineering schools from the 1880s. The consulting relationships noted earlier, and the inauguration of cooperative education at the University of Cincinnati in 1907 in which students alternated semesters between classes and work in industrial establishments, reflected an increasingly deep-rooted connection. The ties to industry strengthened during the

1920s as industrial sponsorship of research on campuses began, exemplified by MIT's Technology Plan.[18] Further encouragement came from Secretary of Commerce Herbert Hoover, who saw trade associations as the essential component in an "associative" vision of cooperative business/government relations. Cooperative research was made to order for associative efforts, offering one of the few activities that would not bring instant antitrust scrutiny. Moreover, joint investigations that benefited entire industries matched the stations' ideal of public service.[19] The new patrons of station research did not alter, however, the basic pattern of practical, applied research.

Researchers at Illinois began their first cooperative investigation in 1916 for the Association of Manufacturers of Chilled Car Wheels.[20] Over the next three decades, cooperative research projects at Illinois grew steadily in number, size, and importance and by 1928 brought more than $100,000 to the station. Once started, projects often continued for decades. For example, in 1919 Illinois engineer (and later university president) Arthur C. Willard first attracted support from the National Warm Air Heating and Ventilating Association for studies of furnaces. The project eventually grew to include two "research houses" on the Illinois campus used to test home furnaces, and work continued into the 1950s. Related projects begun in the 1920s included tests on hot water furnaces for the Illinois Master Plumbers' Association (1924) and the National Boiler and Radiator Manufacturing Association (1926) and on air conditioning for the American Society of Heating and Ventilating Engineers (1932). Other long-lived research programs included a study of fatigue in metal for the Engineering Foundation (1919–30) and a series of investigations for the Utilities Research Committee (URC), an arm of Samuel

[18]Technology Plan sponsors paid a membership fee for access to work in progress at the institute; they also could hire faculty researchers to solve small problems. See David F. Noble, *America by Design: Science, Technology, and the Rise of Corporate Capitalism* (New York, 1977), pp. 184–97; John Servos, "Industrial Relations of Science: Chemical Engineering at MIT, 1900–1939," *Isis* 71 (1980): 531–49; John M. de Bell, "Cooperation with Industries by the Massachusetts Institute of Technology," in Association of Land-Grant Colleges, *Proceedings of the Thirty-fourth Annual Convention,* October 22, 1920, pp. 169–72; Geiger (n. 6 above), pp. 175–80; and Carlson (n. 8 above).

[19]Hoover's ideas are summarized in Ellis Hawley, "Herbert Hoover, the Commerce Secretariat, and the Vision of an 'Associative State,'" *Journal of American History* 61 (June 1974): 116–40. A contemporary statement about cooperative research is William E. Wickenden, "Research in the Engineering College," *Mechanical Engineering* 51 (August 1929): 585–88. For an example of this philosophy in operation, see Bruce E. Seely, "Engineers and Government-Business Cooperation: Highway Standards and the Bureau of Public Roads, 1900–1940," *Business History Review* 58 (Spring 1984): 51–77.

[20]"Chilled car wheels" were cast-iron railroad wheels, so named because the surface of the casting was cooled quickly to produce a very hard iron.

Insull's utility empire. The first studies were made in 1924 on boiler feedwater treatments, aging of porcelain insulators, and refractories. By 1950, the URC had funded more than one hundred projects related to utility operations.[21]

While individual contracts usually generated less than $5,000 annually, industrial sponsorship doubled the budget of the Illinois station in the 1920s. Other schools followed Illinois's lead. Engineers at Ohio State, for example, strengthened ties to the state's ceramic industries with a new facility to test glazes and clays. Engineers at Kansas State University studied rural electrification for the Kansas Public Utilities Commission, while fellowships from corporations and associations supported projects at many stations. Researchers studying culvert pipe at Iowa State were funded by the American Concrete Institute, the American Concrete Pipe Association, the American Railway Engineering Association, and other engineering groups. Even the Southeastern Missouri Sunflower Growers' Association supported research into uses of sunflower oil at the station at the University of Missouri.[22]

Dean A. A. Potter of Purdue, builder of one of the largest cooperative education programs in the country, used external research funds to transform Purdue's station into the nation's largest in the late 1920s. In 1922, Purdue drew only $8,750 from the American Railway Association and the Indiana Quarrymen's Association as a supplement to an internal budget of $25,000. But at a 1926 conference on industrial research, Potter announced a desire for Purdue "to become the Indiana Bureau of Standards and Indiana's Mellon Institute," as well as "the technical laboratory of our smaller industries." Industrial

[21]Information from project files in Subject Files (SF), 1909–61, ser. 11/2/1, Engineering Experiment Station Records (EESR), University of Illinois Archives, Urbana; University of Illinois, *Transactions of the Board of Trustees* (annual statement of expendable gifts to the college of engineering) (1912–60); bulletins and circulars published by the engineering experiment station at the University of Illinois (1919–40); R. A. Seaton, "How Industry Can Cooperate with Engineering Colleges in Furthering Research," *Journal of Engineering Education* 16 (November 1925): 209; and A. A. Potter, "Cooperative Engineering Research," *Journal of Engineering Education* 22 (October 1931): 92–94.

[22]Information obtained from E. A. Hitchcock and J. M. Weed, *A Description of the Engineering Experiment Station, Ohio State University,* Ohio State University Engineering Experiment Station Bulletin no. 50 (Columbus, 1929); articles in various issues of *Engineering Experiment Station News* (1929–40), published quarterly after 1929 by the station at Ohio State; various issues of *Engineering Experiment Station Record* (1929–40); A. A. Potter and G. A. Young, "Tendencies in Research at Engineering Colleges," *Society of Automotive Engineers Journal* 20 (May 1927): 623–32; A. A. Potter, "Research Relations between Colleges and Industry," *American Institute of Electrical Engineers Journal* 45 (1926): 1272–76; and Seaton, pp. 203–14.

support followed, totaling $275,000 in 1929, a sum accounting for more than half the external research funds at all stations combined. Typical projects included studies of road materials and construction techniques, automobile engines, electrical generating machinery, sewage collection and treatment, and radio and television.[23]

Few stations enjoyed such success, however, and one study found only ten had adequate resources for real research. Indeed, only a handful of engineering schools attracted much research. MIT, a land-grant school without an engineering experiment station, was the only one to match Purdue in the late 1920s. Illinois came next, while the University of Michigan, which created a Department of Engineering Research in 1920, attracted $60,000 from industrial sources in 1930. But large academic engineering research programs were exceptions in 1930, and research expenditures at engineering experiment stations totaled only $1.3 million. Just two firms in the electrical industry spent more than twice as much on research as all engineering schools combined.[24]

The Depression ended this period of growth as budgets at most stations declined sharply. Purdue's external research income plummeted from $240,000 in 1931 to $15,000 in 1933, while the number of research workers fell from 100 in 1927 to about 10 in 1931. Research at Texas A&M was crippled when the research budget fell from $15,000 in 1932 to $1,900 a year later; Kansas State absorbed a decline from $63,000 in 1931 to $17,000 in 1934. The station at Illinois weathered the Depression better than most, as many projects continued on a reduced scale, but the budget fell by 50 percent from 1931 to 1934.[25]

[23]Figures from the Association of Land-Grant Colleges, *Proceedings of the Annual Convention* (1920–31); "Purdue as an Engineering Center," *Domestic Engineering* 124 (July 7, 1928): 18–20; and Purdue University, Engineering Extension Department, *Proceedings of Industrial Conference Held at Purdue University, June 1, 1926,* Bulletin no. 15 (1926), pp. 22, 24. On Potter, see Noble, pp. 180–81.

[24]Information obtained from various issues of *Engineering Experiment Station Record* (1921–30); files of the Department of Engineering Research, Records of the Engineering College, Box 20, University of Michigan Archives, Bentley Historical Library, Ann Arbor; University of Michigan, *University Extension Service to Manufacturers, Technical, and Civil Interests Offered through the Department of Engineering Research, University of Michigan, Ann Arbor, Michigan* (Ann Arbor, ca. 1921); Benjamin Bailey, "Can the University Aid Industry?" *American Institute of Electrical Engineers Journal* 45 (1926): 742–45; and Walter Donnelly, ed., *The University of Michigan: An Encyclopedic Survey,* 4 vols. (Ann Arbor, 1953), 3:1243–48. The comparison is drawn from George Wise, *Willis Whitney, General Electric, and the Origin of U.S. Industrial Research* (New York, 1985), pp. 262–71.

[25]Data from statistical tables in *Engineering Experiment Station Record,* January issues (1930–40). The assessment of the Illinois station is based on the research files in SF,

Engineering researchers used several strategies to attract funds for research, all connected to the traditional ideal of practical service. Most stations depended on state and local government agencies, small state appropriations, or university funds. The station at Texas A&M developed a laboratory for the Texas Game, Fish and Oyster Commission to determine the effect of salt water discharges from East Texas oil fields on marine life. The station also studied treatment of dairy, citrus, and packing wastes in conjunction with several cities and the state health department. Other research grew out of the station director's appointment as college architect, which permitted him to use campus construction projects to study foundation and roofing techniques; the design and operation of air conditioning, heating, and ventilation systems; and swimming pool construction.[26]

Another source of revenue during the 1930s, often for the first time, was federal agencies. From 1929 to 1932, and again in 1937–38, land-grant engineers renewed efforts to win federal appropriations. The obvious disparity in resources at agriculture and engineering experiment stations rankled the engineers.[27] Congress again rejected the engineers' claim, but some stations received federal funds through New Deal programs. The Texas A&M station administered Rural Electrification Administration funds that opened several research opportunities, while the Federal-State Cotton Research Program of the Department of Agriculture led to a cotton fibers laboratory. Other projects were carried on with research assistants paid by Works Projects Administration and National Youth Administration funds.[28] The Illinois station also paid staff with relief funds, including about $4,500 in 1934 from the Civil Works Administration.

1909–16, ser. 11/2/1, EESR; and reports of gifts and research contracts in University of Illinois, *Annual Report of the Board of Trustees* (1930–39).

[26]Information from statistical tables in *Engineering Experiment Station Record* 13 (April 1933): 4–5; 14 (January 1934): 5; 15 (January 1935): 6; C. E. Sandstedt and C. H. Samson, "Civil Engineering," in C. W. Crawford et al., *One Hundred Years of Engineering at Texas A&M, 1876–1976* (College Station, Texas, 1976), p. 135; Texas Agricultural and Mechanical College, *Biennial Report of the Agricultural and Mechanical College of Texas* (1930/31–1937/38), in Texas A&M University Archives, College Station; and Dethloff, 2:422.

[27]Engineering stations employed 960 researchers and spent $1.3 million in 1936, while agriculture stations had 3,818 research personnel and budgets totaling $16.4 million; they had received $78 million in federal funds from 1888 to 1936; see various issues of *Engineering Experiment Station Record* (1930–40); also see note 16 above.

[28]Texas Committee on the Relation of Electricity to Agriculture, *Progress Report* (September 1944), pp. 86–93; Texas Agriculture and Mechanical College, *Budgets, 1935–47,* in Texas A&M University Archives; and *Engineering Experiment Station Record* 15 (July 1935): 5.

Substantial assistance from the Bureau of Public Roads (BPR) for studies of bridges, concrete beams, construction materials, and soil mechanics proved especially important to researchers at Purdue, Illinois, Ohio State, and Iowa State. The inauguration at Texas A&M of a cooperative investigation with the BPR on bridge types was an important step for the station in the late 1930s.[29]

The last means of supporting research explored by the largest universities during the 1930s was the creation of private, incorporated research foundations. In 1925, the University of Wisconsin created the Wisconsin Alumni Research Foundation (WARF) to administer agricultural chemist Harry Steenbock's patented process for synthesizing vitamin D. After the university's regents refused to administer the patent, Steenbock formed a nonprofit corporation "capitalized by privately subscribed funds, managed by 'friends' of the University but operated independently." This patent quickly enriched the foundation, which returned the profits to the university beginning in 1928. By 1950, WARF had given $4.43 million to the University of Wisconsin in support of research and for construction of research facilities.[30]

Several schools copied Wisconsin, but none had a windfall like Steenbock's patent. At Purdue, industrialist and board president David Ross and Potter promoted a private foundation that could solicit funds, negotiate with individual companies, and sign contracts protecting proprietary information. As one participant explained, "The Purdue Research Foundation was organized . . . to include functions not clearly provided for in the Federal and State laws governing the organization and operation of Purdue University, and so that the rights of each group might be securely protected and an equitable distribution of the profits resulting from cooperative effort guaranteed." The board created the Purdue Research Foundation in 1930, linking it to a new Department of Research Relations with Industry. The Depression crimped plans to build a multi-million-

[29]The BPR projects brought $6,000 to Illinois in 1929, $15,000 in 1931, and $12,300 in 1932. Large sums also went to Purdue, Ohio State, and Iowa State. See University of Illinois, *Annual Report of the Board of Trustees* (annual statement of expendable gift funds) (1927–34); U.S. Department of Agriculture, Bureau of Public Roads, *Annual Report* (1924–39); and Texas Agricultural and Mechanical College, *Biennial Report of the Agricultural and Mechanical College of Texas* (1940–41), p. 48. Several bulletins were published by the station on this project, including one on the 1940 collapse of the Tacoma Narrows bridge in 1950.

[30]For the period of 1950–73, WARF provided $50 million to the university. See Allen G. Bogue and Robert Taylor, eds., *The University of Wisconsin: One Hundred and Twenty-five Years* (Madison, Wisc., 1975). Quotation from University of Wisconsin, "WARF Plays Vital Role," *University-Industry Relations (UIR) Memo* 1, no. 3 (1965): 7.

dollar endowment, but Ross and pharmaceutical manufacturer Josiah Lilly each donated $25,000. In addition, contracts with industrial firms soon generated projects for departments in engineering and across the university.[31]

In 1936, Ohio State also formed a research foundation. Prominent industrial leaders, including James F. Lincoln of the Lincoln Electric Company and Charles F. Kettering, head of research at General Motors, again played key roles. Purdue's research foundation, which they visited in 1934, was their model. By 1940, Minnesota, Cornell, Virginia Polytechnic Institute (VPI), Delaware, and Washington State had created their own research foundations, and Maine was about to. Georgia Tech and Texas A&M acted after the war.[32]

Foundations were attractive because they offered a means of bypassing restrictions on sponsored research within universities and experiment stations. For the first time, individual companies—not just associations—could fund research on campus, using the foundations as middlemen to accept corporate funds and write contracts with university faculty or departments. These contracts specified the degree of corporate control over results, with sponsors always getting the first look at results, while researchers usually (but not always) retained rights to publish. MIT's Technology Plan of the 1920s had worked in this way, but state universities believed they could not copy the private institute. After Ohio State formed its research foundation, however, the new organization signed contracts to conduct confidential investigations for the Drackett Chemical Company (maker of Windex and Drāno), and the Cambria Clay Products Company. Such initiatives constituted a major shift in academic research policy whose full ramifications were felt on campus only after World War II.[33] In

[31]Information from Knoll (n. 9 above), pp. 94–96; Purdue University, *Annual Report of the President,* vols. 60–65 (1932/33–1939/40); Purdue Research Foundation, *Purdue Research Foundation: Its Organization and Purpose,* Bulletin no. 1 (1931). Quotation from G. S. Meikle, "The Department of Research Relations with Industry," in Purdue University, *Annual Report of the President* 61 (1935): 164. Meikle was director of the Department of Research Relations with Industry.

[32]On Ohio State, see Robert C. Thompson, "A History of the Ohio State University Research Foundation, 1936–1969" (September 1969), pp. 1–21, prepared for the Ohio State University centennial, copy in OSU Archives. On developments at other schools, consult A. A. Potter, "Research and Invention in Engineering Colleges," *Science* 91 (January 5, 1940): 1–7; and Cornelia Weil, "UDRF: Seed Money for Science," *University of Delaware Magazine* (Fall 1988), pp. 12–15.

[33]OSU Research Foundation, *Annual Report,* nos. 1–3 (1937–39), copies in Research Foundation, *Annual Reports,* Box 1, Record Group (RG) 38/c/1, OSU Archives; also see Thompson, pp. 18–21. The acceptance of individual corporate sponsorship through research foundations provided a crucial precedent for justifying military research on campus after 1945. Moreover, one might trace the disappearance of the public service

the 1930s, the projects conducted through the foundations remained practical studies of immediate utility.

Indeed, right up to 1940, research remained a highly practical matter at most engineering schools. When the president of the University of California at Berkeley announced a major commitment to research in the 1930s, his rhetoric stressed public service and practical results.[34] Equally symbolic of the service ethos in land-grant engineering schools was the creation of a Public Service Engineering program at Purdue in 1935. Open to the best engineering students, the program prepared engineers for government service, an orientation congenial with the original goals of engineering experiment station supporters.[35] Land-grant engineers continued to define service to the public, local industries, and government in terms of practical testing and highly applied research.

An Alternative Pattern: Engineering Science

Through 1940, the one constant in academic engineering research was this emphasis on solidly practical subjects. Yet even before the appearance of engineering experiment stations as the home of such work, an alternative style of engineering education and research was emerging. The pressure for change was first felt in the new fields of chemical and electrical engineering, where engineers found traditional rules of thumb no longer sufficed; an understanding of scientific principles and greater facility in mathematics were needed. The push for change came when key figures in electrical engineering were trained in physics, not engineering.[36] Palmer C. Ricketts, president of Rensselaer Polytechnic Institute from 1893 to 1934, was not alone when he complained in 1893 that the curriculum imparted "a smattering of so called practical knowledge." Engineering colleges, he charged, turned out " 'surveyors, and those of mechanical engineering, mechanics, rather than engineers.' Too much attention was being given to the machine shop, and 'too little to head work.' "[37]

---

research ideal to the emergence of research foundations. Yet foundations remain little-studied elements in the history of academic research. On MIT's Technology Plan, see Servos (n. 18 above); and Carlson (n. 8 above).

[34]Verne A. Stadtman, *The University of California, 1868–1968* (New York, 1970), p. 509.

[35]Knoll, p. 99; and "Public Service Engineering at Purdue," *Engineering Experiment Station News* (OSU) 7 (February 1935): 8–9.

[36]See Robert Rosenberg, "American Physics and the Origins of Electrical Engineering," *Physics Today* 36 (October 1983): 48–54.

[37]Samuel Reznick, *Education for a Technological Society: A Sesquicentennial History of Rensselaer Polytechnic Institute* (Troy, N.Y., 1967), p. 256.

By head work, Ricketts referred to the methods and information of science. But Edwin Layton has shown that leading engineers who shared this concern and acted on it, including W. J. M. Rankine and Robert Thurston, did not borrow blindly but created a synthesis between science and engineering by linking scientific methods to engineering values. The crucial engineering value was the emphasis on problem solving and design—on doing, not knowing. The result was engineering science.[38] Yet if engineering sciences differed in intent from traditional sciences, engineering students still needed a foundation of science and math. This may explain why many engineers, even those creating engineering sciences, assumed that engineering was simply applied science. This belief led many engineering educators to propose curricula resting on "fundamentals," by which they meant science courses. Thurston tried to build such a program at Cornell's Sibley College of Engineering in the 1880s and 1890s, and after 1893 the newly formed Society for the Promotion of Engineering Education (SPEE) provided a forum for other proposals. One member wrote in 1898 that "the aim of teaching is . . . thorough instruction in underlying principles, especially those theoretical and scientific principles which cannot be correctly estimated by the layman."[39] Science courses were always part of engineering curricula, but now science began to enter engineering education more systematically.

By 1900, the need to incorporate science had become orthodoxy among academic engineers. A SPEE survey of American engineers that year found that 70 percent endorsed the need to teach fundamentals in engineering schools. General Electric's Charles Steinmetz advocated more mathematics in engineering education and even wrote a text. Dugald Jackson built electrical engineering programs at

[38]The best of Layton's essays are "Mirror-Image Twins" and "American Ideologies of Science and Engineering" (both in n. 3 above); see also Channell (n. 3 above); and Walter Vincenti, *What Engineers Know and How They Know It* (Baltimore, 1990). Hunter Rouse and Simon Ince, in *History of Hydraulics* (Iowa City, Iowa, 1957), trace the emergence of one of the engineering sciences.

[39]Quotation from Henry T. Eddy, quoted in Jeffrey K. Stine, "Professionalism vs. Special Interest: The Debate over Engineering Education in Nineteenth Century America," *Potomac Review,* no. 26/27 (1984–85), p. 85. On Thurston, see Lawrence P. Grayson, "A Brief History of Engineering Education in the United States," *Engineering Education* 67 (December 1977): 253; and Calvert (n. 2 above), pp. 47, 97–104. Authors of articles in *SPEE Proceedings* (n. 5 above) during the period of 1893–1905 were not in universal agreement about what constituted the "fundamentals" of engineering education. Nor did all engineers endorse this argument. But the basic intent of those supporting this educational shift was clear—give engineering students a better grounding in mathematics and in scientific principles relevant to engineering work.

Wisconsin and MIT that reflected his belief that "the engineering college must . . . rigorously hold itself to the fundamentals."[40] Demands for change were more evident immediately after World War I, yet Maurice Caullery, a visitor from France, in 1919 found that American engineers needed "more solid scientific instruction at the foundation."[41] A decade later, another SPEE survey, this time of directors of corporate research laboratories, produced similar recommendations. One respondent observed: "I find it universally true that the American is lacking in a good solid grounding in the elements of engineering mechanics, physics, chemistry, and the natural laws by which the world goes around. . . . I feel it is primarily the duty of the faculties of our American universities to eliminate as far as possible the shop courses, the so-called research courses, and in fact, all so-called practical work and concentrate all effort in preparing the foundation to better advantage."[42] Writers in professional journals echoed this call, as did five major studies of engineering education conducted from 1918 through 1944, although they did not condemn the practical side of engineering education so stridently. By 1935, the dean of engineering at Virginia Polytechnic Institute concluded, "The scientist is pretty much of an engineer and the engineer must be a fair sort of scientist."[43]

Clearly, many engineers favored emphasizing scientific fundamentals, but the repetition of calls for changes suggests that engineering schools moved slowly to adjust curricula, if at all. One historian of technology has noted that suggestions by Steinmetz and in the studies of engineering education were "virtually ignored."[44] The pattern was not uniform, for electrical and chemical engineering were always

[40]Robert Fletcher, "The Present Status of Engineering Education in the United States," *SPEE Proceedings* 8 (1900): 186; Ronald R. Kline, "Origins of the Issues," *IEEE Spectrum* 21 (November 1984): 38–39.

[41]Maurice Caullery, *University and Scientific Life in the United States,* trans. James Haughton and Emmet Russell (Cambridge, Mass., 1922), p. 11.

[42]H. H. Higbie, "Research in Engineering Colleges of Interest to Industry," *Journal of Engineering Education* 23 (October 1932): 154.

[43]Earle B. Norris, "Research as Applied to Engineering," *Civil Engineering* 5 (May 1935): 408. The reports on engineering education are C. R. Mann, *A Study of Engineering Education,* Carnegie Foundation for the Advancement of Teaching, Bulletin no. 11 (1918); Society for the Promotion of Engineering Education, *Report of the Investigation of Engineering Education, 1923–1929,* 2 vols. (Pittsburgh, 1930, 1934); Dugald C. Jackson, *Present Status and Trends of Engineering Education in the United States* (New York, ca. 1939); and H. P. Hammond, "Aim and Scope of Engineering Curricula," *Journal of Engineering Education* 30 (March 1940): 555–66, "Engineering Education after the War," *Journal of Engineering Education* 34 (May 1944): 589–613.

[44]See comment of A. Michal McMahon in Effie G. Bryson, "Frederick E. Terman," *IEEE Spectrum* 21 (March 1984): 73.

linked to fundamentals, and many newer areas such as radio and aeronautical engineering also developed in accord with the newer thinking. Courses in mechanics and strength of materials also seemed closer to the ideals of engineering science.[45]

Yet closer examination suggests that, even in courses on materials, older patterns were firmly entrenched. Strength-of-materials courses were service courses and not well integrated into specific fields. A common arrangement was found at the University of Illinois, where a separate Department of Theoretical and Applied Mechanics taught such classes. Often, courses in these departments were designed to improve skills in the use of apparatus and to teach formulas, not to develop an understanding of fundamental principles of materials behavior.[46] Stephen Timoshenko, a Russian immigrant who built a reputation in this field, was stunned at the poor preparation of American engineering students. As an engineer at Westinghouse in the 1920s, he offered an informal course on strength of materials and asked recent graduates for their class notebooks: "I was amazed at the complete divorce of strength-of-materials theory from experimental research. Most of my students had done no work whatever in mechanical testing of materials with measurements of their elastic properties. The newer methods of calculating beam deflection and investigating flexure in statically indeterminate cases had not been taught them at all." Timoshenko felt obliged to offer a course given to sophomores in Russia. Teaching at the University of Michigan in 1927 only confirmed Timoshenko's complaints, for those students knew nothing about the physical properties of building materials, allowable stresses, and fatigue in metals.[47]

Few engineering colleges in the 1920s provided what Timoshenko considered a good education. Perhaps only the California Institute of Technology (Caltech), the creation of scientists, embraced the new vision of scientific engineering. Physicist Robert Millikan became the

[45]Work in the strength of materials may provide the best demonstration of engineering science emerging in the fashion described by Layton in "Mirror-Image Twins." I am indebted to Walter Vincenti for the comment that different areas of engineering developed more scientific approaches at different times, with the newer subspecialty fields often proving more responsive.

[46]Over time, the changing goals stated for the Department of Theoretical and Applied Mechanics, organized at the University of Illinois in about 1890, reflect the changing place of science in engineering education. The stated purposes always mentioned problem solving, with knowledge of the behavior of materials paramount. Mathematical analysis was being added to statements of purpose by 1930, along with knowledge of the laws of mechanics, but application of physics was not specifically stated until 1940. Baker and King (n. 1 above), pp. 394–96.

[47]Stephen P. Timoshenko, *As I Remember* (Princeton, N.J. 1968), pp. 253, 280.

school's president in 1922, and told astronomer George Ellery Hale that, if the school "is to make a new contribution to the educational development of the country, it has to be done, I think, through making engineering grow out of physics and chemistry."[48] Hale and Millikan hired another European immigrant, the "strongly theoretical" Theodore von Kármán, to bring a scientific orientation to aeronautical engineering at Caltech.[49]

Even MIT would not have satisfied Timoshenko in the 1920s. Until physicist Karl Compton became president in 1931, MIT emphasized things practical, as evidenced by its Technology Plan and by the number and variety of faculty consulting projects. Dugald Jackson, the head of electrical engineering and a proponent of fundamentals, emphasized application. Arthur A. Noyes, a physical chemist engaged in fundamental research, departed for Caltech, unhappy with MIT's emphasis on applied research. Historian Roger Geiger has argued that Frank Jewett, president of Bell Labs, persuaded Compton to come to MIT and change "the obsolescence of traditional engineering education, locked into teaching practical skills. As the leading institution of engineering education, MIT had a responsibility, according to Jewett, to introduce fundamental science into engineering." Compton limited industrial consulting by faculty and decreed that "only research touching upon important scientific questions" would be performed at the institute.[50]

Only a few other schools showed similar willingness to implement a more scientific approach to engineering education. Harvard was one, according to Eric Walker, a student there in the 1930s who later became dean of engineering and president of Penn State. Walker was highly critical of Harvard—and other schools as well—for failing to link theory and practice, but Walker felt that Harvard (unlike most schools) placed too much stress on theory. Many of his professors were European engineers, who "taught us elegant theory: vector diagrams for rotating machinery, hyperbolic functions for transmis-

[48]Millikan quoted in Wise (n. 24 above), p. 262.

[49]See Robert H. Kargon, "Temple to Science: Cooperative Research and the Birth of the California Institute of Technology," *Historical Studies in the Physical Sciences* 8 (1977): 3–31; Theodore von Kármán, with Lee Edson, *The Wind and Beyond: Theodore von Kármán, Pioneer in Aviation and Pathfinder in Space* (Boston, 1967), pp. 122–26, 146–59. The description of von Kármán is from Clayton R. Koppes, *JPL and the American Space Program: A History of the Jet Propulsion Laboratory* (New Haven, Conn., 1982), p. 2.

[50]Quotations from Geiger (n. 6 above), p. 181. Geiger argues that MIT's board also acted because of unfavorable comparisons in the number of graduate students and research fellowships at Caltech, Berkeley, and MIT. Nonetheless, MIT's researchers did not completely abandon practical research activities. See Noble (n. 18 above), pp. 184–97; Servos (n. 18 above); and Carlson (n. 8 above).

sion lines and even triple integrals" but paid little attention to applications.[51]

Significantly, Walker's description of Harvard contains one feature common to MIT and Caltech: European engineers were crucial proponents and promoters of a more scientific approach in American engineering schools. Caltech's von Kármán, for example, brought to America the German engineer Ludwig Prandtl's work in fluid dynamics and boundary-layer theory, crucial elements in the development of aerodynamics.[52] And he trained a generation of aeronautical engineers. Equally important though less flamboyant was Timoshenko, who transformed strength of materials, structural mechanics, and dynamics—especially vibration—by placing work on a mathematical footing. Although he disparaged the poor preparation of American engineers in mathematics and theory, he also helped remedy those problems. His training course at Westinghouse grew into a graduate program at the University of Pittsburgh; he inaugurated a summer school in mechanics at the University of Michigan in 1929. He, too, trained a generation of engineers comfortable in a scientifically based, mathematical approach to engineering at Michigan and Stanford.[53]

Other less visible Europeans also helped push American engineering toward theoretical and mathematical expressions of engineering principles. A memorial volume to von Kármán issued in 1941 had a contributors' list that was a Who's Who of leading European-born engineers active in the United States, including Boris Bakhmeteff, Max Munk, A. L. Nádai, Richard von Mises, even Albert Einstein and colleagues from the Institute for Advanced Study, and Harald M. Westergaard. Westergaard came from Denmark in about 1914 to do graduate work at Illinois and taught structural theory and the theory of elasticity there until moving in 1936 to Harvard where he became dean of engineering in 1937. Karl Terzaghi was another important European immigrant. A civil engineer who traveled the world consulting on dam foundations, he was a key figure in the development of soil mechanics. Terzaghi also taught at MIT from 1925 to 1929 and at Harvard after 1938. Max Jakob, who fled Nazi Germany in 1937, brought crucial understandings of the theory of heat transfer to the

[51]Eric A. Walker, *Now It's My Turn: Engineering My Way* (New York, 1989), pp. 54–70, 105–8. Quotation is from p. 107.

[52]See Paul Hanle, *Bringing Aerodynamics to America* (Cambridge, Mass., 1982); and von Kármán.

[53]See Timoshenko; also see "Stephen Prokofievitch Timoshenko, 1878–1972," *Stanford Engineering News,* no. 82 (May 1972). I wish to thank Walter Vincenti for sharing his recollections and thoughts about Timoshenko.

Armour Research Institute.[54] Only a few Americans helped transfer the European style of engineering, all of them after studying in Europe. Morrough P. O'Brien, later dean of engineering at Berkeley in the 1950s, introduced fluid dynamics, especially as applied to coastal engineering, after spending more than a year in Europe in the late 1920s. George Swain, professor of engineering at Harvard from 1881 to 1927, brought advanced methods of structural analysis to this country from Germany. L. M. K. Boelter, who taught at Berkeley from 1918 until 1941, played a similar role in heat transfer by introducing the newest European work through a set of highly influential lecture notes.[55]

Taken together, these European engineers and their American students played the crucial role in supporting and encouraging the emphasis on fundamentals that many American engineering educators had advocated since the 1890s. Nonetheless, the pace of change in American engineering remained deliberate. Although many Europeans consulted for industry, the new approach generally was confined to academia. Even there, most American engineers remained tied to older approaches. Another story from Timoshenko illuminates the continuing resistance to change. In 1924, Timoshenko presented a paper on stress concentration that Swain of Harvard critiqued as "a useless fantasy of theoreticians." Timoshenko, who strongly defended

[54]I encountered Terzaghi and Westergaard, whom Timoshenko mentions in his autobiography, in the course of earlier research on highway research in this country. See also *Who's Who in Engineering,* 5th ed. (New York, 1941); 7th ed. (New York, 1954). On heat transfer, see Elizabeth Jakob, "Max Jakob: Fifty Years of His Life and Work," pp. 87–116, and Virginia Dawson, "From Braunschweig to Ohio: Ernest Eckart and Government Heat Transfer Research," pp. 125–37, both in *History of Heat Transfer: Essays in Honor of the Fiftieth Anniversary of the ASME Heat Transfer Division,* ed. Edwin T. Layton, Jr., and John Lienhard (New York, 1988); Edwin T. Layton, Jr., "Innovation and Engineering Design: Max Jakob and Heat Transfer as a Case Study," *Innovation at the Crossroads between Science and Technology,* ed. Melvin Kranzberg, Y. Elkana, and Z. Tadmar (Haifa, 1989), pp. 132–52; "Theodore von Kármán Anniversary Issue," *Applied Mechanics* (1941), copy at Engineering Library, University of California at Berkeley.

[55]See "Morrough P. O'Brien: Dean of the College of Engineering, Pioneer in Coastal Engineering, and Consultant to General Electric," College of Engineering Oral History Series, University of California at Berkeley (1988), pp. 13–16, copy in Bancroft Library, University of California at Berkeley; and Frank Kreith, "Dean L. M. K. Boelter's Contribution to Heat Transfer as Seen through the Eyes of His Former Students," in Layton and Lienhard, eds., pp. 117–24. It is significant that almost all of the key figures in Edwin Layton's discussion of changes in engineering science during this period were European. See Edwin T. Layton, Jr., "The Dimensional Revolution: The New Relations between Theory and Experiment in Engineering in the Age of Michelson," in *The Michelson Era in American Science, 1870–1930,* ed. Stanley Goldberg and Roger Stuewer (New York, 1988), pp. 23–41.

mathematical analysis, seemed to consider Swain a symbol of backward American engineering practice. Yet Swain had adopted European approaches to structural analysis. According to noted engineer Hardy Cross, Swain was not opposed to these mathematical methods but to "the emphasis on methodology that they too frequently represented. He emphasized training in interpreting the data of science as distinguished from drill in the technique of assembling it." Swain and Timoshenko actually had much in common in the way they practiced engineering, but not all European theorists shared their interest in real-world problems. Eric Walker's experience at Harvard in the 1930s seems to support Swain's critique, for Walker considered the European immigrants on Harvard's engineering faculty theoreticians who disdained questions of practice or application.[56]

This dichotomy helps explain the Timoshenko/Swain confrontation and reflects the extreme poles of European and American engineering education. Significantly, most of the engineers who occupied a middle ground and used the new methods to improve engineering in the real world—men like Timoshenko, von Kármán, Boelter, and others—were European or trained in Europe. The most influential European engineering theorists remained committed to solving problems, to doing, however scientific their approach. They proved the utility of scientific engineering. But the deeply entrenched practical emphasis of American engineering colleges meant that new approach was adopted very slowly.

Change was especially slow on land-grant campuses. Ironically, most land-grant researchers would not have opposed an emphasis on fundamentals in either education or research. Indeed, the 1903 charter for the Illinois station read, "The Station is conducted as an institution of scientific research rather than as a commercial testing laboratory."[57] But rarely did research live up to this description. Few projects at stations equaled the quality of the aircraft propeller research of Frederick Durand and E. P. Lesley at Stanford from 1917 through 1930.[58] Not until the 1930s were a few stations pursuing investigations that involved more than routine tests. At Purdue,

[56]Timoshenko, p. 255. Hardy Cross authored the *Dictionary of American Biography* entry on Swain. Hardly a relic of the past, Walker became dean of engineering and then president of Penn State after 1945. See Walker, p. 107.

[57]Paine (n. 10 above); Marston Papers (n. 11 above), no. 172. See also the bulletins and circulars published by the two stations.

[58]This work was a model of engineering research. Durand and Lesley were less interested in theory than in helping aircraft designers choose efficient propellers, so results appeared in tabular form, although Durand was comfortable with theory. See Vincenti (n. 38 above), pp. 137–69; also Frederick E. Terman, "William Frederick Durand," *Biographical Memoirs* (Washington, D.C., 1976), 48:153–93.

researchers were working on television, while studies of metal fatigue, structural strength, material behavior, and structural design were conducted by Illinois's Department of Theoretical and Applied Mechanics. The shifting norms of research were evident by 1939, however, when outside reviewers severely criticized the College of Engineering at Texas A&M for the almost complete absence of basic research.[59]

Yet Texas A&M was hardly unique. Even at Illinois, one could question how much research in the Theoretical and Applied Mechanics Department (TAM) was really fundamental. The most noted TAM project during the 1930s was a study of failures in railroad rails. H. F. Moore began by placing rails in fatigue-testing and drop-testing machines. The results proved confusing; only the chance reading of a metallurgy report on the damaging effects of hydrogen in steel enabled Illinois engineers to move beyond *how* rail fatigue occurred to *why* it happened.[60]

Moore's approach to research was typical before 1940. According to Frederick Terman, later dean of engineering at Stanford, "what passed in academic institutions for 'research' was typically advanced testing . . . [and] was in most cases superficial, although here and there some significant work was carried on by an unusual professor."[61] Thus engineers at Texas A&M tested attic fans and those at Kansas State road materials, while researchers at Illinois tested furnaces, radiators, and concrete beams. Critics such as D. B. Keyes lamented this practical orientation: "The primary mistake that has been made in the past has been that industries have forced upon the universities the type of problem which is not fundamental in nature and which may be classified under the head of commercial research."[62] In short, even though it was widely advocated after 1890, greater attention to scientific fundamentals did not dominate American engineering campuses before 1940. There was some change, as European engineers helped prove the importance of engineering science to American engineers. But most American engineering colleges remained wedded to the traditional approach of practical education and research.

[59]Dethloff (n. 17 above), 2:478, 482. The report itself and related correspondence are available in Texas A&M file, Aubrey A. Potter Papers, box 4, Special Collections, Purdue University Archives, West Lafayette, Ind.

[60]See R. A. Kingery, R. D. Berg, and E. H. Schillinger, *Men and Ideas in Engineering; Twelve Histories from Illinois* (Chicago, 1967), pp. 54–65; also Baker and King (n. 1 above), pp. 408–9, 424–25.

[61]Frederick E. Terman, "A Brief History of Electrical Engineering Education," *Proceedings of the Institute of Electrical and Electronic Engineers* 64 (September 1976): 1399.

[62]D. B. Keyes, "Co-operative Chemical Engineering Research in the University," *Industrial Engineering and Chemistry* 24 (August 1932): 947–49, quoted in *Research in Industry*, ed. C. C. Furnas (New York, 1948), p. 484.

After World War II, however, practical engineering education finally gave way to the newer scientific engineering, even at land-grant colleges. Several factors help explain why the change occurred when it did, including the preparatory efforts of European theorists and the increasing sophistication of technology that required better scientific foundations. Another important element was the wartime experience of engineers who were chagrined at their limited role in advanced engineering projects. Stanford's Terman was acutely sensitive to this situation, complaining that engineers "had neither the fundamental knowledge required to think creatively about these new concepts, nor the research experience to carry through. Thus most of the great electrical developments were produced . . . by scientists, particularly physicists who had turned engineers for the duration." Terman urged renewed emphasis on fundamentals in engineering curricula.[63] There were other reasons to be sure, but of central importance to the adoption of the new style of engineering by academia was a massive volume of federal funding. The changes in research that followed proved crucial to the adoption of fundamental research and scientific engineering in mainstream American engineering schools after 1945.

Scientific Engineering Enters the Mainstream: Academic Research after World War II

During World War II, dramatic changes began in academic engineering research, initiated by the first appearance of significant federal funds. These were not evenly distributed since only the largest stations attracted military projects. At many land-grant campuses engineering faculty taught in the military training programs. At Texas A&M, more than 45,000 men passed through training programs, while only five research projects remained by 1943. This experience was typical, and even at Illinois training contracts in 1943 and 1944 were larger than research contracts.[64]

At a few stations, however, research continued, and Illinois was again the leader. But there were changes, for while a few industrial projects continued, all new studies were sponsored by the military.

[63]Terman.

[64]The Navy V-12 and Army Engineering-Science-Management War Training programs were the most important such efforts. See Dethloff, 2:456; also Texas Agricultural and Mechanical College, *Biennial Report* (1942–43), pp. 38–43. On the experience of other schools, see James E. Pollard, *The Bevis Administration,* pt. 1, *The University in a World at War, 1940–1945,* History of the Ohio State University, vol. 7 (Columbus, Ohio, 1967), pp. 86–116; Bezilla (n. 7 above), pp. 139–43; Knoll (n. 9 above), pp. 107–17; and University of Illinois, *Annual Report of the Board of Trustees* (annual statement of expendable gifts) (1942–45).

The Office of Scientific Research and Development (OSRD) alone provided $250,000, while contracts with the Office of Emergency Management, the War Production Board, the army, navy, and army air force, the surgeon general, and the National Advisory Committee for Aeronautics supported studies ranging from cathode-ray indicators and propeller blades to counter-sunk screws for aircraft and ceramic coatings for disposal systems. In 1944, OSRD contracts totaled $350,000 at Illinois, more than the total research budget of the station during any year in its history.[65]

While impressive by station standards, research at Illinois paled beside the programs at MIT, Johns Hopkins, Chicago, or Berkeley.[66] But station engineers liked what they saw, for even small military contracts were larger than most industrial projects. At Illinois, on average, industrial projects had totaled about $5,000 a year before 1940 while military contracts averaged $12,000–$25,000 or more. When federal funds were continued after the war, all stations wanted a share. But as in the past, not all stations gained access to these lucrative projects and Illinois moved first.

The Office of Naval Research (ONR) proved to be the most important research patron in the postwar era. An Illini alumnus with the ONR informed his alma mater of the navy's intentions, and engineering administrators spent 1945 preparing proposals that were funded beginning in 1946. Several continued for almost fifteen years.[67] Contracts with other federal agencies followed, bringing an avalanche of federal money to Illinois—$1,100,000 in 1946, compared to $150,000 in industrial contracts.

Illinois had found a strategy for growth that other stations attempted to follow.[68] Stanford was perhaps the most successful in tapping federal funding. More amazing, perhaps, was the growth of the Georgia Tech station from one of the smallest to one of the largest, thanks to such funds. Wartime studies of helicopter blades and microwaves opened the door for additional military contracts, and since 1950 Georgia Tech has been one of the largest military

[65]Information from project files in SF, 1909–61, ser. 11/2/1, EESR (n. 21 above); and from University of Illinois, *Report of the Board of Trustees* (1942–45).

[66]The MIT Radiation Laboratory alone received 35 percent of the $450 million allocated to the OSRD during the war. See Daniel Kevles, *The Physicists: The History of a Scientific Community in Modern America* (New York, 1978), p. 342.

[67]See correspondence in various files related to the ONR for 1945–46 but especially in ONR General Correspondence, 1945–54, in SF, 1909–61, ser. 11/2/1, EESR, Box 7.

[68]The following statistics are drawn from *Engineering Experiment Station Record,* vols. 20–31 (1940–51); and Engineering College Research Council of the American Society for Engineering Education, *Engineering College Research Review* (1951–59).

TABLE 2
SOURCES OF RESEARCH FUNDS FOR SELECTED ENGINEERING COLLEGES, 1945–59 ($ IN THOUSANDS)

Station	University Funds	Industry	Federal, Military	Federal, Nonmilitary	Other	Total
Illinois:						
1946	172	N.A.	N.A.	N.A.	1,150	1,322
1952	N.A.	N.A.	N.A.	N.A.	3,080	3,080
1955	N.A.	473	3,156	150	627	4,406
1959	N.A.	403	4,384	1,128	1,447	7,362
Purdue:						
1945	17	N.A.	N.A.	N.A.	499	516
1951	N.A.	75	550	125	130	880
1955	N.A.	200	600	50	225	1,075
1959	N.A.	N.A.	350	880	470	1,700
Georgia Tech:						
1945	70	N.A.	N.A.	N.A.	161	231
1951	N.A.	N.A.	1,221*	N.A.	161	1,382
1955	N.A.	N.A.	2,100*	N.A.	150	2,250
1959	N.A.	240	1,900	230	500	2,870
Ohio State:						
1945	N.A.	N.A.	N.A.	N.A.	53	53
1951	69†	113	149	10	...	341
1955	N.A.	112	130	48	143	433
1959	N.A.	260	93	139	219	711
Texas A&M:						
1946	80	N.A.	N.A.	N.A.	144	224
1950	245	N.A.	N.A.	N.A.	181	426
1955	249†	32	10	13	...	304
1959	1,000	427	N.A.	128	...	1,555

SOURCES.—*Engineering Experiment Station Record,* vol. 25 (January 1945); and Engineering College Research Council, *Research Review* (1951, 1955, 1959).
NOTE.—N.A. = not available; ellipses indicate that there are no funds in that category.
*Includes government and industrial funds.
†Legislative appropriation.

research contractors among American universities.[69] By 1955–56, $8.2 million of the $14.4 million spent by thirty-two engineering experiment stations came from military sources, with another $947,000 from nonmilitary federal agencies. (See table 2.)

In short, federal monies rapidly eclipsed traditional sources of experiment station research funds—state government, university, and trade associations. Few engineers seemed concerned about the implications of this change. A few raised questions about conducting so much classified research on campus; others worried about overreli-

[69]Stuart W. Leslie, "Playing the Education Game to Win: The Military and Interdisciplinary Research at Stanford," *Historical Studies in the Physical Sciences* 18 (January 1987): 56–88; also McMath et al. (n. 2 above), pp. 212–17, 256–70.

ance on a single source of research revenue. But perhaps the most far-reaching effect of government money was the way it transformed the goals of academic researchers. Federal contracts favored fundamental research, encouraging university engineers to move away from their practical emphasis. Once that shift began, engineering education followed suit. The greater emphasis on basic research and then on scientific fundamentals that leading engineering educators had long advocated finally came to pass.

Vannevar Bush, wartime head of the OSRD, is often considered the architect of the change in federal research policy. But the steps he urged were not original; indeed his views of the relationship between research and social progress contained elements of traditional Progressive assumptions about engineering research. Bush adopted the view of engineering-as-applied-science long used by many academic engineers to buttress calls for engineering curricula stressing fundamentals.[70] Bush's main contribution was his study, *Science: The Endless Frontier,* which laid out a rationale and a mechanism for federal funding of fundamental studies. Moreover, he influenced government officials concerned with postwar research, including Admiral Harold G. Bowen, head of the ONR. As the assistant secretary of the navy put it at the dedication of a laboratory at Penn State in 1949, "The last war proved conclusively that it is not possible to conduct basic research during hostilities and to convert knowledge gained thereby into weapons soon enough to have a decisive effect." By August 1946, the ONR had signed 177 contracts with eighty-one universities or private laboratories, and most involved fundamental research.[71]

Few engineering researchers, even those who had conducted practical research before the war, opposed the shift to fundamental studies. For many engineers, the new research strategy conformed to long-held beliefs. W. D. Coolidge of General Electric, for example, worried during the war about the increasingly applied focus of university research. Boris Bakhmeteff, chair of the Panel on a National Research Foundation for the Joint Engineer's Council, echoed Bush's fears about lapses in basic knowledge and added in 1946, "In no branch of fundamental research is the void more acute and pressing than in the realm of basic engineering science." And in

[70]I appreciate the suggestion of a referee about clarifying this, especially the point that a study by the National Resources Planning Board, *Research: A National Resource* (Washington, D.C., 1940), demonstrated the antecedents of Bush's proposal.

[71]Quotation from Bezilla, p. 155. See Vannevar Bush, *Science: The Endless Frontier* (1945; reprint, Washington, D.C., 1960), pp. 5–22; Kevles, pp. 348–355; and Harold Sapolsky, *The History of the Office of Naval Research* (Princeton, N.J., 1990).

1948, F. M. Dawson, dean of engineering at the University of Iowa, wrote that "we must use extreme caution before we enter into projects which do not involve 'fundamental' studies. We must give every emphasis we can . . . to the achievements of fundamental engineering research."[72]

Such sentiments were not new. The difference this time was that academic engineers had access to funding for exactly this type of scientific engineering research. Moreover, changes occurred in both land-grant colleges and the elite schools. At Illinois the flagship projects of the 1930s—furnace studies, rail and car-wheel behavior, and materials testing for utilities—received no notice after the war. Instead, studies funded by the ONR took precedence, including a betatron in the physics department, advanced studies of structural behavior in the Department of Theoretical and Applied Mechanics, the behavior of fluid streams in chemical engineering, and jet propulsion in aeronautical engineering. Other engineers applied theory to welding or studied traveling wave tubes or computers. Federal funds supported all of this work. Indeed, military dollars built the facilities for basic research, including the Control Systems Laboratory opened in 1951 at the request of the Defense Department. The head of the physics department oversaw a classified research program that totaled $2.7 million during its first two years of operation. Reorganized as the Coordinated Sciences Laboratory in 1959, the lab remained heavily oriented toward government contracts in computers, plasma, atmospheric and surface physics, and space science. Other research programs at Illinois included a radio telescope, funded in 1958 by the ONR ($323,000), and a nuclear reactor, partially supported by the Atomic Energy Commission in 1958 ($150,000).[73]

Other stations also sought large contracts for fundamental investigations. At Georgia Tech, the key projects no longer concerned turpentine and textiles, but high-speed aerodynamics, microwave acoustics, ultra-high-frequency interference, underwater acoustics,

[72]W. D. Coolidge, "The Role of Science Institutions in Our Civilization," *Science* 96 (November 6, 1942): 412–17, quoted in Furnas, ed. (n. 62 above), p. 484; Boris Bakhmeteff, "Science and Engineering," *Civil Engineering* 16 (March 1946): 99; F. M. Dawson, "The Role of Fundamental Research," *Journal of Engineering Education* 39 (December 1948): 195.

[73]Information from files on the ONR in SF, 1909–61, ser. 11/2/1, EESR, Box 7; and University of Illinois, *Report of the Board of Trustees,* vol. 44 (1946–48), see esp. (May 29, 1946), pp. 1103–4; 47 (September 23, 1953): 1087–89; 49 (1956–58): 1240; 50 (1958–60): 190, 572; and "The Coordinated Science Laboratory: Some Historical Highlights," March 10, 1964, in Engineering—Coordinated Science Laboratory, file 1964–67, ser. 11/30/05, University of Illinois Archives.

and cosmic radiation. Not all of this research was as theoretical as it sounded, for engineers knew how to label their work to enhance success with funding agencies.[74] But it was also true that research with obvious applications relied far more on science and required a much stronger theoretical bent than most prewar studies.[75] The author of a history of engineering at Purdue noted the change almost plaintively: "The tendency seemed to be toward such far-out matters as 'a theoretical study of the scattering of electrical waves by perfectly reflecting bodies,' and when . . . the engineering editor tried to get pictures to illustrate reports on research in progress, he was sometimes told there was nothing to photograph unless he was willing to photograph an equation."[76]

Federal funding also encouraged a transformation of fields that had not previously built a basis in scientific methodologies. It was not surprising that Illinois's aeronautical engineers launched a series of theoretical investigations for the air force's experimental center at Wright-Patterson Air Base; American aviation began moving that way under von Kármán's influence in the 1930s. But ceramic engineers also became heavily involved in theoretical studies for the air force on high-temperature materials, while electrical engineers worked on antenna projects.[77] Not every school moved this quickly; indeed, the pace of change depended on a variety of local considerations at every university. But Illinois set a standard that others hoped to emulate.

The connection between the shift to scientific engineering and federal funding stands out when examining how research grew in engineering schools and where. A pattern appears: Only schools that embraced scientific engineering received large federal projects, and only schools with federal funding developed large research programs. Engineers at Texas A&M, for example, who still conducted practical studies, performed very little federally sponsored research. The same pattern held at many similar land-grant schools, including Kansas

[74]I thank Edwin Layton for stressing this point in comments about an earlier version of this article.

[75]See McMath et al., pp. 256–70; and also correspondence with August Giebelhaus, one of the coauthors.

[76]Knoll (n. 9 above), pp. 137–38. Leslie (n. 69 above) stresses that at Stanford the new style of engineering research actually focused on areas in the gaps between science and engineering, so that the distinctions between basic and applied research were increasingly blurred. This observation describes developments at many stations as well, but for this article, it is convenient to retain the terms applied and basic to indicate the contrasts between prewar and postwar research.

[77]Information from files of government research contracts in SF, 1909–61, ser. 11/2/1, EESR, Box 7 (n. 21 above).

State and Washington State at Pullman. Without federal contracts, research budgets were limited.[78]

By the mid-1950s, academic engineering administrators understood this linkage. Some engineering colleges abolished or renamed their experiment stations to shed the stigma of a history of applied research. Forty-six stations existed in 1945, but there were only about thirty in 1960. Several had become research institutes (Georgia Tech and Berkeley); Maine's had become a Technology Experiment Station; several colleges turned stations into research administration offices; and a few downgraded them to departments or divisions of research. Penn State followed this last path because "the term 'experiment station' connoted testing activities, when in reality testing of materials had taken on secondary importance in favor of more scientific studies."[79] Indeed, the pendulum swung so far that the chair of the Engineering College Research Council became concerned when the Aircraft Industries Association proposed to the secretary of defense in 1954 that universities should conduct only basic research, with aircraft builders handling all applied studies.[80]

As this transformation in engineering research took place, parallel changes followed in the classroom. As Frederick Terman put it in a 1955 essay in the Institution of Radio Engineers *Student Quarterly*, "Electrical Engineers Are Going Back to Science!"[81] The critical point here is the timing of the changes. Many engineers had endorsed an education with a heavy foundation in science and math, but only after research had installed the new directions did most colleges implement such curricula. More was at work than the desire of engineers like Terman, who wanted electrical engineers to compete with physicists. To account for a shift that swept through every engineering school in the two decades after the war, one also must recognize the impetus provided by federal research funds.

That curricula changed after World War II is not a matter of dispute. But it is difficult to generalize about the experience of

[78]See table 2 for Texas A&M's budget; information largely from the author's interviews with Calhoun and Fred J. Benson, one of Calhoun's successors; budget information drawn from Engineering College Research Council, *Research Review* (1951–61).

[79]Bezilla (n. 7 above), p. 193; other information from Engineering College Research Council, *Research Review* (1951–61); and John J. Desmond, "The Engineering Experiment Station and the Community," in *Engineering Colleges, Legislatures and the Community*, ed. George Bugliarello (New York, ca. 1978), p. 18.

[80]See correspondence between Eric A. Walker and the secretary of defense and Engineering College Research Council members, December 12, 1953, and March 1, 1954, Engineering College Research Council, SF, 1909–61, ser. 11/2/1, EESR, Box 4.

[81]Bezilla, pp. 172–74; comment of McMahon in Bryson (n. 44 above), p. 73.

American engineering colleges. The Engineers Council for Professional Development, an accrediting agency organized in the late 1930s, produced some comparability, not uniformity, among curricula for similar degrees. With that caveat in mind, an examination of the civil engineering curricula at Texas A&M demonstrates the transformation in the classroom. Importantly, Texas A&M was not considered a progressive institution in 1945. Perhaps for that reason, several points emerge clearly in tables 3–8.

TABLE 3
TEXAS A&M UNIVERSITY TECHNICAL CURRICULA, SOPHOMORE THROUGH SENIOR YEARS
CIVIL ENGINEERING, 1920

Classes	Hours Theory/Practice
Fundamental science courses:	
Sophomore year:	
Physics	6, 6
Calculus	10, 0
Junior year:	
General Geology	3, 3
Other technical courses:	
Sophomore year:	
Elementary Steam Engineering	2, 0
Junior year:	
Electrical Machinery	5, 3
Mechanics of Materials	3, 2
Drawing courses:	
Sophomore year:	
Mechanical Drawing	0, 4
Junior year:	
Civil Engineering Graphics	0, 3
Topographical Drawing	0, 2
Senior year:	
Railroad Drafting	0, 4
Civil engineering courses:	
Sophomore year:	
Surveying	3, 3
Railroad Engineering	2, 3
Analytical Mechanics	3, 0
Summer Field Practice (three weeks)	...
Junior year:	
Railroad Engineering	3, 3
Mechanics of Materials	3, 2
Hydraulics	3, 2
Railroad Construction	2, 0
Roofs and Bridges	3, 0
Strength of Materials	2, 0
Summer Field Practice (three weeks)	...

TABLE 3 (*continued*)

Classes	Hours Theory/Practice
General civil engineering courses:	
Senior year:	
Roofs and Bridges	3, 6
Roads and Pavements	3, 0
Materials on Construction	0, 3
Elements of Reinforced Concrete	2, 0
Bridge Design	0, 6
Contracts and Specifications	2, 0
Masonry	2, 0
Water Supply and Sewerage	3, 0
Irrigation and Drainage	2, 0
Electives (one is selected):	
Engineering Geology	2, 2
Highway Laws and Economics	3, 0
Water Bacteriology	2, 4
Highway and municipal engineering courses:	
Senior year:	
Highway Construction and Maintenance	5, 0
Highway Materials	2, 6
Elements of Reinforced Concrete	2, 0
Bridge Design	1, 3
Contracts and Specifications	2, 0
Masonry	2, 0
Water Supply and Sewage	3, 0
Bacteriology	2, 4
Electives (one is selected):	
Engineering Geology	2, 2
Highway Laws and Economics	3, 0
Water Treatment	1, 3

SOURCE.—Texas A&M College, *Course Catalog, 1920–21* (College Station, Tex., 1920).

First, although the most obvious changes became apparent in 1960 and 1970, movement could be seen earlier. Between 1940 and 1960, calculus and introductory engineering mechanics were slowly moved forward in the curricula. While not shown in the tables, calculus moved to the freshman year, while mechanics moved to the sophomore year. Second, engineering science courses displaced several traditional courses. Classes in specific subjects, such as bridge structures and roads and pavements, started disappearing, as did practical laboratory classes. Engineering Mechanics became a three- and four-hour course, not two three-hour classes. Thermodynamics replaced Steam and Gas Power; Electric Circuits and Machinery had appeared

TABLE 4

Texas A&M University Technical Curricula, Sophomore through Senior Years
Civil Engineering, 1930

Classes	Hours Theory/Practice
Fundamental science courses:	
Sophomore year:	
Physics	6, 6
Calculus	10, 0
Junior year:	
General Geology	3, 2
Other technical courses:	
Sophomore year:	
Elementary Steam Engineering	2, 0
Junior year:	
Electrical Machinery	3, 3
Steam and Gas Power	3, 0
Senior year:	
Engineering Administration	2, 0
Drawing courses:	
Sophomore year:	
Mechanical Drawing	0, 4
Junior year:	
Estimating and Drafting	0, 4
Structural Drafting	0, 4
Civil engineering courses:	
Sophomore year:	
Surveying	3, 3
Railroad Engineering	3, 3
Analytical Mechanics	3, 0
Summer Field Practice (six weeks)	...
Junior year:	
Railroad Surveying	0, 3
Mechanics of Materials	3, 0
Material Laboratory	0, 2
Hydraulics	3, 0
Hydraulics Laboratory	0, 2
Analytic Mechanics	3, 0
Railroad Construction	2, 0
Elementary Structural Analysis	3, 0
Masonry	2, 0
Senior year:	
Roads and Pavements	3, 0
Materials of Construction	0, 4
Elements of Reinforced Concrete	2, 0
Water Supply and Purification	3, 0
Sewerage and Sewage Disposal	3, 0
Structures	2, 4
Reinforced Concrete Design	2, 3

TABLE 4 (*continued*)

Classes	Hours Theory/Practice
Twelve hours technical electives chosen from:	
Chemical Testing, Water and Sewage	2, 3
Bituminous Materials	2, 3
Engineering Economics	3, 0
Structural Design	0, 6
Sanitary Design	0, 6
Water Bacteriology	1, 4
Irrigation and Drainage	3, 0
Highway Administration	3, 0
Structural Engineering	3, 0
Sanitation and Public Health	3, 0
Municipal Administration	3, 0

SOURCE.—Texas A&M College, *Course Catalog, 1930–31* (College Station, Tex., 1930).

by 1960. Soil Mechanics and Foundations replaced Masonry and Foundations. Theory of Structures and Hydraulics of Drainage Structures also appeared. Even when content remained similar, the course name now reflected the emphasis on engineering science. By 1970, the shift was completed. Upper-level drafting courses were gone and fluid dynamics, differential equations, numerical methods, materials science, indeterminate structures, and systems design were in the curriculum.

But more than course titles were changed. At many schools, new academic programs appeared. Several colleges created degrees in engineering science, almost always after the war. Illinois started a degree in engineering physics in the early 1940s, taking advantage of the fact that the physics department was in the college of engineering. Cornell created a program in engineering science in 1946, as did Stanford. By 1959, at least three additional schools offered such curricula, while seven schools offered engineering physics. All limited these programs to their brightest students. Penn State, for example, created an honors course in engineering science in 1953–54.[82]

In 1955, another study of engineering education appeared that both confirmed and accelerated the diffusion of scientific tendencies within American engineering schools. The Grinter Report strongly urged greater attention to fundamental science and math courses, arguing that such courses prepared engineers for rapid technological

[82]Bezilla, pp. 172–74.

TABLE 5

TEXAS A&M UNIVERSITY TECHNICAL CURRICULA, SOPHOMORE THROUGH SENIOR YEARS
CIVIL ENGINEERING, 1940

Classes	Credit Hours	Lecture-Laboratory Hours
Fundamental science courses:		
Sophomore year:		
Physics	10	4-3, 4-3
Calculus	8	4-0, 4-0
Junior year:		
Geology for Civil Engineers	4	3-3
Other technical courses:		
Sophomore year:		
Engineering Mechanics	4	4-0
Junior year:		
Electrical Machinery	4	3-3
Steam and Gas Power	3	3-0
Drawing courses:		
Junior year:		
Estimating and Drafting	1	0-4
Structural Drafting	1	0-3
Civil engineering courses:		
Sophomore year:		
Surveying	4	3-3
Railroad and Highway Engineering	4	2-6
Summer Field Practice (six weeks)	...	...
Junior year:		
Mechanics of Materials	4	4-0
Materials Laboratory	1	0-2
Hydraulics	3	3-0
Hydraulics Laboratory	1	0-2
Structural Analysis and Design	4	3-3
Masonry and Foundations	3	2-2
Senior year:		
Roads and Pavements	3	3-0
Materials of Construction	1	0-4
Reinforced Concrete	2	2-0
Reinforced Concrete Design	3	2-3
Steel Bridge Structures	3	2-3
Sewerage and Sewage Disposal	3	3-0
Professional Relations	2	2-0
Water Supply and Purification	3	3-0
Twelve hours technical electives chosen from:		
Bituminous Materials	3	2-3
Engineering Economy	3	3-0
Steel Building Structures	3	2-3
Sanitary Design	3	1-5
Sanitation and Public Health	3	3-0
Highway Administration and Design	3	2-3

TABLE 5 (*continued*)

Classes	Credit Hours	Lecture-Laboratory Hours
Hydraulic Engineering	3	3-0
Statically Indeterminate Structures	3	3-0
Sanitary Laboratory	3	1-5
Municipal Administration	3	3-0

SOURCE.—Texas A&M College, *Course Catalog, 1940–41* (College Station, Tex., 1940).

change better than courses in specific areas of engineering and design. This time the recommendations were not ignored. By the mid-1960s, engineering science curricula were in place in engineering schools across the country.[83] In fact, the report's proposal for a two-tiered system of engineering education, in which most schools would train engineers for industry, while a few produced graduate engineering scientists for government research, was stillborn. No college wanted to be labeled "second class" or to miss out on federal research funds. Symbolizing the embrace of change was the abolition in 1960 of Purdue's summer surveying camp, a hallmark of its civil engineering program since 1914.[84] Utilizing the lever of federal funding for basic research, leading academic engineers finally transformed engineering education in ways they had long advocated.

The Social Dimension of the Rise of Engineering Science

Several broad conclusions emerge from this examination of the historical development of engineering research on American college campuses. First, although engineering schools usually have been viewed as the home of a theoretical and scientific "school culture" that replaced an apprentice-style "shop culture" in the late 19th century, engineering schools remained decidedly practical in orientation well into the 20th century. It was not that engineers opposed emphasizing theoretical and general studies of engineering and scientific principles, especially after 1900. But movement was slow, and when change came it appeared from the newer fields, beginning with electrical and

[83]On the 1955 report and its impact, see Grinter (n. 4 above); American Society for Engineering Education, *Goals of Engineering Education: Final Report of the Goals Committee* (Washington, D.C., 1968); Nathan W. Dougherty, "Foundation for Our Future," *Journal of Engineering Education* 58 (May 1968): 1019–31; Bryson; Kline (n. 40 above), pp. 38–39. Eric Walker claimed that not until this report "was enough emphasis put on science in the engineering curriculum." See Walker (n. 51 above), p. 106.

[84]Knoll, pp. 251–54, 260–61.

TABLE 6

Texas A&M University Technical Curricula, Sophomore through Senior Years

Civil Engineering, 1950

Classes	Credit Hours	Lecture-Laboratory Hours
Fundamental science courses:		
Sophomore year:		
Physics	10	4-3, 4-3
Calculus	8	4-0, 4-0
Junior year:		
Geology for Civil Engineers	3	2-2
Other technical courses:		
Sophomore year:		
Engineering Mechanics	3	3-0
Junior year:		
Steam and Gas Power	3	3-0
Senior year:		
Electrical Machinery	4	3-3
Civil engineering courses:		
Sophomore year:		
Plane Surveying	4	3-3
Advanced Surveying	4	2-6
Summer Field Practice (six weeks)	...	...
Junior year:		
Mechanics of Materials	4	4-0
Strength of Materials Laboratory	1	0-2
Hydraulics	3	3-0
Hydraulics Laboratory	1	0-2
Elementary Hydrology	2	2-0
Plain and Reinforced Concrete	3	2-3
Design of Members and Connections	3	2-3
Water and Sewage Treatment	3	2-3
Soil Mechanics and Foundations	3	3-0
Senior year:		
Roads and Pavements	3	3-0
Materials of Construction	1	0-4
Analysis and Design of Structures	3	2-3
Cost Estimating	3	3-0
Water Supply and Sewage Practice	3	3-0
Engineering Economy	2	2-0
Contracts and Specifications	2	2-0
Seminar	1	1-0
Six hours technical electives chosen from:		
Bituminous Materials	3	2-3
Bridge Design	3	2-3
Sanitary Design	3	2-4
Sanitation and Public Health	3	3-0
Sanitary Laboratory	3	2-4
Municipal Administration	3	3-0

TABLE 6 (*continued*)

Classes	Credit Hours	Lecture-Laboratory Hours
Hydraulic Engineering	3	3-0
Construction Plant and Methods (a sanitary option was available)	3	3-0
Statically Indeterminate Structures	3	2-3
Highway Design	3	2-3
Hydrology	3	3-0
Building Design	3	2-3
Aerial Photogrammetry	3	2-3
Advanced Engineering Geology	4	4-4

SOURCE.—Texas A&M College, *Course Catalog, 1950–51* (College Station, Tex., 1950).

chemical engineering and continuing through aeronautical and radio engineering. Only at a handful of private colleges and state universities had emphasis on the scientific and mathematical fundamentals become a central concern by the 1930s. More general change came to most schools and to the traditional fields after World War II.

One consequence of the new style of academic research, funded by a new patron, was the disruption of the pattern of practical research and testing for industry. The connections between American industries and engineering colleges had a long history, and, more than any other factor, this probably accounts for the slow pace of change in both engineering education and research from 1900 through 1940. At many schools, industrial leaders were consulted about curricular reforms, while alumni in industry also had input on programs. Engineering educators felt steady pressure to produce engineers ready to step into technical positions. This connection with industry has continued into Accreditation Board for Engineering and Technology accreditation programs. Certainly this feature largely distinguished American schools from the European institutions attended by Timoshenko and von Kármán. Facing a much weaker industrial demand for practicality, German and French engineering schools had always been more theoretical in approach. In the United States, however, only federal patronage overwhelmed the traditional sources of research funding and opened the possibility of an education similar to that available in Europe.

The results of this change soon became apparent in the American engineering profession. For the first time, American academic engineers and those in industry had significantly different expectations about education and research. No longer were engineering faculty expected to

TABLE 7

Texas A&M University Technical Curricula, Sophomore through Senior Years
Civil Engineering, 1960

Classes	Credit Hours	Lecture-Laboratory Hours
Fundamental science courses:		
Sophomore year:		
Physics	8	3-3, 3-3
Calculus	6	3-0, 3-0
Junior year:		
Geology for Civil Engineers	3	2-3
Other technical courses:		
Sophomore year:		
Engineering Mechanics	3	3-0
Junior year:		
Engineering Mechanics	3	3-0
Thermodynamics	3	3-0
Senior year:		
Electrical Circuits and Machinery	4	3-3
Civil engineering courses:		
Sophomore year:		
Plane Surveying	4	3-3
Mechanics of Materials	3	3-0
Strength of Materials Laboratory	1	0-2
Summer Field Practice (six weeks)	...	...
Junior year:		
Mechanics of Materials	2	2-0
Hydraulics	3	3-0
Hydraulics Laboratory	1	0-2
Theory of Structures	3	2-3
Hydraulics of Drainage Structures	2	2-0
Plain and Reinforced Concrete	4	3-3
Design of Members and Connections	3	2-3
Soil Mechanics and Foundations	3	2-2
Senior year:		
Water and Sewage Treatment	3	2-2
Highway Engineering	3	3-0
Materials of Construction	2	1-3
Analysis and Design of Structures	3	2-3
Cost Estimating	3	3-0
Water Supply and Sewage Practice	3	2-2
Engineering Economy	2	2-0
Seminar	1	1-0
Six hours technical electives chosen from:		
Bituminous Materials	3	2-3
Sanitation and Public Health	3	3-0
Sanitary Design	3	2-3
Hydrology	3	3-0
Aerial Photogrammetry	3	2-3

TABLE 7 (*continued*)

Classes	Credit Hours	Lecture-Laboratory Hours
Construction Plant and Methods	3	3-0
Statically Indeterminate Structures	3	2-3
Traffic Engineering	3	3-0
Municipal Administration	3	3-0
Hydraulic Engineering	3	3-0
Advanced Engineering Geology	4	4-4

SOURCE.—Texas A&M University, *Course Catalog, 1960–61* (College Station, Tex., 1960).

have practical experience in industry. And the most valued faculty were not designers but theoretically oriented researchers, who published papers in academically oriented research journals. Eventually, academic and industrial engineers developed different conceptions of engineering, and they almost stopped talking to each other. By the 1960s, practicing engineers routinely complained to professional societies about the limited utility of much academic research. As a result of changes in research funding and practice, two subcultures had appeared in engineering.[85]

Perhaps the most important consequence of this bifurcation in the engineering profession was its effect on the relationship of engineering and science. Historians have come to understand engineering science as linking the spectrum of knowledge that ranges from knowing (science) to doing (engineering). While the methods may be largely the same as science, the values place stress on design and doing. But the changes in research encouraged academic engineers to place more emphasis on the former and shifted their values toward those of scientists. This does not mean that they became scientists, for most engineering scientists continued to pursue their work differently. Sometimes the difference was in the approach taken to the work; other times the field itself indicated the distinction. Usually one can at least glimpse a connection to the world of things, of doing and making—in other words, a link to engineering. This is the point made by aeronautical engineer Walter Vincenti in his recent book, *What*

[85]Terry Reynolds helped me see this point, based on his work with the American Institute of Chemical Engineers. See Terry Reynolds, *Seventy-five Years of Progress: A History of the American Institute of Chemical Engineers* (New York, 1983), pp. 62–63. Examples of complaints from practicing engineers about academic research and publications can be found in every professional engineering society after 1960. Just one example, delightfully titled, is John Huston, "Stop the World—I Want to Get Off!" *Civil Engineering* 51 (December 1981): 66–67.

TABLE 8

TEXAS A&M UNIVERSITY TECHNICAL CURRICULA, SOPHOMORE THROUGH SENIOR YEARS
CIVIL ENGINEERING, 1970

Classes	Credit Hours	Lecture-Laboratory Hours
Fundamental science courses:		
Sophomore year:		
Physics	7	3-0, 4-3
Physics Laboratory	1	0-3
Calculus	3	3-0
Differential Equations	3	3-0
Junior year:		
Geology for Civil Engineers	3	2-3
Other technical courses:		
Sophomore year:		
Engineering Mechanics 2	3	3-0
Junior year:		
Materials Science (Aero)	3	2-3
Numerical Methods (Aero)	3	2-3
Senior year:		
Thermodynamics	3	3-0
Electrical Circuits and Machinery	4	3-3
Civil engineering courses:		
Sophomore year:		
Plane Surveying	4	3-3
Mechanics of Materials	3	3-0
Summer Field Practice	...	...
Junior year:		
Engineering Mechanics of Materials	2	2-0
Fluid Dynamics	3	3-0
Fluid Dynamics Laboratory	1	0-2
Theory of Structures	3	3-0
Hydraulics of Drainage Structures	2	2-0
Design of Members and Connections	3	2-3
Soil Mechanics and Foundations	3	2-2
Water and Sewage Treatment	3	2-2
Highway Engineering	3	3-0
Water Resources Engineering	2	2-0
Senior year:		
Materials of Construction	3	2-3
Water Supply and Sewage Practice	3	2-2
Reinforced Concrete Structures	4	3-3
Systems Design	3	2-3
Seminar	1	1-0
Technical electives*	3	

SOURCE.—Texas A&M University, *Course Catalog, 1970–71* (College Station, Tex., 1970).

*During senior year, three hours of technical electives are chosen from almost any 300- and 400-level course in civil engineering.

Engineers Know and How They Know It. Vincenti notes that, while theoretical research in engineering can be almost indistinguishable from science, distinctions may still be seen: "Theoretical research as an activity in engineering differs from that in science probably less than does the resulting knowledge, which is noticeably more slanted toward the requirements of design. The differences in process, however, are not insignificant."[86]

Yet the difference between engineering and science is very tenuous at this level. Eda Kranakis has taken an approach that examines, not the distinctions, but the overlap between the two areas, finding that often the only differences between engineers and scientists are what they call themselves.[87] This situation emerged when academic engineering researchers embraced not just the methods but also the orientation and goals of science. Some became far more concerned with knowing and were eager to be judged, as were their scientific colleagues, by the elegance of their theories, not by applicability to practice. Eric Walker found this situation at Harvard even in the 1930s among his European professors: "Their [his European professors'] sacred cow was 'engineering science'—meaning theoretical analysis regardless of whether it could be applied. . . . As for applications, the general attitude among these European superstars was 'That's not our department.' "[88] More recently, academic engineers have adopted the reward structures of science, seeing the publication of scientific papers (i.e., knowledge) as the basis of professional advancement. All this suggests that changes in the nature of academic engineering research have erased any clear boundaries between science and engineering.[89]

One final observation concerns the development of engineering science. Usually, historical scholarship has presented the growth of this area of knowledge as an internalist phenomenon, a response to

[86]Edwin Layton, personal correspondence and discussion with the author; Vincenti (n. 38 above), p. 232.

[87]Eda Kranakis, "Science and Technology as Intersecting Socio-cognitive Worlds" (paper presented at the conference on Technological Development and Science in the 19th and 20th Centuries, University of Technology, Eindhoven, the Netherlands, November 6–9, 1991).

[88]Walker, p. 107.

[89]Layton might not go this far, but he does note that the movement of engineering scientists toward science has had an impact on engineering: "There is now a tension between science and design in modern engineering that would have been unimaginable a century ago." See Layton, "The Dimensional Revolution" (n. 55 above), p. 29.

changing technology. Layton stated the argument clearly in a discussion of the development of chemical engineering: "One of the most critical things driving chemical technology to this more scientific and theoretical approach was the changing needs of chemical technology."[90] To be sure, this factor is among the most important, but those internal demands by themselves do not seem sufficient to explain all the changes discussed here. If they were, one might have expected the arguments of leading academic engineers to have started the transition much sooner. It is the lag that is disconcerting.

A fuller explanation emerges by considering the various social factors that are involved in this story. For example, experiment stations were first shaped by the land-grant ideal of service and the Progressive ideal of disinterested expertise, a combination that created a pattern of public service and industrially oriented applied research. Only after World War II did another rationale that harnessed science and engineering to the service of the military reverse those older goals for research. Similarly, both the resistance to change because of the traditional close links between engineering schools and industry and the impetus to change from European engineers hinge more on social factors than on technical questions; they have little to do with the internal demands of engineering. Another important social consideration was the status concerns of engineers vis-à-vis scientists. From the formation of engineering experiment stations through the efforts of Frederick Terman, engineers often seemed driven to achieve status and funding equal to that given scientists. In short, looking beyond the laboratory provides a far more complicated view of the development of engineering research and education in this country and reveals extensive interaction between technology and culture.

[90]Edwin T. Layton, Jr., "Through the Looking Glass, or News from Lake Mirror Image," *Technology and Culture* 28 (July 1987): 606; see also James Kip Finch, *The Story of Engineering* (Garden City, N.Y., 1960), pp. 387–88.

Technology in the Seamless Web: "Success" and "Failure" in the History of the Electron Microscope

GREGORY C. KUNKLE

The microscope, rivaled only perhaps by a white lab coat, has assumed a status as *the* symbol of science in the modern world. The mere image of a generic microscope conjures up mental pictures of a scientist hard at work in a laboratory uncovering the mysteries of the universe. Indeed, increasing the power of the microscope has, by implication, been tantamount to increasing the power and knowledge of scientists. In this scenario, the most powerful microscope, the state-of-the-art electron microscope, unlocks nature's secrets by enabling scientists to view erstwhile hidden parts of the universe through magnification levels on the order of 200,000 times an object's actual size. Such microscopes, it could be argued, are the leading tools with which scientists push forward the frontiers of knowledge and reveal the secrets of the molecular world. In such a view, the objective scientist is empowered by his or her instrument, which has been technically determined by factors quite distant and distinct from social forces with which, say, an office manager or a teacher has to contend.

Yet the history of the electron microscope, particularly its commercial development, reveals quite a different and more complex story—one that supports the idea that, while technology obviously has an impact on society, social factors, in their turn, have as forceful an impact on technology. In examining the early history of commercial development of the electron microscope at RCA and General Electric (GE), one can glean general insights into the relationship between social forces and the pathways of technological development.[1]

DR. KUNKLE is an assistant professor of Science, Technology and Society at Rochester Institute of Technology. He wishes to thank Dr. Charles Lyman and the Electron Microscopy Society of America for their support and also Professor Stephen Cutcliffe, John Smith, and Roger Simon for their helpful suggestions and criticisms.

[1]For other insights into the role of social factors in the development of technological artifacts, see David Noble, "Social Choice and Machine Design: The Case of Automatically Controlled Machine Tools," in *Case Studies on the Labor Process*, ed. A. Zimbalist (New York,

Early in the development of the electron microscope, a question arose concerning two different types of lens systems. It was not at all obvious whether the *electromagnetic* lens, on the one hand, or an *electrostatic* design, on the other, would prove to be superior. Buried within this choice between two technologies are lessons in how technology and science function within the larger contexts of industry, communities of practitioners, and society at large. The question of technological choice[2] is instructively revealed on examining the development of the electromagnetic and electrostatic electron microscopes in the international arena. Such a comparative analysis is suggestive of those factors that are most significant in shaping the course of technological development. Further, the efforts of two American firms, RCA and GE, to develop and market an instrument provide an interesting look at the relationship between instrument makers and the research community.

In the United States during the 1940s and 1950s, GE made two unsuccessful attempts to market electron microscopes with electrostatic lenses. In the same period, RCA achieved success in producing and selling electron microscopes with electromagnetic lenses. It has generally been accepted by scientists involved in electron microscopy that GE's lack of success was due to the inherent technical inferiority of the electrostatic lens.[3] A closer examination of the electron microscope, however, including developments in the international setting—notably in Germany and Japan—suggests that this assumed technical inferiority was not necessarily the cause for its commercial failure. Contemporary

1979); and Ruth Schwartz Cowan's discussion of the refrigerator in *More Work for Mother: The Ironies of Household Technology from the Open Hearth to the Microwave* (New York, 1983), pp. 127–50.

[2]The word "selection" could just as easily be employed here in the sense that George Basalla uses the term in his evolutionary model of technological development. Basalla argues that decisions regarding the path of technology are somewhat arbitrary, based on social as well as technical factors. I have used the word "choice," however, in order to employ the connotations of social factors inherent in that term. See George Basalla, *The Evolution of Technology* (New York, 1988), chap. 7, "Selection: Social and Cultural," esp. pp. 189–90.

[3]Representative of the predominating retrospective view is that offered by V. E. Cosslett, "50 Years of the Electron Microscope," *Advances in Optical and Electron Microscopy* 10 (1987): 224–26. In this account, Cosslett speaks of the magnetic electron microscope having "won the competition" based on its technical superiority. When we compare this 1987 assessment with Cosslett's more contemporaneous observation of 1951, however, the results of the "competition" are less preordained. In his earlier work, *Practical Electron Microscopy* (London, 1951), he speaks of the advantages of the electrostatic system without dismissing it as a possibility for the future of electron microscopy, noting its potential for "production as a compact, cheaper instrument of limited range of performance for routine work." See pp. 25–27 and esp. 271–74.

evaluations of lens systems conducted by firms in other countries, as well as successes in marketing electrostatic microscopes abroad, indicate that much more was involved in the successes and failures of particular electron microscopes made by RCA and GE.[4]

Thus, an examination that includes a comparative perspective supports the contention that the success of the electromagnetic microscope was not due to its technological superiority. A look at this machine's early history also reveals the sort of effects that institutional structures and approaches can have on the development of technology.

Because of the electron microscope's special position in advancing the frontiers of science, its development may have ramifications not only for technology but also for the subsequent path of science as well. For instance, the dynamics of technological development in this case also provide insights into the development of scientific disciplines closely related to, if not wholly contingent on, advances in electron microscopy—such as materials science, microanalysis, and a host of biological sciences concerned with the submicroscopic world.[5] Inasmuch as these fields of study are dependent on a technical instrument for their advance and indeed their very existence, the history of the device that

[4]Wiebe Bijker, Thomas Hughes, and Trevor Pinch et al. argue for an evaluation of technology that considers it as developing and functioning in, as well as acting on, such a societal context, or "web." When viewing technology from such a vantage point, we see that various social factors impinge on technical development, and we must thereby dismiss our erroneous assumption of a "linear structure of technical development." Instead of technology progressing in a straight-ahead fashion according to what is technically possible, it proceeds along socially determined paths characterized by an "interpretive flexibility." In this model, more than one possibility exists regarding what is accepted as valid, and this validity is continually subject to social influence as it gains "rigidity" through social mechanisms and finally reaches "closure" in its acceptance in the social-cultural milieu. Thus, what is deemed a "success" is only such because it has, for some reason or combination of reasons, found acceptance in the contextual web. Conversely, what is considered a technological "failure" in retrospect must be seen not necessarily for its intrinsic technical deficiencies but, rather, for the particular impediments it has met in the larger contextual setting. We can, and must, therefore, view the history of technology "symmetrically"—that is, taking into account and examining the so-called failures as carefully as the successes in order to garner the fullest insights into the processes of technological development. See Wiebe E. Bijker, Thomas P. Hughes, and Trevor Pinch, eds., *The Social Construction of Technological Systems* (Cambridge, Mass., 1987), passim; and also Thomas P. Hughes, "The Seamless Web: Technology, Science, Etcetera, Etcetera," *Social Studies of Science* 16 (1986): 281–92.

[5]The electron microscope, in this aspect, reflects Nathan Rosenberg's characterization of scientific instruments in general as leading to "interdisciplinary research," the rise of "entirely new subdisciplines," and the "migration" of scientists from one field of study to another. See Nathan Rosenberg, "Scientific Instrumentation and University Research," *Research Policy* 21 (1992): 381–90.

commands the threshold of scientific endeavor in these areas takes on added significance. Simply put, if social factors are important in shaping the technology that has given rise to new fields of science, then these areas of science are, by extension, socially configured to a proportionate extent.

Electron "Lenses" and the First "Choice"

A transmission electron microscope operates in a fashion analogous to a conventional light microscope. Instead of using light as the medium and optical lenses to magnify and focus the image, however, the electron microscope utilizes a "beam" of electrons magnified and focused by either magnetic or electric lenses. An electron moving in a magnetic field changes direction as it moves along a trajectory at right angles with respect to the direction of the magnetic field. The degree to which its direction is changed is inversely proportional to the speed with which it is moving and directly proportional to the strength of the field and the charge on the electron, hence, degree of shift = (constant × charge × field strength)/velocity. Similarly, an electron moving in an electric field changes direction. As a result of the attraction between a positive plate and the negatively charged electron, the electron is drawn toward the plate, hence, degree of shift = (constant × charge × electric field)/velocity.

These, then, are two fundamental ways of "bending" electrons in order to utilize them analogously to light and thus magnify an object.[6] While an electromagnetic electron microscope utilizes a magnetic field, the electrostatic design employs an electric field as its "lens." This choice between lens types presented itself to the first builder of an electron microscope, Ernst Ruska, who, in the late 1920s and early 1930s as a graduate student at the Technological University of Berlin, was studying electron lenses. As Ruska later recognized, however, his choice of the magnetic design was predicated on an incorrect assumption.

When Ruska began examining the relative merits of different electron lenses, he initially misunderstood the properties of an "electrostatic einzel lens," and so he opted in favor of the magnetic design. On the basis of his understanding of how electrons would act in the field of an electrostatic lens, he concluded that they "would not be appreciably altered on passage through the lens because of the symmetrical field distribution about the mid-plane of the lens." Having "overlooked that

[6]In both lens systems, electric currents are used: in the latter case, obviously, to create the electric field, while in the former case, to create an electromagnet, which in turn supplies the magnetic field that acts as the "lens."

as a consequence of the changing electron velocity a strong focussing of the ray bundles occurs," he thus "suggested another arrangement ... *with spherically-shaped grids.*"[7] As a result, "the images were appreciably *distorted by the two meshes immersed in the beam.* This none-too-pleasing result of the investigation made it seem ... at the time more fruitful to concentrate on the properties of magnetic lenses."[8] As these remarks reveal, it was the combination of misunderstandings that induced Ruska to abandon the electrostatic system in favor of the magnetic type. Hence, the first choice concerning the electromagnetic and electrostatic lenses was not a result of any real deficiency in the latter.[9]

As a result of what happened in Berlin, however, the electromagnetic lens received a "head start" and, consequently, a measure of technological momentum that made it the front-running technology in the early commercial development of electron microscopes.[10] Because the electron microscope was intended to surpass the resolving power of the light microscope, the momentum that the electromagnetic type received as a result of this episode established it as the leading design in the quest to "see the atom"—an overarching goal of microscopy virtually since the birth of atomic theory.[11]

The lead that the electromagnetic design received as a result of Ruska's choice relegated the electrostatic instrument to playing catch-up in both Germany and America. Because both RCA in America and Siemens in Germany were among the first to initiate electron microscope development, each had an advantage in securing patent positions for the electromagnetic-type design.[12] Siemens began manufacture of electron microscopes in 1939, and RCA began marketing an

[7]Ernst Ruska, *Early Development of Electron Lenses and the Electron Microscope* (Stuttgart, 1980), p. 21, emphasis added.

[8]Ibid.

[9]Ruska also related this episode to his audience in his acceptance speech for the 1986 Nobel Prize in physics. See Ernst Ruska, "The Development of the Electron Microscope and of Electron Microscopy," *Reviews of Modern Physics* 59 (1987): 629.

[10]By "technological momentum" I am referring to the various social and institutional forces in which the electromagnetic technology became embedded from this early point onward. That is, I am using the word "technological" to denote a socially constructed phenomenon as opposed to a "technical" characteristic.

[11]The editors of *Scientific American,* in a piece entitled "Our Point of View," reflected the sentiment of this quest to see smaller and smaller structures of matter. See *Scientific American* 163 (1940): 9. This aim of the microscope is also revealed in S. Bradbury, *Evolution of the Microscope* (New York, 1967). The extent to which this remains a driving goal for scientific instruments is demonstrated by a recent article by Ronald Hoffman, "For the First Time, You Can See Atoms," *American Scientist* 81 (1993): 11.

[12]Reinhold Rudenberg addresses the early patents in Germany in "The Early History of the Electron Microscope" (letter to the editor), *Journal of Applied Physics* 14 (1943): 434.

electron microscope in America in 1941.[13] While unable to surmount the technical advantages RCA held with the electromagnetic microscope, GE adopted a strategy of targeting what its engineers perceived as a need for a "practical commercial instrument."[14]

The question of what results in technological success or failure can be better understood if one explores GE's quest to develop and market such a commercial instrument. Central to such an exploration is determining specifically what factors affected GE's attempt to build and market this instrument. Here, the story turns to the work of C. H. Bachman and Simon Ramo, GE research scientists pursuing the development of the electrostatic electron microscope at the company's Electronics Laboratory in Schenectady, New York.

The Case for the Electrostatic Design

Bachman and Ramo first published word of their intention to build an electrostatic electron microscope in a three-part article in the *Journal of Applied Physics* in 1943. One year earlier and in the same journal, E. G. Ramberg of RCA Labs had published results of research that found that greater aberration effects resulted from the use of electrostatic lenses compared to electromagnetic lenses.[15] Nevertheless, Bachman and Ramo believed that electrostatic lenses would prove to be a more viable system for a "practical commercial instrument."[16] In order to understand fully the factors that went into this decision, familiarity with some of the basic physical properties of electron lenses is essential.

There are two major sources of distortion of the image in electron microscopy: chromatic aberration and spherical aberration. In light optics, aberration refers to an image that is blurred because not all of the light rays from the object are focused at the same distance from the lens. Similarly, in electron optics aberrations are caused by variations in the point at which the electrons, after passing through the lens, converge to form the image. In light optics, aberrations are the result of imperfectly ground lenses and the varying frequencies (which are directly related to velocity in a given medium, e.g., glass) of constituent

[13]Several summations of RCA's development of the electron microscope are available. For the most complete, see Jerome H. Reisner, "An Early History of the Electron Microscope," *Advances in Electronics and Electron Physics* 73 (1989): 163–215.

[14]C. H. Bachman and Simon Ramo, "Electrostatic Electron Microscopy I," *Journal of Applied Physics* 14 (1943): 8.

[15]E. G. Ramberg, "Variation of the Axial Aberration of Electron Lenses with Lens Strength," *Journal of Applied Physics* 13 (1942): 582.

[16]Bachman and Ramo (n. 14 above), p. 8.

colors of the spectrum. Analogously, in electron optics, blurred images, or aberrations, are the result of imperfections in the field lines of the lenses, known as spherical aberration, and variations in electron velocity, or chromatic aberration.[17] Both types of aberration can present difficulties that severely distort the image.[18]

As was explained earlier, the degree to which an electron's direction is changed is a function of its velocity. The velocity of the electron is, in turn, directly related to the voltage of the "electron gun"—as voltage increases, the velocity of the electron increases. If the voltage in the gun varies, the velocity of the electron will increase or decrease accordingly. Consequently, as electrons of varying velocity pass through the lens, the degree to which they are "bent" by the lens will differ. This introduces an aberration effect whereby electrons that represent the object are focused at varying distances from the lens. This effect, chromatic aberration, is illustrated in figure 1. Spherical aberration, the other major source of focusing problems, is caused by variations in the voltage that cause the strength of the fields that constitute the lenses to waver. Similar to the case with imperfectly ground light-optical lenses, electron lenses of varying field strength will cause electron "beams" to be focused at varying distances from the lens. Spherical aberration is illustrated in figure 2.

In order to prevent aberration in an electron-focusing system, the voltage supply must be carefully regulated. Bachman and Ramo recognized, however, that, in the electrostatic lens system, variations of voltage in the "gun" would be offset by proportional variations in the lens system. Consequently, any potential chromatic aberration introduced by the electron gun would be offset by counteracting variations in the lens system, and the need for elaborate voltage regulation would be eliminated. This inherent simplicity of the electrostatic design led Bachman and Ramo to regard this system as well suited for the practical commercial instrument they had in mind.

In outlining their plan, Bachman and Ramo described a microscope characterized by "simplicity of design, operation, and maintenance," having a resolving power ten times greater than the light microscope, and possessing a "size, weight, and complexity less than previously

[17]D. Gabor's *The Electron Microscope* (New York, 1948) provides a good introduction to these concepts.

[18]In addition, there also exists *relativistic* aberration—a more arcane phenomenon that results from the changing mass of the electron as it approaches the speed of light—with which we need not be as concerned because neither system holds an advantage in this regard. For a discussion of the contemporaneous understanding of relativistic aberration, see V. Zworykin et al., *Electron Optics and the Electron Microscope* (New York, 1946), pp. 650–51.

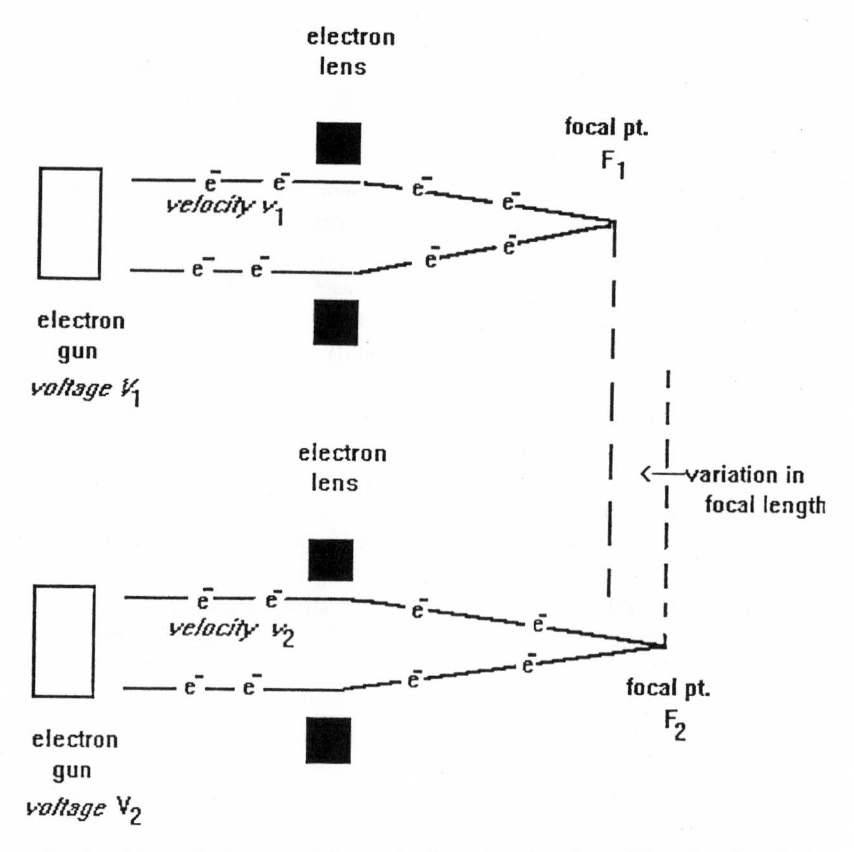

FIG. 1.—Schematic diagram of chromatic aberration. As voltage (V) varies, the velocity of electron (v) varies; thus, the focal point wavers, causing a "blurring" effect.

described instruments."[19] In formulating the design for their machine, they were willing to achieve simplification at the expense of forgoing the state-of-the-art resolving power then possible for electron microscopes. As a trade-off, they provided a compact and relatively mobile instrument—having a source-to-image distance of 11 inches and mounted on casters in order that it could be rolled from place to place—aimed at offering "care-free use by the operator."[20]

General Electric's attempts in 1944 to produce and market an electron microscope based on this design did not meet with much

[19] C. H. Bachman and S. Ramo, "Electrostatic Electron Microscopy III," *Journal of Applied Physics* 14 (1943): 155.

[20] Ibid.

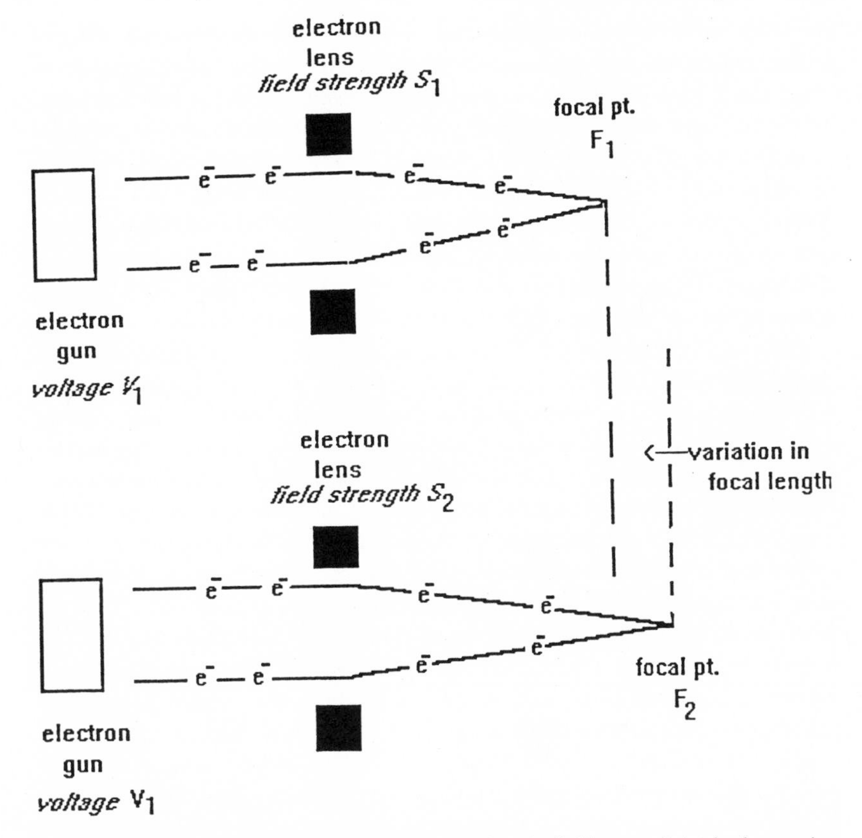

FIG. 2.—Schematic diagram of spherical aberration. As field strength in the lens varies, the focal point wavers, causing a "blurring" effect.

commercial success.[21] Yet it does not appear that its failure was due simply to an inherent inferiority in the performance capabilities of this lens design.[22] Successes achieved internationally demonstrate that the electrostatic design was technically viable and that a market for such machines *did* exist. Before looking at successful ventures in Japan and Germany, however, let us turn to some contemporaneous evaluations of the needs of electron microscopists.

It was not only those at GE who suggested needs other than high resolving power. James Hillier and R. F. Baker, two scientists involved in

[21]Reisner; and Sterling Newberry, "Electron Microscopy, the Early Years: Part I," *EMSA Bulletin* 15, no. 1 (1985): 39.

[22]Keiji Yada, "History of Electron Microscopes, Tohuku University," in *History of Electron Microscopes,* ed. Hiroshi Fujita (Kyoto, 1986), p. 29.

electron microscope development at RCA, criticized evaluating the performance of an instrument by "measuring the least separation observable in a single micrograph [i.e., resolution]," remarking that "such measurements bear little, if any, relationship to the *every-day* performance of a particular instrument."[23] Further, they criticized this criterion of evaluation because it "do[es] not take into account spoilage of micrographs caused by defects of specimen technique, by inaccuracies in the adjustment, and by contamination of the instrument."[24] Although never explicitly stated as such, these latter concerns directly relate to the relative ease of operation of the instrument.

The supposition that the electrostatic instrument was rejected out of hand because it was designed with a lower resolving power is also dubious in light of a 1946 survey conducted by the Electron Microscopy Society of America (EMSA). Responses to a questionnaire indicated that, in over two-thirds of the work in contemporaneous electron microscopy, resolution "could be neglected as a limiting factor."[25] This suggests that microscope users were not constantly pushing the frontiers of resolution and, consequently, were not so dependent on this state-of-the-art aspect of the electron microscope. Even discounting the third of the work that presumably demanded ultimate resolution, a practical instrument offering ease of use and a lower price in lieu of optimal resolving power should have been adequate for a sizable portion of microscope applications. Had it been marketed effectively, such a practical instrument should have been appropriate for tasks such as routine quality-control testing in industry.[26] Such a hypothesis is supported by events on the international scene.

[23]James Hillier and R. F. Baker, "A Discussion of the Illuminating System of the Electron Microscope," *Journal of Applied Physics* 16 (1945): 469 (emphasis in original).

[24]W. G. Kinsinger, J. Hillier, R. G. Picard, and H. W. Zieler, "Report of the Electron Microscope Society of America's Committee on Resolution," *Journal of Applied Physics* 17 (1946): 989.

[25]Ibid.

[26]The need for a simplified machine can be inferred from two letters written to Thomas F. Anderson, RCA electron microscope fellow of the National Research Council working at the RCA labs in Camden, New Jersey, in the early 1940s. John L. Magee of the B. F. Goodrich Corporation wrote to his friend Anderson about problems relating to the installation of a complex machine in a corporate environment: "La Rue of [the] engineering dept. is apparently in charge of installation but the work must be done by the plumbing and electrical departments with equipment which is bought by the purchasing dept., etc., etc." Related to the manpower demands of the complex machine that would be obviated with an easy-to-use instrument, a J. H. Matthews of the University of Wisconsin wrote to Anderson expressing doubt that the university would allocate funds "to get a trained man to work with it," feeling that, with such a complex tool, "the technique is often fully as important as the instrument itself." See Magee to Anderson, April 7, 1942, and Matthews to Anderson, February 25, 1941, Thomas F. Anderson Papers, American Philosophical Society, Philadelphia (hereafter cited as Anderson Papers).

Electrostatic Successes in the International Setting

After World War II, firms in Germany and Japan successfully manufactured and marketed electron microscopes with electrostatic lenses. In Japan, an electrostatic electron microscope was first built by Yasumasa Tani, a researcher at the University of Tokyo, in 1939.[27] Although development was hampered by the aerial bombings and material shortages of World War II, this instrument eventually attained a resolution on an order of magnitude equaling the best of the contemporaneous electromagnetic microscopes and remained in use until 1950.[28]

A development at the Japan Electron Optics Laboratory (JEOL) in Tokyo also demonstrates that contemporary researchers perceived the electrostatic design as technically viable. In September 1946, Kenji Kazato headed a JEOL team that evaluated current electron microscope technology. A group of "first-class scientists from physics, electronics and vacuum technology" chose the electrostatic machine as the most propitious design.[29] Although JEOL would switch to the electromagnetic design years later, the fact that its initial study found the electrostatic design favorable suggests that in the mid-1940s the choice between electrostatic and electromagnetic lenses was not plainly obvious.

Further evidence in support of the viability of electrostatic microscopes is offered by the success the Toshiba Corporation achieved with the electrostatic design. Toshiba successfully built and marketed several electrostatic electron microscopes from 1941 through 1943, including the Toshiba model numbers 1, 3, and 6.[30] These instruments were designed with a maximum resolution of 80 angstroms (Å), or 80×10^{-10} meters. Development and manufacturing continued through the 1940s into the 1950s. In addition to achieving gains in resolving power (the EUL-1B model in 1947 provided 60Å resolution), Toshiba continually implemented new features such as the capability of viewing three different types of images on a single instrument. So equipped, the EUL-1B functioned as a transmission electron microscope and a shadow microscope, as well as offering electron diffraction analysis.[31] In addition to the electrostatic models, Toshiba also produced electromagnetic

[27]Akira Fukami, Koichi Adachi, and Kentaro Asakura, "Development of Electron Microscope in Tokyo Imperial University," in Fujita, ed. (n. 22 above), p. 30.

[28]This microscope's resolution was better than 100Å. An angstrom (Å) is equal to 1×10^{-10} meters. Contemporary electromagnetic microscopes as of December 1943 were performing at about 50Å. See Vladimir Zworykin and James Hillier, "A Compact High Resolving Power Electron Microscope," *Journal of Applied Physics* 14 (1943): 661.

[29]Kazuo Ito, "Development of Electron Microscopes in JEOL," in Fujita, ed., p. 54.

[30]Hiroshi Kamogawa, "Electron Microscope Research in Toshiba Corporation," in Fujita, ed., pp. 64–79.

[31]Ibid., pp. 74–75.

microscopes.[32] Apparently the company felt that neither lens design was necessarily superior for all purposes and that the electrostatic models were clearly suited to existing market demands.

The successes achieved in Japan were echoed in Germany, and perhaps most convincing was the pursuit of the electrostatic electron microscope by the Carl Zeiss Company, a firm already possessing "a great name and international reputation as [a] manufacturer of superior optical microscopes." In 1950 Zeiss performed an evaluation of electron microscope capabilities. Its findings, in combination with its subsequent achievements with the electrostatic design, almost unequivocally demonstrate that this design was perceived as, and indeed was, a practical possibility.[33]

Zeiss researchers cited "lower demands of [the] electrostatic [design] on the stability of the high voltage [electron 'gun'] and the lens current supply," in addition to "cost-saving lens designs" that need "not be water-cooled" (as did electromagnetic lenses). They also noted that "the image is not rotated when the magnification is changed" and concluded that "according to the technical knowledge of around 1950 electron microscopes with electrostatic lenses were certainly up over instruments with electromagnetic lenses in terms of the price/performance ratio." Both independently and in joint ventures with AEG and Suddeutche Laboratorien, Zeiss successfully produced electrostatic electron microscopes from 1942 until 1962.[34]

In the face of these undeniable successes with the electrostatic design on the international scene, GE's hardships in its attempted development of this design require explanation. Clearly it is insufficient to conclude that its failure resulted from the technical inferiorities of the electrostatic lens as compared to the electromagnetic design—no matter what the present (1990s) state of the technology reveals. Therefore, one must delve deeper to reveal the processes involved in this instance of technological choice. Ultimately, the respective successes and failures of the RCA and GE electron microscopes are best explained by taking account of nontechnical factors. This can best be done by viewing how each technological system was situated in what social constructionists have termed the "seamless web" of human activity.[35]

"Success" for RCA and "Failure" at GE

One critical factor at RCA was its special relationship with scientists and researchers either actively or potentially involved with electron

[32]Ibid.

[33]"The History of Electron Microscopy at Carl Zeiss," in Fujita, ed., p. 209.

[34]Ibid., pp. 210–11; and Cecil E. Hall, "Commercial Electron Microscopes," in his *Introduction to Electron Microscopy* (New York, 1953), pp. 201–25.

[35]Bijker et al., eds. (n. 4 above), p. 3 and passim.

microscopy. This special relationship with the newly emerging community of practitioners is evidenced by the role RCA's leading scientist in the field, James Hillier, played at early EMSA meetings. Another critical factor was the presence of Thomas Anderson, a National Research Council (NRC) fellow, for whom a specific laboratory for electron microscopy was established at RCA.[36] The roles of Hillier and Anderson served two purposes that have relevance to this story. First, Hillier's visible presence at gatherings of electron microscopists provided market linkages to potential customers. Second, Anderson's lab in combination with Hillier's visibility helped establish the credibility of the technology and, more specifically, RCA's instrument, in the eyes of the scientific community.

RCA fostered and openly encouraged relations with users and potential users of the electron microscope. As evidenced by Anderson's work, establishing the usefulness of the instrument to science was a large part of the practical as well as the philosophical aims of electron microscope development at RCA.[37] In 1940, the NRC awarded Anderson a research fellowship to "explor[e] the possibility that the newly developed electron microscope might have applications in biology."[38] In addition to the basic-research value of the NRC fellowship, RCA, not surprisingly, sought to reap commercial rewards from the arrangement. This is demonstrated in the explicit terms under which the fellowship was set up, as outlined in an agreement between Ross Harrison, director of the NRC, and Vladimir Zworykin, director of electronics research at RCA. The agreement provided that RCA would contribute $3,000 for Anderson's salary, while it was "understood that technical credit will be given to the RCA Manufacturing Company in publications." Further, RCA and the NRC agreed that "patentable discoveries, developments in the nature of physical improvement of the instrument, its mechanical adaptation for biological work and methods relating to the mounting of specimens are to be the property of the RCA Manufacturing Company."[39]

[36]See Sterling Newberry, "Electron Microscopy, the Early Years: Part II," *EMSA Bulletin* 15, no. 2 (1985): 39, for the importance of this in the context of competition between RCA and GE.

[37]Hillier recalled that the feeling that the electron microscope should be pursued for philanthropic reasons was always present, and this emphasis, he remembered rather fondly, was specifically encouraged by David Sarnoff, president of RCA (personal interview, September 1991).

[38]Thomas F. Anderson, "Memories of Research," *Annual Review of Microbiology* 29 (1975): 7. For more information on the fellowship, see Reisner (n. 13 above), esp. pp. 225–27; and also Thomas Anderson, "Electron Microscopy of Phages," in *Phage and the Origins of Molecular Biology,* ed. John Cairns, Gunther Stent, and James Watson (Cold Spring Harbor, N.Y., 1966), pp. 63–78.

[39]Ross Harrison to Vladimir Zworykin, June 17, 1940, Anderson Papers.

Anderson's lab at RCA was, from its inception in early 1941, involved in scientific research and publication in matters related to electron microscopy. The first images of viruses viewed with RCA's developmental model received much attention in both the scientific and popular literature.[40] Thus, as the work at RCA was made known to outside groups, RCA became identified with state-of-the-art microscopy. The linkages to relevant scientific circles established a social milieu, or "technological frame," that increased the likelihood of commercial success with the instrument.[41] RCA's consciousness of the public-relations value of Anderson's work is made quite clear in a letter from M. C. Banca, of the company's Engineering Products Division, requesting Anderson's signature to release use of pictures he made on the electron microscope. Wondering if Anderson "had forgotten about it," Banca explained that "we are rather anxious to use this for publicity."[42]

People affiliated with the applications of the RCA electron microscope, including the directors of the fellowship and others from industry and academia, came from all over the country.[43] The prestige of these scientists and their institutions reflects the institutional linkages and personal interactions that were undeniably important in fostering a respected position for RCA in the nascent field of electron microscopy.[44]

[40]For example, S. Luria and T. F. Anderson, "The Identification and Characterization of Bacteriophages with the Electron Microscope," *Proceedings of the National Academy of Sciences* 28 (1942): 127–30; S. Luria, M. Delbruck, and T. F. Anderson, "Electron Microscope Studies of Bacterial Viruses," *Journal of Bacteriology* 46 (1943): 57–77; Stuart Mudd, Katherine Polevitzky, Thomas Anderson, and Leslie Chambers, "Bacterial Morphology as Shown by the Electron Microscope," *Journal of Bacteriology* 42 (1941): 251–64; Wendell Stanley and Thomas Anderson, "A Study of Purified Viruses with the Electron Microscope," *Journal of Biological Chemistry* 139 (1941): 325–38; Thomas Anderson, "The Study of Colloids with the Electron Microscope," *Advances in Colloid Science* 1 (1942): 353–90; Thomas Anderson and Wendell Stanley, "A Study by Means of the Electron Microscope of the Reaction between Tobacco Mosaic Virus and Its Antiserum," *Journal of Biological Chemistry* 139 (1941): 339–44; Glenn Richards, Thomas Anderson, and Robert Hance, "A Microtome Sectioning Technique for Electron Microscopy . . . ," *Proceedings of the Society for Experimental and Biological Medicine* 51 (1942): 148–52; and "Never Seen Before: EM Reveals Viruses for First Time," *Scientific American* 164 (1941): 358.

[41]"Technological frame" is the term Wiebe Bijker employs to denote the combination of social and technical factors that constitute a given "technology." Its applicability to the case of the electron microscope will be elaborated below. See Wiebe Bijker, "The Social Construction of Bakelite: Toward a Theory of Invention," in Bijker et al., eds. (n. 4 above).

[42]M. C. Banca to Anderson, November 17, 1942, Anderson Papers.

[43]Anderson's correspondence from the period of his fellowship, as well as names listed on a guest register, reveal that those interested in the electron microscope and those who sent or delivered material to be examined on the machine were from places as diverse as New York City; Pasadena, Calif.; Dallas; Madison, Wis.; and Portland, Maine, to name just a fw.

[44]Diana Crane, who has explored and revealed the significance of networks of scientists, i.e., "invisible colleges," has found that communications among scientists are influential in

Those individuals with whom Anderson recalls being involved in the biological applications of the electron microscope at RCA include Stuart Mudd, concurrently a professor of bacteriology at the Henry Phipps Institute in Philadelphia; Charles W. Metz, a professor of zoology at the University of Pennsylvania and the director of its Zoological Labs from 1940 to 1945; the preeminent Ross G. Harrison, chairman of the NRC from 1938 to 1946, a professor emeritus of biology at Yale and also concurrently an emeritus trustee at Woods Hole; Wendell M. Stanley, an associate member of the Rockefeller Institute and contemporaneously a visiting lecturer of virology at the University of California (1940), Cornell (1942), and Princeton (1942) and a Nobel Prize recipient in 1946; David B. Lackman, an associate instructor of bacteriology at the Medical School of Pennsylvania from 1939 to 1941 and an assistant bacteriologist for the U.S. Public Health Service from 1941 to 1946 who would later go on to become the senior scientist and scientific director of that agency in 1946; S. E. Luria, a resident assistant surgeon at the College of Physicians and Surgeons, Columbia University (1940–42) and later a Guggenheim fellow in bacteriology at Princeton (1942–43); Harry Morton, a professor of bacteriology at the Medical School of Pennsylvania; renowned physicist Max Delbruck, then a professor at the Vanderbilt Institute of Physics; and also Leslie Chambers, a biophysicist at the Medical School of Pennsylvania, who in 1946 would become the chief physical scientist of the Defense Division Biological Labs.[45]

People at RCA were not only cognizant of the commercial contacts established by Anderson's lab, they were also concerned with utilizing Anderson's experimental microscope specifically for demonstrations to prospective customers. In November 1941, at the end of Anderson's first year at RCA, the RCA project engineer and the sales department arranged a new schedule of microscope allocation to "better accommodate" the demonstration of the machine to "prospective customers" in coordination with the time required for Anderson's research. Not only was one week out of each month set aside, an hour was reserved for

the growth of scientific knowledge in newly emerging fields of study. Crane's findings with respect to scientific developments seem to be paralleled by the technological development of the electron microscope—particularly insofar as this instrument served, and continues to serve, as a technological frontier to scientific advance and, indeed, as we may judge electron microscopy as a field of scientific study. The applicability of Crane's analysis is especially appropriate in light of the existing circle of practitioners centered about the RCA instrument. See Diana Crane, *Invisible Colleges* (Chicago, 1972).

[45]Anderson, "Memories of Research" (n. 38 above). Affiliations of individuals are gleaned from Jacques Cattell Press, ed., *American Men and Women of Science* (New York, 1971–86).

demonstrating the instrument to prospective customers even on days established for Anderson's research.[46]

The network of individuals important for RCA's success also reached beyond Anderson's lab. The involvement of many other RCA scientists with the EMSA provides further evidence of the company's close-knit relationship with the inchoate electron microscope community.[47] Furthermore, Vladimir Zworykin, the director of electronics research who assembled the team to develop the microscope at RCA, also took a keen interest in the success of electron microscope development, having been personally involved in electron optics research at least since the early 1930s.[48] Thus, Zworykin, Banca, and Hillier were all enthusiastically involved in securing success for the RCA efforts.

General Electric, in contrast to RCA, did not cultivate such a congenial relationship with relevant scientific circles. In his recollections of the early days of electron microscopy, Sterling Newberry, head of electron microscope development at GE in the late 1940s (after Bachman and Ramo left), gives a description of his company that is quite different from the approach of RCA. According to Newberry, GE's personnel were much less visible in these early electron microscopy gatherings. For instance, at the second meeting of EMSA in Chicago, Newberry recalls that no GE scientist gave any formal paper.[49] From this, it can be inferred that GE was not utilizing this assemblage of practitioners as effectively as was RCA. Indeed, as Newberry remembers, EMSA's third meeting, held at Princeton University in late 1945, was virtually run by RCA's leading electron microscope scientist, James Hillier.

[46]Memorandum from Perry C. Smith, project engineer, to J. P. Taylor, engineering products sales, with carbon copy sent to Anderson, November 17, 1941, Anderson Papers.

[47]Newberry, "Electron Microscopy, the Early Years: Part II" (n. 36 above), p. 39. The first meeting of EMSA was held in Chicago in 1941, sponsored by the American Chemical Society.

[48]The importance of Zworykin in serving as a conduit between management and the electronics lab is attested by George Morton, a contemporary scientist at RCA labs in the 1940s. Morton describes Zworykin as "one of those rare and invaluable men who can command the trust and aid of top management (i.e. for funding, etc.) and at the same time have the respect, loyalty and cooperation of those under him." Correspondence between Morton and Eric Weiss, a scientist involved in electron microscope development at RCA, February 8, 1993. I am indebted to Mr. Weiss for a copy of this letter and other insights into "laboratory life" at RCA. In addition to serving as a conduit within the company, Zworykin also championed the microscope outside RCA, notably in a lecture entitled "The Electron Microscope in Relation to Chemical Research," delivered at the first EMSA meeting. Being that, as noted above, the first EMSA meeting was sponsored by the American Chemical Society, Zworykin's choice of topic suggests that he fully realized the significance of this gathering. See Newberry, "Electron Microscopy, the Early Years: Part II," p. 39.

[49]Newberry, p. 43.

Hillier's role in explaining the functioning of his company's instrument and in relating his laboratory's latest findings was undoubtedly important in establishing and buttressing the electron microscope's credibility. As Ian Hacking argues, understanding the physical properties that allow an instrument to work leads one to find its data credible.[50] More important, perhaps, is the image RCA was able to present by having its man, Hillier, recognized as leading the vanguard into this new area of scientific inquiry.

Here, then, is a picture of a network of practitioners, and potential customers, that contributed to the development of a positive image for RCA. Undeniably there were direct effects, as these researchers were affiliated with many scientific institutions and were also potential buyers of the new technology. Too, there were indirect benefits of having members of the scientific elite utilize one's technology. As discoveries with the instrument were published in the 1940s, RCA became increasingly identified with electron microscope technology. And as the scientific literature in various fields proliferated, the burgeoning field of electron microscopy became increasingly identified with RCA. Although these effects cannot be directly measured other than by citing the number of publications mentioning RCA instruments and suggesting the positive correlation to RCA's business success vis-à-vis GE, it is nevertheless self-evident that the establishment of a "good name" is essential for commercial success.[51] This RCA achieved by ensconcing its technology in colleges of practitioners—both visible colleges, as outlined by the institutional affiliations of the individuals listed above, with direct and tangible results, and invisible colleges, as their relations and communications fanned out with more indirect, but ultimately just as important, effects.

The more indirect influences of RCA's involvement with the relevant community of practitioners in electron microscopy are best explained utilizing the notion of "inclusion" as articulated by Wiebe Bijker; that is, the extent to which relevant players who might contribute to the acceptance of a given technology are literally included in the social network in which that artifact is to find its place.[52] As RCA attempted to develop an electron microscope, there were various social groups,

[50]Conversely, if an instrument remains an opaque "black box," its data are likely to be interpreted as artifacts of the machine. See Ian Hacking, *Representing and Intervening* (New York, 1983), p. 209.

[51]For the dangers inherent in trying to reduce technological successes and failures to a single factor in relation to the electric car, see Michel Callon, "Society in the Making: The Study of Technology as a Tool for Sociological Analysis," in Bijker et al., eds. (n. 4 above), p. 95.

[52]Bijker (n. 41 above).

scientific theories, and technological artifacts that constituted the "technological frame" in which the instrument existed—really, this technological frame *is* the electron microscope in the broadest sense of its existence in the sociotechnical world. As such, the human actors in this setting were indeed a significant element in determining the "success" of the instrument, and RCA, by literally "including" so many important actors in its technological frame, helped provide a viable subculture in which the artifact could flourish.

The actors involved here varied both in their orientation and in their level of inclusion. Obviously RCA's own scientists, men like Hillier, were directly grounded in electron microscope science; this GE had as well, with Bachman and Ramo and, later, Sterling Newberry. But RCA's relative prowess lay in the level of inclusion of actors with other groundings, most notably actors from the life sciences tied to RCA via Anderson's lab, including Anderson himself. Especially important was the extent to which RCA's relationship to the scientific community resulted in users becoming increasingly familiar and dependent on the RCA machine.[53]

Another key element in RCA's technological frame was the high level of inclusion of the business side of the corporation, most notably and convincingly demonstrated by the financial and emotive support offered by David Sarnoff, president of the company.[54] Sarnoff's support emanated directly from a very high opinion of electron research that went above and beyond concern for the commercial success or failure of the electron microscope. In the words of his biographer, Kenneth Bilby, Sarnoff envisioned "a total system approach to a new industry—'the whole ball of wax' he called it—and at the time it was unique in the industrial landscape. . . . From this concept, Sarnoff moved on to an ever more dynamic gestalt for the management of technology, which he began articulating during the mid-thirties in speeches and at stockholder meetings. *Fortune* would later call it his 'missionary approach to the science of electronics.' RCA would muster all its research resources behind the electron."[55]

Reflecting on RCA's accomplishments in the mid-1950s, Sarnoff echoed this sentiment, remarking, "we are an organization founded upon science. We made our living by the tiniest thing known in the

[53]Direct evidence of this occurring is attested to by a 1942 letter to Anderson from G. C. Clark of the Chemistry Department at the University of Illinois. Clark, who was involved in setting up a national meeting of electron microscope users, admitted to Anderson that "most of us [electron microscope users] obviously depend on the RCA instrument" (G. C. Clark to Anderson, November 2, 1942, Anderson Papers).

[54]See n. 37 above.

[55]Kenneth Bilby, *The General: David Sarnoff and the Rise of the Communications Industry* (New York, 1986), pp. 124–25.

world, the tiniest particle that scientists know about—the electron."[56] That Sarnoff's emphasis on the importance of electron research contributed directly to the development of the electron microscope is evidenced by the personal interest he took in the project and by specific references to the microscope in public addresses. For instance, in a 1943 speech outlining RCA's accomplishments, Sarnoff noted, "we have the electron microscope, one of the most important new scientific tools of the twentieth century."[57] It was thus the relative advantages of the technological frame about the electron microscope at RCA that proved most advantageous in its competition with GE and consequently determined the "success" of the electromagnetic configuration.

Evidence of direct competition between GE and RCA is noted by Newberry, who was a pioneer in early electron microscope development at Washington University in St. Louis in the 1930s and would be hired by GE in 1947 to head its "second attempt" at developing an electrostatic microscope. Newberry, on recalling the 1943 meeting of EMSA, relates that each company was "obviously partisan" with respect to its own machine and, further, suggests that competition at that time was especially marked regarding GE's development of a portable electrostatic microscope.[58] While both companies were vying for recognition of their technologies, RCA clearly utilized such occasions more advantageously. That the RCA instrument should have reaped greater commercial success should perhaps come as no surprise.

It is important to note here that neither at this 1943 meeting nor at any of the other early meetings was there a situation where one company's instrument outperformed the other and thereby won support. On the contrary, as Newberry states regarding the 1943 EMSA meeting, "there were difficulties for both [RCA and GE] instrument displays . . . [consequently] there was no attempt to display instruments for several meetings after this one."[59] Thus, it is insofar as scientific, business, and professional relationships were developed that these meetings translated into commercial success.

In addition to RCA's advantageous relationship with the scientific community, the company's internal approach to electron microscope

[56]David Sarnoff, "Remarks at a Dinner Honoring Dr. Vladimir K. Zworykin," in his *Looking Ahead: The Papers of David Sarnoff* (New York, 1968), p. 254.

[57]David Sarnoff, "Address before the American Association for the Advancement of Science," Lancaster, Pa., November 11, 1943, in ibid. Hillier also spoke of Sarnoff's personal interest in the course of an interview with the author in September 1991. It is also noteworthy that at a dinner honoring Vladimir Zworykin, the acclaimed television scientist, Sarnoff noted Zworykin's work on the electron microscope as well as his efforts in television. See ibid., p. 252.

[58]Newberry, "Electron Microscopy, the Early Years: Part II" (n. 36 above), p. 40.

[59]Ibid.

development differed from GE's. Hillier recalls that Sarnoff engendered an environment at RCA that was quite responsive and amenable to the needs of developing the electron microscope.[60] This approach contrasts sharply with Newberry's recollections regarding electron microscope undertakings at GE. In 1944, GE began production of an electrostatic electron microscope based on the design of Bachman and Ramo. While the first ten instruments sold quickly, trouble resulting from hurried design and inadequate machining adversely affected sales of the next "batch" of twenty microscopes. With this sour turn of events, Ramo moved to the West Coast because of his wife's health, and Bachman became discouraged and left for a teaching job at Syracuse University. Thus, the first attempt at commercial production of an electron microscope at GE came to an end. Rather than a technical deficiency of the electrostatic lens design, however, Newberry points to the lack of corporate support.[61]

On his arrival at GE in May 1947, Newberry was told by a former salesman for the electronics lab "that some minor adjustments should have been possible to make them perform to specification."[62] A month later, Newberry was able to uncover the flaws in Bachman and Ramo's design. Examining their notebooks, he found that they had decided, imprudently, to go ahead with a smaller, 2-inch lens design, which introduced a distortion into the image. Newberry concluded from the sketchiness of their notes that Bachman and Ramo were apparently hurrying "under pressure" and had thus performed "no experimental verification . . . before a large program was launched."[63] That Bachman and Ramo would have made such an error in the absence of some outside influence seems highly unlikely, for in outlining their research in February 1943 they specifically acknowledged negative effects that limit "the reduction in diameter of the microscope body for any given lens design."[64]

Investigating further two years later, Newberry found that "critical parts" subcontracted for the production models had "not received the workmanship required by the design."[65] In GE's second attempt to develop and market an electron microscope, moreover, Newberry's efforts would also suffer from a lack of support. In a retrospective appraisal of his years at GE, he alluded to the lack of financial resources

[60]Hillier interview (n. 37 above).

[61]Newberry, "Electron Microscopy, the Early Years: Part II," p. 44.

[62]Ibid.

[63]Ibid.

[64]C. H. Bachman and Simon Ramo, "Electrostatic Electron Microscopy II," *Journal of Applied Physics* 14 (1943): 70.

[65]Newberry, "Electron Microscopy, the Early Years: Part II," p. 45.

made available to electron microscope development, noting specifically how he and his colleagues "used to jokingly say 'as soon as we purchase a proper metal name plate, we have exceeded the allowable manufacturing cost.' " He also referred to difficulty in obtaining a suitable metal cabinet to house the microscope: "The current model of the GE Electric Ironer had an ideal frame for the purpose. . . . However, the Factory would not even consider selling them to us in small numbers because of possible production foulups." He then punctuated this remembrance by noting, "There were many more such disappointments."[66]

Newberry went on to present additional evidence of GE's lack of adequate support, as he related being "sent back to the drawing board with essentially no funds." And perhaps most convincing are his remembrances of having to "devise a self-guiding boring bar similar to those made by *early cannon makers*" and boring the tube for holding the lenses *by hand* for twelve hours because the factory would not provide the requisite machining, all in order to acquire a "not bountiful" extension of funds.[67]

Despite Newberry's formidable effort, GE scrapped the second attempt at electron microscope production due to a corporate reorganization that entailed a "decentralization of special products" which, in turn, brought about the abandonment of "even well established products such as the Analytical Mass Spectrometer."[68] This restructuring is vital to understanding the role the electron microscope played in the broader corporate strategy prevailing at GE in the postwar years. Unlike the convivial environment in which the technology was situated at RCA, at GE the microscope was seen merely as another product, one which either had to yield immediate profits or be eliminated.

After Charles E. Wilson succeeded Gerard Swope as president of GE in the 1940s, he hired Ralph Cordiner to implement a new corporate organization. Although not publicized until the plan had been almost fully implemented with Cordiner's ascension to the presidency at the end of 1950, the restructuring of GE began as early as 1944 and thus had a direct bearing on the fate of the electron microscope there.[69] Decentralization was the essence of Cordiner's reconfiguration. He divided GE into fifty virtually independent divisions or "profit centers." A man with a calculating, "bottom-line" approach, Cordiner was inclined toward bold and sweeping action, as is illustrated by an episode in his brief experience as vice chairman of the War Production Board.

[66]Ibid.

[67]Ibid., emphasis added.

[68]Ibid., p. 46.

[69]"Mr. Wilson at Work," *Fortune* 35 (1947): 121; "Cordiner of General Electric: Reorganization by Pure Reason," *Fortune* 45 (1952): 132.

Apparently miffed by the inefficiency of government, he sought to fire 3,000 civil servants, and when he "found that he couldn't, the fact seemed to induce in him a sort of cold horror." On instituting the reorganization at GE, Cordiner placed stringent profit requirements on each of the divisions, emphasizing immediate and high rates of return on investments. Divisions either had to produce or be cut back.

This proclivity for sweeping action combined with Cordiner's business outlook—shaped by his own experience in appliance sales and merchandising—led to critical changes at GE: decentralization, greater market responsiveness, and a corporate strategy configured around the production and sale of domestic appliances. This new approach was epitomized, on the one hand, by the construction of the massive "Appliance Park" production facility on a 942-acre site near Louisville, Kentucky, in the 1950s.[70] Production for industry, on the other hand, was concentrated on building heavy power equipment. In 1948 the company launched a multibillion-dollar program to build and market apparatus such as turbine generators.[71] As a consequence of this dual focus, specialized electronic products for industry such as leak detectors, X-ray photometers, and mass spectrometers were left out of GE's corporate strategy during just the period when the attempts at electron microscope development were being undertaken.

General Electric's departure from specialized electronic equipment is apparent in the changing nature and emphasis of the company's advertising from the late 1940s into the early 1950s. Trade journals such as *Chemical and Engineering News,* which had included many GE ads in the late 1940s, became conspicuously devoid of such ads by the early 1950s.[72] And the later ads began to differ in tone. The 1940s' ads had emphasized GE's prowess at helping "industry to solve thousands of problems" with "new testing and measuring equipment." They even held out the hope that some particular problem "may justify a development program to create a new product."[73] By early 1950, however, GE's ads were no longer mentioning new development programs, and, rather than urging those in industry to write to the research labs, they instead instructed readers

[70]Robert Slater, *The New GE: How Jack Welch Revived an American Institution* (Homewood, Ill., 1993), pp. 11–12.

[71]Reported in *Fortune* 38 (1948): 8–9.

[72]*Chemical and Engineering News* is mentioned specifically as an example because it had the largest circulation of any periodical in the chemical industry. It kept abreast of new developments in industrial instruments technology and carried rather extensive advertising for industrial electronic equipment, including some ads for RCA's electron microscope. Other journals, such as *Rubber Age* and *Chemical Engineering,* in which GE originally advertised scientific instruments also reflect this trend.

[73]See, e.g., the two-page spread in *Chemical and Engineering News* 26 (1948): 782–83.

to "call your nearest GE sales office."[74] Completing the de-emphasis on electronic products for industry, the ads featured fewer and fewer products until finally disappearing by 1952.

At RCA, Anderson's lab served as an excellent conduit whereby the company could not only make its technology responsive to the needs of users but through its position could also make users dependent on its technology, all the while furthering its reputation in the field. RCA's relationship to the scientific user-community centered around the electron microscope also provided what might be described as a de facto feedback loop for ongoing development of the instrument. An outline of this can be gleaned from a sampling of articles published in journals specializing in scientific instruments, which reveals that electron microscopists were publishing descriptions of modifications and adaptations of their RCA machines.[75] This ongoing improvement taking place *in the field* suggests that the electron microscope was not a product amenable to a strategy of merely being dumped on the market in the hope of generating immediate profits. RCA not only offered accessories to improve performance of instruments previously sold but also published information to enable users themselves to make improvements.[76] It carved out a market niche not simply by producing a superior technical artifact but by setting up and maintaining a frame where users and the producer remained interactive as well.[77] RCA's role as a scientific research leader in electron microscopy, a reputation established in great measure by Anderson's lab, was undoubtedly of great significance in achieving commercial success with the instrument.

[74]See *Chemical and Engineering News* 28 (1950): 1733.

[75]A few examples, with their affiliations, are H. Crane, University of Michigan, "Additional Stabilization for the Beam Current in the RCA Type B Electron Microscope," *Review of Scientific Instruments* 16 (1945): 58; John T. Quynn, Camp Detrick, Md., "Adjustable Aperture for the Electron Microscope—RCA Type EMU," *Review of Scientific Instruments* 19 (1948): 472–73; J. A. Simpson and Alan Van Bronkhorst, National Bureau of Standards, "Modification of the Electron Microscope for Electron Optical Shadow Method," *Review of Scientific Instruments* 21 (1950): 669; and B. O. Heston and P. R. Cutter, University of Oklahoma, "Molecular Diffraction Attachment for RCA Microscope," *Review of Scientific Instruments* 21 (1950): 608.

[76]For instance, RCA Laboratories, "Laboratory Modifications in the RCA Model EMC Electron Microscope," *Review of Scientific Instruments* 21 (1950): 255; and accessories such as "charge neutralizers," "focusing magnifiers," and "self-bias gun kits," announced in the new products section of the *Review of Scientific Instruments* 20 (1949): 844.

[77]Direct interaction between Anderson and users in the field is evident in Anderson's correspondence and his calendar of "informal talks" during his years at the RCA laboratory. For example, in the early 1940s Anderson made appearances throughout the country at various colleges and organizations such as the American Medical Association and American Chemical Society.

In contrast to the situation at RCA, the electron microscope's prospects at GE did not bode well in a company refocusing its strategy toward the popular consumption of domestic appliances. And the electron microscope held no particular fascination with those higher up the corporate ladder at GE. Rather, a scientific instrument like the electron microscope, not in line with broader corporate goals, faced a greatly diminished chance of commercial success.

Conclusion

As this accumulated evidence demonstrates, the fate of the electron microscope at GE must be understood in full context. It cannot be sufficiently explained by simply stating that the electrostatic instrument was technically inferior to the electromagnetic microscope. While this may be the case as the technics of electron microscopy is currently understood half a century later, it does little to elucidate the dynamics of technological development as it was occurring in the nascent days of electron microscopy.

The teleological retrospective argument of "superior technology" does not stand up in the face of international successes with the electrostatic lens design. Equipped with the knowledge that the electrostatic lens was technically viable, we are better able to understand "successes" and "failures" by viewing RCA's relative success as a product of the technological frame about its electromagnetic lens design. A significant factor was the inclusion of relevant practitioners from without and sufficient direct and indirect support from within the corporation. And in contrast to GE, support for the electron microscope at RCA resulted from a corporate approach that was predicated on electron research and thereby saw the electron microscope as a logical and essential component of its corporate strategy.

Momentum Shifts in the American Electric Utility System: Catastrophic Change—or No Change at All?

RICHARD F. HIRSH AND ADAM H. SERCHUK

Electric utility executives and several business analysts claim that the American electric utility system began to change in the 1970s. They argue that the system, which binds together massive turbines, transmission lines, nuclear reactors, human decision makers, millions of customers, and countless other components, has somehow been transformed, even though its physical nature remains much the same. As evidence of the mutation, they point to the loss of control over the business of electricity generation and to the growing influence of scores of entrepreneurs, some of whom produce power with technologies once dismissed as fancies of the "counterculture." At the same time, environmental activists and interventionist regulators have realigned corporate business policies dramatically in some states by forcing utilities to enter the energy-efficiency business. Contrary to their business instincts, utility managers in these states must now "unsell" their product after nearly a century of aggressively promoting its use.

While we acknowledge the major changes under way in the electric utility industry, we remain critical of the view, often accepted unanalytically within the industry itself, that the American electric utility system has been profoundly reshaped. Certainly, we agree with those participants who cite evidence of permanent restructuring. Yet we interpret many recent events as ways for established utilities to retain essential social and economic control over their industry. In

DR. HIRSH is professor of the history of technology and science and technology studies at Virginia Polytechnic Institute and State University (Virginia Tech). He is the author of *Technology and Transformation in the American Electric Utility Industry* (Cambridge, 1989) and several shorter pieces on the recent history and management of the utility industry. DR. SERCHUK is research coordinator of the Renewable Energy Project at the University of Maryland. He received his Ph.D. from Virginia Tech in 1995, with a dissertation titled "Federal Giants and Wind Energy Entrepreneurs: Utility-Scale Windpower in America, 1970–1990." Drs. Hirsh and Serchuk continue to collaborate on scholarly analyses, consulting projects and training seminars that use academic tools to explore the evolution of the American electric system.

other words, though we concede the presence of change, we classify it as conservative (at least up to the early 1990s), intended to maintain vital aspects of the current system, rather than as a radical deconstruction of the way the nation supplies, distributes, and consumes its electricity. Although both utility managers and their critics in the 1970s cast the system as rigid and endangered, it has proven able to assimilate numerous potential threats. Current policy makers and business analysts who warn of impending disintegration of the system may likewise underestimate its resilience and the ability of its human elements to defend their basic positions of control.

In this article, we employ Thomas Hughes's ideas about large technological systems to frame our discussion of the possibility of radical system change. After considering Hughes's work, we address two stresses that have challenged the traditional actors' control of the electric utility system: the emergence in a deregulated industry of new players who generate electricity, using California wind-power entrepreneurs as an example, and the impact of environmental activists, also primarily in California, who have encouraged utility companies to pursue conservation efforts. We describe two responses to those stresses: the incorporation of small-scale technologies into the electric supply grid, and the development of so-called demand-side management programs intended to increase energy efficiency. We then ask whether these responses signify radical structural changes or conservative modification of an essentially unchanged system. Finally, we offer a way to extend the conceptual value of Hughes's notions of technological change.

Systems and Change

As a framework for this discussion of systems change, we refer to the work of Thomas Hughes, whose *Networks of Power: Electrification in Western Society* and other works provide analytic tools for talking about change.[1] Perhaps most important, Hughes demonstrates that the generation, transmission, and distribution of electricity occurs within a technological *system.* The continued existence of any such large technological production system depends on system builders' abilities to bind into a "seamless web" considerations that a casual observer might categorize as economic, educational, legal, administrative, and technical. Large modern systems knit these elements into a whole, with system builders constantly striving, in Hughes's words, to "construct or . . . to force unity from diversity, centraliza-

[1] Thomas P. Hughes, *Networks of Power: Electrification in Western Society, 1880–1930* (Baltimore, 1983).

tion in the face of pluralism, and coherence from chaos."[2] If the managers succeed, the system thrives and expands. As the system matures, it may effectively close itself. That is, the influence on it of the outside environment may ebb, perhaps because the system has expanded to encompass factors to which it might be vulnerable.

Hughes believes that humans play a special role in systems and act to control feedback between system performance and goals. And, of course, the capacity of the system to grow depends largely on the talents of the managers. Nevertheless, Hughes argues that modern developments tend to minimize the voluntary role of humans in technological systems and that managerial control decreases within an individual system as it matures.

Systems reach closure in various ways. Some companies try to ensure their dominance in a system by pursuing vertical integration, by which they manage all phases of a given commodity from raw material to final consumer product. In the electric utility system, the urge to integrate vertically can be seen in the career of Thomas Edison, who aimed not only to produce and deliver electricity but also to manufacture the equipment used in generating and consuming his product.

System closure also can be achieved by "capturing" the regulatory apparatus, by which institutions of bureaucratic oversight come to serve the interests of the controlling stakeholders. In the utility system, power company managers appeared to have successfully captured regulators, at least until the 1970s. During the 1930s, regulators in some states blocked creation of proposed cooperatives, sponsored by the Rural Electrification Administration (REA), so that established utility companies could skim off the best rural customers for their own.[3] Regulators also gave favorable treatment to construction plans, valuations of the "rate base" on which utilities earn a profit, and rate reduction requests (as a form of competition against other fuels) after World War II and until the 1960s.[4] While scholars differ as to the causes, they agree that regulators did indeed

[2]Thomas P. Hughes, "The Evolution of Large Technological Systems," in *The Social Construction of Technological Systems: New Directions in the Sociology and History of Technology*, ed. Wiebe E. Bijker, Thomas P. Hughes, and Trevor J. Pinch (Cambridge, Mass., 1987), p. 52.

[3]For charges of regulatory capture in the electric utility industry during the 1930s and 1940s, see Marquis Childs, *The Farmer Takes a Hand: The Electric Power Revolution in Rural America* (New York, 1952), pp. 76–77.

[4]According to some critics, regulatory capture continued to occur through the 1980s. See Richard Rudolph and Scott Ridley, *Power Struggle: The Hundred-Year War over Electricity* (New York, 1986), pp. 187–92.

help utility company management efforts to close the system and preempt interference by outsiders.[5]

According to Hughes, stakeholders also attempted to close the utility system by encouraging conservative invention—new technology that preserves the existing institutional arrangements. At the same time, participants tried to stifle radical invention, such as often originates outside a system and which might initiate competing systems. After establishing steam turbines and generators as the core generation technology in the early 1900s, system builders made good use of academic institutions and research and development laboratories owned by equipment manufacturers to maintain the dominance of that core technology for another fifty years. With almost no alternatives to fossil-fuel–burning equipment, utilities bought and helped develop increasingly sophisticated steam-turbine generators, whose incrementally improving thermal efficiencies and economies of scale contributed so much to the industry's growth and improved productivity until the 1960s.[6] Managers viewed radical inventions outside this engineering realm as inimical to established financial and intellectual interests. Consequently, the industry came

[5]According to one school of thought, public utility commissioners struck an implicit bargain with power company executives because the legislators and civic leaders who had pushed for creation of regulatory bodies in the early 1900s became apathetic soon thereafter. As the public outrage over industry abuses subsided, newly created commissions found they had little political power. State politicians gave regulators few resources for building their staffs or for broadening their oversight activities, and the commissions often became "dumping grounds for political hacks and cronies of the governor" (Thomas K. McCraw, *Prophets of Regulation* [Cambridge, Mass., 1984], p. 243). By making decisions amenable to the industry managers they supposedly oversaw, utility commissioners aligned themselves with a strong source of political and financial power, thus ensuring their survival. A second school of thought suggests that, in general, regulators suffer capture because they realize that, after their term of office, they can market only the specialized knowledge gained during their tenure to the industry they formerly supervised. Hence, to ensure future employability, they shape policy so as not to offend industry. A third model of capture seeks to discredit the popular notion that regulation arose as an effort by a benevolent government to police unscrupulous industries. Some economists and historians argue that the utility industry *sought* regulation, either to bar market entry by new players, or to obtain legitimacy and forestall government prosecution. Regulation, therefore, constituted part of a strategic plan to maintain control over part of the market. The various schools of thought regarding regulatory capture are explored in Douglas D. Anderson, *Regulatory Politics and Electric Utilities: A Case Study in Political Economy* (Boston, 1981), pp. 1–4; and George J. Stigler, "The Theory of Economic Regulation," *Bell Journal of Economics and Management Science* 2 (Spring 1971): 3–21.

[6]Richard F. Hirsh, *Technology and Transformation in the Electric Utility Industry* (Cambridge, 1989), p. 83.

to rely more heavily on the conservative output of corporate engineers, who were perhaps more fettered by established ways of seeing problems, than on freewheeling individual inventors.[7] Utility managers appeared eminently successful, dispelling scrutiny as they oversaw price declines of more than 90 percent in real, inflation-adjusted terms and an annual growth rate in consumption averaging 7 percent from World War II until 1973.[8]

Despite tendencies to mitigate novelties, systems can and do change. Though he rejects the idea that systems direct themselves,[9] Hughes asserts that technological systems, "even after prolonged growth and consolidation, do not become autonomous; they acquire momentum."[10] Elsewhere Hughes defines momentum as a "mass of technological, organizational and attitudinal components [that tend] to maintain their steady growth and direction."[11] The system's inertia, or tendency to continue along a given path, results from the synergy of educational and regulatory organizations, thousands of skilled individuals, millions of dollars worth of industry-specific equipment, and ingrained ways of looking at the world; together, all encourage the continuation of business as usual. Investment, Hughes points out, refers to more than just money. Large technological systems may appear autonomous, but for Hughes this autonomy reflects human managers' careful construction of the system so as to exploit the existing social environment.

Faced with the threat of radical change, "the people and investors in technological systems construct a bulwark of organizational structures, ideological commitments, and political power to protect themselves and the system."[12] In time, system participants may come to mistake unhindered momentum for system autonomy. Changes in the business environment that alter system momentum may then

[7] Hughes, "The Evolution of Large Technological Systems," pp. 56–62. For the development of utility management culture, see Hirsh, pp. 26–35.

[8] Hirsh, pp. 82–83.

[9] Some observers of technological change posit some principle within the system itself that motivates its growth. Social theorist Jacques Ellul, for example, is usually read as a believer in the existence of technological autonomy; Ellul argues that "technique is autonomous with respect to economics and politics. . . . Its progress is likewise independent of the social situation." Rather, Ellul contends that technology represents a prime mover that conditions all other species of social change "in spite of all appearances to the contrary, and in spite of all human pride." See Jacques Ellul, *The Technological Society* (New York, 1984), pp. 133–34.

[10] Hughes, "The Evolution of Large Technological Systems," p. 76.

[11] Thomas P. Hughes, *American Genesis: A Century of Invention and Technological Enthusiasm, 1870–1970* (New York, 1989), p. 460.

[12] Ibid., pp. 460–61.

provide a shock for managers and clients who believe in the inexorable forward progress of their supposedly autonomous system. As we will discuss, such a shock certainly occurred in the electric utility industry in the 1970s, when managers had taken as gospel the notion of steady "progressive" growth of electricity consumption and continuously improving power technologies.

Hughes's work focuses on the social construction of momentum in the electric utility industry, and it does not fully explore the conditions under which momentum might change. His most provocative statements on that subject come at the end of his *American Genesis.* "In the face of this conservatism and momentum," he asks, "what might bring the displacement of large-scale, centralized, hierarchically controlled production systems? What forces might counter the tendency of large technological systems to determine social change, even history?"[13] Hughes suggests three possibilities.

Given that the market forms such an important part of the environment of modern systems, a rapid shift in consumer inclinations might cause change. Hughes considers the example of the 1973 oil embargo, when increased fuel costs expanded the American market for smaller and more energy-efficient vehicles (especially those made in Japan), to the dismay of American manufacturers, who preferred selling bigger, more profitable cars. Ultimately, American automakers altered their production strategies to accommodate market preferences.

Second, system catastrophes, such as the accident at Three Mile Island, the explosion of the space shuttle *Challenger,* and the Chernobyl nuclear power plant disaster have become increasingly common. Not only highly interconnected and complex, these systems are embedded in a social milieu that demands, in the case of Chernobyl, increasing amounts of power or, in the case of the *Challenger,* frequent newsworthy launches. Catastrophic system failure indicts not only faulty machines but social values of technological progress as well. Hughes suggests that such failures may alert the public to the dangers inherent in pushing tightly coupled complex systems too hard or too far.

Third, Hughes finds radical change stemming from "a change in belief, attitudes and intentions comparable to a religious upsurge, or a religious conversion."[14] Hughes has in mind mass acceptance of values today considered counterculture, rejecting the materialism and the concentration of power embodied in large technological

[13] Ibid., p. 461.
[14] Ibid., p. 466.

systems. He sees evidence of such change in new styles of small-scale, decentralized, and flexible management systems that appear better suited to today's uncertain economic growth and global marketplace. He concludes *American Genesis* by wondering whether the coming age of multinational corporations will demonstrate increasing decentralization or further imposition of hierarchical power.

To summarize Hughes's views, systems develop through human management of internal and external factors. They acquire momentum that is often—misleadingly—presented as autonomous motion and growth. As systems mature, they increasingly resist social construction. The most likely possibility for a radical change in momentum seems to be, in Hughes's phrase, "a confluence of contingency, catastrophe and conversion."[15]

What are the chances of such a confluence in the American electric utility system, whose accretion of momentum Hughes has done so much to explicate? We begin by examining the stressful environment in which the electric utility system found itself in the early 1970s and the nest of related technological, public policy, and cultural developments that provide superficial evidence of a greatly transformed utility system, especially as it operated in California.

First Stress: Deregulation and Introduction of New Actors

After earning a generally positive reputation in the aftermath of World War II as a well-regulated natural monopoly that extended near-universal service at steadily declining prices, electric utility companies suffered greatly in the 1970s.[16] The energy crises of 1973 and 1979 sparked oil price hikes of more than 400 percent (coal prices increased by about 200 percent) during the decade, causing electricity rates to rise as well. From 1973 to 1983, residential customers saw rates jump by an average of 11 percent annually, which spurred sharply curtailed usage growth of only 2.5 percent per year. High price inflation, reaching more than 13 percent in 1980, also meant skyrocketing borrowing costs for the most capital-intensive industry in the country.[17] As prices to consumers soared, regulatory commissions shed their role as captives to utility companies. Becoming more interventionist, they frustrated power company managers by denying the requested levels of "rate relief"—a euphemism meaning higher

[15] Ibid., pp. 470–71.

[16] By 1956, more than 99 percent of American homes enjoyed electrical service; Bureau of the Census, *Historical Statistics of the United States: Colonial Times to 1970, Part 2* (Washington, D.C., 1973), p. S-111.

[17] Hirsh, *Technology and Transformation* (n. 6 above) pp. 110–13.

rates for consumers—and began working to protect customers' interests.

Though disconcerted by the new regulatory activism, utility managers, often trained as engineers, found more disheartening the deceleration of technical improvements in generating equipment, which had previously mitigated cost increases. Cramped by inflation, managers sought greater economies of scale from successively larger turbine-generator sets. In response to a frenzy of orders for new equipment in the 1960s and early 1970s, manufacturers abandoned their practice of refining their designs through small incremental steps, which had allowed the incorporation of gradually accruing operating experience into each successive model. Instead, as utilities' orders for ever-larger equipment came more rapidly, manufacturers designed new units without benefit of operators' experiences with previous models. By the 1970s, this new "design by extrapolation" process yielded huge machines—up to 1,300 megawatts (MW) each—which often proved unreliable and costly to maintain.[18] Moreover, metallurgical problems prevented larger units from boosting the efficiency of fuel-to-electricity conversion, eliminating yet another means by which utilities traditionally reduced their cost of generating power. This socially constructed end of progress, which has been dubbed "technological stasis," helped create an unstable environment that challenged previous business assumptions and practices.[19]

One such assumption was the rationale long used by utility managers to justify their natural monopoly status for generating electricity. Early in the century, utilities had won the right from state governments to exclusive retail franchise areas partly on the basis of their ability to exploit economies of scale and to make the best use of natural resources. That is, because of the high ratio of fixed costs to operating costs in the utility business and the apparent wastefulness of duplicated resources, policy makers accepted the argument that multifirm competition would lead to higher rates for customers. In return for their protected status, utilities agreed to regulatory oversight of rates and service—a concept trumpeted not only by progressive reformers before World War I but also by industry leaders who saw regulation as legitimation for a system that appeared poised for profitable expansion. Technological stasis eroded the notion of

[18] A megawatt, which equals 1,000 kilowatts (kW), is a measure of the instantaneous demand for electrical power. For purposes of comparison, a toaster draws 1 kW of power, a stereo receiver 0.1 kW, and a 60-watt light bulb 0.06 kW.

[19] Hirsh, *Technology and Transformation*, pp. 2–3.

scale economies because bigger units did not necessarily provide the expected benefits. Moreover, some critics suggested that utilities' power plants, which converted about 35–40 percent of raw fuel into electricity, did not maximize resource efficiency.

This last criticism found a supporter in President Jimmy Carter, who made energy efficiency a major element of his National Energy Plan, passed by Congress in 1978. One of the laws making up the plan, the Public Utility Regulatory Policies Act (PURPA), included a provision that sought to optimize overall energy efficiency by encouraging "cogeneration" technology in papermaking, oil refining, and other industries that used large quantities of process heat. Cogeneration exploits waste heat from industrial processes to run generators that produce electricity for local use. An old technique, cogeneration had lost favor as power companies took advantage of larger-scale plants and attractive rates to entice industrial producers of electricity to subscribe to utility service.[20] The 1978 law sought to reverse this trend by encouraging companies outside the utility industry to produce steam and electricity in smaller and more efficient plants. By exempting the companies from federal and state laws that would have classified them as electric utilities, the law freed cogenerating firms from burdensome regulation.[21] It also required utilities to interconnect with them through existing transmission lines and to purchase the excess electricity cogenerators produced at a rate equal to the power companies' own marginal cost of producing power.[22]

Though intended only to reduce American vulnerability to disruptions of the foreign fuel market by increasing energy efficiency, PURPA unexpectedly started a process of deregulating the electric utility industry. It did so by creating a special class of independent generators (known as "qualifying facilities" or "QFs") and guaranteeing them a market of regulated utilities required to purchase the QFs' power. Moreover, the QFs could enter and withdraw from markets at will, something utilities could not do because of their legal

[20] Cogenerators in 1922 produced almost 30 percent of electricity in the United States, much of it, along with steam, for industrial use. In 1960, however, industrial plants delivered 10.5 percent of the nation's electric power by using cogeneration technology, while in 1978, they produced only 3.5 percent. See Edison Electric Institute, *EEI Pocketbook of Electric Utility Statistics,* 34th ed. (Washington, D.C., 1988), p. 6.

[21] In particular, the law exempted cogenerators from portions of the Federal Power Act and the Public Utility Holding Company Act dealing with business organization and state regulation of rates and service (Public Law 95-617, *Public Utility Regulatory Policies Act,* sec. 210(e)(1)).

[22] Ibid., sec. 210(b).

responsibility to serve all customers within their franchise areas. Utility managers generally opposed the special treatment that QFs obtained, especially after the independents gained an increasing share of the generation market. (By 1992, cogenerators produced 15 percent of American electricity.)[23] To overcome the threat posed by the unregulated entities, some utilities unsuccessfully petitioned the Supreme Court to overturn PURPA.[24] These actions suggest that PURPA-inspired deregulation quickly became a major stress for power company managers—the people previously in control of the electric utility system.

Aside from cogenerators, PURPA encouraged development of new technologies for producing power, such as those using biomass, waste, and other renewable resources (such as solar- and wind power) as primary energy sources.[25] These technologies previously had applications only where their expensive power output filled specific market niches. (For example, solar power proved useful in isolated areas where no utility transmission or distribution lines reached.) But PURPA provided incentives to expand the scope of these renewable sources of power, which, a few legislators hoped, would relieve dependence on foreign oil.

Though a federal law, PURPA was not implemented uniformly throughout the United States. Congress ordered the Federal Energy Regulatory Commission to set guidelines for instituting PURPA's mandates, but it left the states latitude in interpreting them, which was done in varying ways. In California, regulators applied the law liberally to spur creation of a healthy nonutility generation industry, which in turn helped alter the traditional utility system.

California's aggressive pursuit of "PURPA power" actually began before the 1978 passage of the act, when Jerry Brown won election as governor in 1974. A former Jesuit seminary student and devotee of Zen philosophy, Brown acquired the sobriquet "Governor Moon-

[23] Ann Crittenden, "Generating Competition: Electric Utilities Face a Host of New Rivals," *Barron's*, February 3, 1993, p. 14.

[24] *Federal Energy Regulatory Commission v. Mississippi*, 102 Sup. Ct. 2126 (1982); and *American Paper Institute v. American Electric Power Service Corporation*, 103 Sup. Ct. 1921 (1983). The latter case established the legitimacy of PURPA's requirements that utilities pay QFs for their power and overcame utility objections concerning the requirement to interconnect the producers and transmission grids. For a summary and brief discussion of the case's implications, see "High Court Upholds Utility Rules of United States," *New York Times*, May 17, 1983, p. D5.

[25] *Public Utility Regulatory Policies Act*, sec. 210. Hydroelectric power was also offered special treatment under the law. Cogenerators, however, obtained the greatest benefits under PURPA. While most alternative energy systems could not exceed 30 megawatts to qualify for PURPA exemption, cogenerators could be any size.

beam" because of his unconventional beliefs and sometimes eccentric behavior. Among his interests he counted small-scale, dispersed technologies, such as solar and wind power. By making politically astute appointments to the California Public Utilities Commission (CPUC), as well as the first appointments to the California Energy Commission, created in mid-1974 to shape overall energy policy, Brown changed the course of electricity policy in a state that was the fifth largest energy consumer in the world.[26]

To comply with PURPA, the CPUC in 1983 required utilities to purchase electricity from QFs at rates reflecting the commonly held view that energy prices would continue rising in the future. Most significantly, the commission required utilities to offer standardized contracts specifying fixed long-term purchase prices for QF power. The availability of these contracts, the commissioners believed, would aid potential QFs in obtaining financing by allowing them to anticipate future revenues. The simple contracts also lowered transaction costs for QFs, further facilitating market entry for entrepreneurs touting diverse technologies, ranging from conventional cogeneration plants to solar- and wind-powered projects. Known as Interim Standard Offer #4, the contracts attracted scores of takers, especially when oil prices began declining in the early 1980s and entrepreneurs feared that the lucrative contract would be withdrawn. (From a peak of about $35 per barrel in 1981, oil prices coasted to about $32 in 1983, only to plummet to about $10 in 1986.) Although the CPUC suspended the contract offer in late 1985, many nonutility generation companies had already signed up for the rewarding deals, and the contracts remained valid for a decade, despite grumbling from utilities that had to honor them.[27] When combined with tax credits and accelerated depreciation for the novel power-producing equipment provided by the federal and state gov-

[26] Barbara R. Barkovich, *Regulatory Interventionism in the Utility Industry: Fairness, Efficiency, and the Pursuit of Energy Conservation* (New York, 1989), chap. 4; and California Energy Commission, *Energy Glossary and Guide to Programs, Agencies, and Legislative Committees* (Sacramento, 1990), p. 71.

[27] Author (Hirsh) interview with Margaret E. Rueger, manager, Project Finance, Kenetech Windpower Corporation, February 27, 1991. Rueger represented US Windpower in negotiations with the CPUC and Pacific Gas & Electric Company. See also Janine L. Midgen, "State Policies on Waste-to-Energy Facilities," *Public Utilities Fortnightly* 126 (September 13, 1990): 26–30, for more on the rate packages offered by California and other states. Some complaints about the highly attractive rate packages can be found in Sebastian J. Nola and Fereidoon P. Sioshansi, "The Role of the US Electric Utility Industry in the Commercialization of Renewable Energy Technologies for Power Generation," *Annual Review of Energy* 15 (1990): 99–199, esp. 111–12. The authors were utility managers at Southern California Edison.

ernments,[28] alternative energy took off. Cogeneration in California, for example, provided 6,000 MW of the state's 41,000 MW of demand in 1989.[29] Meanwhile, wind- and solar-powered electricity burgeoned. During the 1980s, for example, California hosted 85 percent of the world's capacity of wind-powered electricity generation,[30] as well as 95 percent of the world's solar capacity.[31]

The story of wind-powered generation illustrates the possibly radical changes in the American electric system in the 1980s.[32] Though unfamiliar to recent generations of Americans, the pedigree of today's electricity-generating wind turbines extends back many centuries, if not millennia. In varying forms, wind-driven machines have enjoyed periods of substantial, though temporary, success. Colonial American millers, for example, erected European-style windmills to grind their grain when they lacked access to fast-running streams for the watermills they preferred. In the late 19th century, cheap, mass-produced, wind-driven pumps drew water for parched homesteaders on the plains of the American Midwest. After World War I, when electric utilities pursued dense urban markets rather than stringing lines to isolated farms, many rural Americans bought or built small individual wind-driven generating sets to charge the batteries that powered their radios and electric lights.[33]

Wind machines met the need for power in a variety of contexts, but wind technology in all its manifestations invariably lost ground to new sources of power. Though it costs nothing, the wind blows capriciously. Dependable coal-burning roller mills, which produced the whiter flour formerly available only to the rich, decisively edged out windmills in the 19th century. Wind-driven pumps and genera-

[28] Public Law 95-618, *Energy Tax Act of 1978;* C. Richard Baker, "Project Financing for Cogeneration Projects," *Public Utilities Fortnightly* 125 (March 15, 1990): 26.

[29] California Energy Commission, *Electricity: 1990 Report,* Publication P106-90-002, (Sacramento, 1990), pp. 3–11.

[30] Carl J. Weinberg and Robert H. Williams, "Energy from the Sun," *Scientific American* (September 1990): 146–55.

[31] Michael Lotker, *Barriers to Commercialization of Large-Scale Solar Electricity: Lessons Learned from the Luz Experience,* Sandia National Laboratories Report SAND91-7014 (Albuquerque, N.M., 1991), p. iii.

[32] For a more complete treatment of wind power, see Adam H. Serchuk, "Federal Giants and Wind Energy Entrepreneurs: Utility-Scale Windpower in America, 1970–1990" (Ph.D. diss., Virginia Polytechnic Institute and State University, 1995).

[33] For colonial windmills, see Walter Minchinton, "Wind Power," *History Today* 30 (March 1980): 31–35; and Volta Torrey, *Wind-Catchers: American Windmills of Yesterday and Tomorrow* (Brattleboro, Vt., 1976). For wind pumps, see Walter P. Webb, *The Great Plains* (Boston, 1931), pp. 333–48. For wind-driven generators in rural America, see Carol Lee, "Wired Help for the Farm: Individual Electric Generating Sets for Farms, 1880–1930" (Ph.D. diss., Pennsylvania State University, 1989).

tors made sense on the American plains when no alternative existed, but the rising tide of cheap, centrally generated electricity, distributed in many cases by REA cooperatives, made wind machines largely a forgotten relic of the frontier, glimpsed occasionally by travelers on the nation's highways.

The first half of the 20th century saw only sporadic attempts to generate electricity from wind power. Most notably, Palmer Putnam and the S. Morgan Smith Company erected a 1,250-kilowatt (kW) turbine at Grandpa's Knob in Vermont during the early 1940s, a unit not surpassed in size until the mid-1970s. The Smith-Putnam machine fed power into the Central Vermont Public Service Corporation grid for about two years, until the innovative machine "threw" a blade. Project planners attributed the failure to design shortcuts they had risked to avoid delays due to wartime material shortages. However, Central Vermont and the Smith Company opted not to pursue the project, as they doubted that they could construct and operate wind turbines at a cost competitive with that of other generating technologies.[34]

Putnam claimed in his final report that "the technical problems of the 1,250-kW wind turbine are understood and have been solved," but he conceded that bringing down the cost of wind power probably required government aid.[35] Beauchamp Smith, former vice president of the Smith Company, told participants at the first federal wind-energy workshop in 1973 that the attempt to generate electricity from wind power succeeded technologically; "what it did not prove," he observed, "is that this can be done on an economically feasible basis!"[36] That both Smith and Putnam themselves described the project as a failure indicates their true goal: not simply to create a machine, but to integrate a generating technology into the profit-oriented utility system.

To observers of the day, the Smith-Putnam experiment provided evidence that wind turbines could not compete economically with larger, centralized generation technology, even as part of a utility's overall generating mix. By the late 1960s and early 1970s, wind power had become a counterculture technology, advocated as an

[34] See, e.g., Frank R. Eldridge, *Wind Machines,* 2d ed. (New York, 1980), p. 24; Torrey, pp. 130–140; Tom Kovarik, Charles Pipher, and John Hurst, *Wind Energy* (Chicago, 1979), pp. 13–15. Our account draws on Palmer C. Putnam, *Power from the Wind* (New York, 1948); and Beauchamp E. Smith, "Smith-Putnam Wind Turbine Experiment" in *Wind Energy Conversion Systems: Workshop Proceedings,* ed. Joseph M. Savino (Washington, D.C., 1973), pp. 5–10.

[35] Putnam, p. 218.

[36] Smith, p. 6.

ideological act of unplugging from the utility grid. In addition to its cleanliness and apparently infinite supply, advocates urged wind power's potential to decentralize power production, erode the power of giant corporations, and encourage individuals and communities to take responsibility for their own energy decisions.

The energy crisis of the 1970s spurred an acrimonious debate over the link between energy and social structure, and over the assertion that widespread adoption of small-scale, renewable-resource energy technology would benefit American society. Much of this debate made reference to physicist Amory Lovins. In a provocative *Foreign Affairs* article published in 1976, Lovins charged that our conventional energy economy had gone bankrupt, and he suggested that a switch to renewable energy sources could help bring about desirable social change. As an alternative to the prevailing energy system, which depended on rapid expansion of centralized high technology to increase energy supplies—a system Lovins called the "hard path"—he outlined a "soft path" utilizing "soft technologies." Soft, he explained, was "intended to mean not vague, mushy, speculative or ephemeral, but rather flexible, resilient, sustainable and benign."[37] Lovins argued passionately that soft technologies relied on renewable resources rather than on depletable energy capital, that they utilized accessible components rather than arcane high technology, and that they remained amenable to decentralized community or family control.

Lovins's prescriptions for social and technological change provoked irate responses. Physicists Aden and Marjorie Meinel, for example, professed themselves "chilled" that a physicist could so "distort physical reality." They interpreted Lovins as advising America to abandon high-technology and large-scale endeavors, which would, they charged, make "mankind once more a slave to the dispatches of a dispassionate environment and his own furies." The Meinels argued that adoption of small-scale dispersed technology such as small solar and wind installations would run counter to the historical trends of centuries, and prove "inconvenient, unreliable and costly" besides. They concluded by advising reliance on the leadership of established electric utilities, and pursuit of the "true option" of nuclear fusion power. Philosopher George Pickering questioned Lovins's suggestion that soft technologies resist the creation of technological dependence and commercial monopoly: "If the soft technologies are as commercially viable as Lovins maintains, they will attract investment and they will become absorbed in the

[37] Amory B. Lovins, "Energy Strategy: The Road Not Taken," *Foreign Affairs* 55 (October 1976): 77.

web of commercial corporations competing in the manufacture, distribution and servicing of them."[38] Still other critics picked at the details of his technological and economic arguments, and maintained that his suggestions, while praiseworthy, failed to stand up to rigorous analysis.[39]

But to those who agreed with Lovins, small wind turbines seemed a logical element of his prescription for an alternative energy future. Indeed, several wind pioneers had explored the path advocated so strikingly by Lovins. Many accepted the proposition that a socially desirable energy future should be judged on grounds other than economic. For instance, in 1981 *Mother Earth News* described an apparently successful attempt by its staff to refurbish an early Jacobs 1.8-kW turbine for household use. Significantly, the crew does not seem to have compared the cost of the turbine to the price of utility power. The partisan author's observation that "the power that is produced by the wind on the hill above will spin no meters, and will be accompanied by no bills" proclaimed not efficiency, but independence.[40]

Meanwhile, the electric utility industry exhibited little interest in such pursuits. As Lovins had frankly proclaimed, the soft energy path entailed not merely a change in technology, but a radical challenge to the prevailing technological system, with its attendant values, skills, ways of life, educational institutions, and so on. In the late 1960s and early 1970s, after decades of steadily increasing scale and decreasing prices, utility managers saw no reason to fix what, in their eyes, did not appear broken. While they may have paid lip service to the concept of renewable resource energy technologies, casting them as the energy sources of the 21st century, utilities made only small investments in renewable energy research.[41]

Even as they reeled from the energy shock of 1973, most utility managers remained resistant to the idea of renewable resource technology. The federal government, however, took modest steps in that

[38] See Aden Meinel and Marjorie Meinel, "'Soft' Energy Paths—Reality and Illusion," and George W. Pickering and Margaret N. Maxey, "The Road Not Taken—and Wisely So: A Path Too Soft to Travel," both in *Soft vs. Hard Energy Paths: 10 Critical Essays on Amory Lovins' "Energy Strategy: The Road Not Taken,"* ed. Charles B. Yulish (New York, 1977), pp. 70–76 and 77–110.

[39] Lovins answered some criticisms in Amory Lovins and his critics, *Energy Controversy: Soft Path Questions and Answers,* ed. Hugh Nash (San Francisco, 1979).

[40] "Wind Power Comes to Mother's Land," *Mother Earth News* 69 (May–June 1981): 181.

[41] The *1978 Annual Report* of Pacific Gas & Electric Company observed that "a significant portion of US electric energy needs by the year 2020 could come from solar cells," while "wind energy may some day become an economical and practical supplemental source of electricity" (pp. 9–10).

direction by establishing the Federal Wind Energy Program. The federal effort explored a variety of wind technologies, mapped the nation's wind resources, and researched institutional barriers to the adoption of wind power. However, the program element that received the most media attention (and funding) was a series of contracts with large aerospace and electrical equipment manufacturers (such as Boeing and General Electric) for design, construction, and testing of very large wind turbines for electricity generation. The largest of these multimegawatt turbines, the 3,200-kW "MOD-5B" machine erected in Oahu, Hawaii, by Boeing, the National Aeronautics and Space Administration, the Department of Energy, and Hawaiian Electric Industries, cost $54 million to design and manufacture;[42] the entire federal program, including research on smaller machines, cost some $460 million.[43]

Private developers lacked the resources to fund turbines on the scale of the Federal Wind Energy Program's giants. Instead, venture capitalists seeking PURPA's incentives and associated tax credits devised an alternate technological variation—the "windfarm." With power pooled from several hundred moderate-size wind turbines, each producing 100–200 kW, the windfarms sold electricity to utilities through direct connections to the local utility grid. Except for a single Hawaiian installation, independent investors undertook all windfarm construction, partly because utilities did not qualify for the incentives under PURPA. Wind entrepreneurs concentrated their efforts in California's Altamont, San Gorgonio, and Tehachapi Passes, long known for high winds, and known today for their fields of spinning turbines.[44]

By the end of the 1980s, the federal and venture-capital paths had

[42]Victor Laniauskas, "Letter from Hawaii," *Far Eastern Economic Review* 139 (January 21, 1988): 76.

[43]"Renewable Energy Budget History (ERDA and DOE)," internal Department of Energy summary of spent appropriations, fiscal year 1975 to fiscal year 1991, courtesy of Daniel F. Ancona (in authors' possession). Another source estimates federal research costs between 1979 and 1987 at $450.9 million; see Fred J. Sissine and Michele Passarelli, "Renewable Energy Technology: A Review of Legislation, Research and Trade," Congressional Research Service, Report no. 87-318 (Washington, D.C., March 1987), p. 29.

[44]For another view of the California wind energy business, see Robert W. Righter, "Wind Energy in California: A New Bonanza," *California History* 73 (1994): 142–55. On the contributions made by Danish companies in wind-turine technology, much of which was used on the California hillsides, see Matthias Heymann, "Why Were the Danes Best? Social Determinants of Wind Technology Development in the 20th Century" (seminar paper presented to the Centre de Recherche en Histoire des Sciences et des Techniques, Centre National de la Recherche Scientifique, Paris, December 20, 1994).

led to markedly different results. The federal program produced about a dozen giant turbines, culminating in the Oahu MOD-5B. But by 1992, that last giant wind turbine had been shut down due to poor economic performance and chronic malfunctions.[45] To the chagrin of the counterculture, household-scale turbines in the range of 1 kW (such as the *Mother Earth News* machine) proved equally unable to capture market share, either in the residential or the windfarm sector, largely because manufacturers failed to develop designs that produced electricity more cheaply than could utilities.

By contrast, the privately financed windfarm concept thrived. Against the expectations of critics such as California's Governor George Deukmejian, who asserted in 1984 that "the windmill industry would disappear into the desert sands without" government-mandated incentives, windfarms survived the expiration of federal tax credits in 1985 and the loss of California state support a year later.[46] As of 1992, the United States had 1,619 MW[47] of installed wind capacity, 99 percent of it in California, where 15,500 turbines met 1.2 percent of the state's electricity demand.[48] The California Energy Commission in the early 1990s estimated the cost of electricity generated by wind turbines at 4.7–7.2 cents per kilowatt-hour (kWh), making utility-financed wind energy projects a cheaper option than nuclear plants, although slightly more expensive than coal boiler plants.[49] In 1994 at least one California firm was willing to sign contracts to supply wind power to utilities at 5 cents per kWh, about half the price of electricity generated by the Pacific Gas & Electric Company's (PG&E) own Diablo Canyon nuclear power plant.[50] In short, many California managers and policy makers viewed wind power as commercially feasible and highly competitive.[51] Because of rapid techno-

[45] "Hawaiian Electric Quits Business," *Wall Street Journal,* October 8, 1992, p. C18.

[46] Ellen Paris, "The Great Windmill Tax Dodge," *Forbes* 133 (March 12, 1984): 40.

[47] Statement of Michael L. Marvin of the American Wind Energy Association; House Committee on Ways and Means, *Comprehensive Energy Policy Act: Hearings,* 102d Cong., 2d sess., April 28, 1992, p. 231.

[48] Lisa Richardson, "Windmill Plan Could Re-energize an Industry," *Los Angeles Times,* October 5, 1992, p. B3.

[49] Nuclear plants produced power at from 5.3 to 9.3 cents/kWh while natural gas plants could sell power for 5.3–7.5 cents/kWh; California Energy Commission figures quoted in American Wind Energy Association, "Wind Energy and Electricity Rates," [early 1990s].

[50] Kenetech Windpower signed contracts for this price; PG&E's cost at its Diablo Canyon plant in 1994 was almost 12 cents per kWh. See PG&E, *1993 Annual Report,* p. 38.

[51] American Wind Energy Association, "Comments by the AWEA for the 1994 Biennial Report on Repowering California's Wind Industry" (Washington, D.C., May 27, 1993), pp. 3–4.

logical advances in wind power, several utilities planned windfarms in the Midwest, which studies indicate has greater wind resources than California.[52] And in Massachusetts, the New England Electric System (NEES) signed a contract with Kenetech Windpower for the first major windfarm in the East.[53]

Deregulation of the generation sector of the utility industry, in a period in which conventional generation technology was static, thus constituted a major stress on the once-stable system. PURPA opened the door for non-utility generators, and suddenly utility companies lost control over one element of their business. At the same time, the success of some small-scale technologies, such as wind turbines, suggested that the system might modify its preference for centrally controlled fossil-fuel and nuclear power plants and begin to incorporate decentralized technologies, some of which had long been applauded by America's counterculture. On the surface, at least, it appeared that the electric utility system had changed markedly in response to the stress of deregulation, despite the desires of the utility managers who were traditionally regarded as the system's primary decision makers.

Second Stress: New Business and Regulatory Policies

The environmental movement created a second stress that altered the electric utility system by empowering a new set of players. Modern environmentalism burst into public discourse with the publication in 1962 of Rachel Carson's *Silent Spring*.[54] The subsequent surge in public enthusiasm for environmental protection, symbolized by events such as 1970's Earth Day, impelled Congress to create the Environmental Protection Agency (EPA).[55] Environmental opposition to the utility industry centered on construction of new fossil-fuel and nuclear power plants. But while new EPA rules requiring

[52] Michael C. Brower, Michael W. Tennis, Eric W. Denzler, and Mark M. Kaplan, *Powering the Midwest: Renewable Electricity for the Economy and the Environment* (Cambridge, 1993).

[53] "East Breaks West's Wind-Power Monopoly," *Wall Street Journal*, August 31, 1993, p. B1.

[54] For a discussion of the birth of the environmental movement, see Samuel P. Hays, *Beauty, Health, and Permanence: Environmental Politics in the United States, 1955–1985* (Cambridge, 1987).

[55] Not all Americans joined the clamor. Many disenfranchised minorities considered the outcry over the environment a distraction from their own struggle to obtain acceptable housing, jobs, and education. See, e.g., "To Blacks, Ecology Is Irrelevant," *Business Week*, November 14, 1970, p. 49. Of course, it served the interests of environmentalism's opponents to note such fractures; in so doing, they trivialized the movement as the fetish of middle-class dilettantes.

submission of environmental impact statements before building power plants may have slowed construction somewhat, the regulations did not require most managers to alter their behavior noticeably.

The picture changed partly as a result of the work of special-interest advocacy groups. In particular, organizations such as the Environmental Defense Fund (EDF) took a novel approach toward challenging utility management practices. Abandoning exclusive reliance on public education as the primary method of winning support for environmental policies, the EDF used litigation against utilities to restrict construction of power plants. In an especially noteworthy development, the organization began in 1976 to employ PG&E's own planning models to argue that a combination of renewable energy technologies and conservation would both satisfy California's electricity needs and increase returns for PG&E's investors.[56]

The new environmental strategy contributed to the stress already building on the utility system. In California, it motivated previously quiescent state regulators to take an active role in negotiations concerning electricity production and delivery. Just as significant, litigious environmentalism struck a serious psychological blow against utility managers. In an unprecedented affront, a group of liberal lawyers had successfully challenged the inner workings of major corporations. They had also publicly questioned the technological competence and business savvy of the managers, who considered themselves hardworking, underappreciated public servants. The humiliation of this process, quite apart from its legal effects, should not be underestimated.

As part of their rhetorical and substantive arsenal in the 1970s and 1980s, the EDF and other environmental groups advocated using technological "fixes" as alternatives to power plants. These fixes included some alternative energy supply systems, such as wind turbines, which rapidly approached maturity in this period. But so did technologies that reduced energy demand by increasing efficiency in appliances and equipment. Through the efforts of entrepreneurial companies, university professors such as Arthur Rosenfeld, a Berkeley physicist turned energy-efficiency expert, and advocates like Lovins, research and development to improve energy efficiency burgeoned. In lighting alone great strides were made. Consuming about 25 percent of all power, electric lighting depended largely on the century-old technology of the incandescent bulb—which has been described as an electric heater that produces light as a by-prod-

[56] On the EDF, see David Roe, *Dynamos and Virgins* (New York, 1984).

uct. New compact fluorescent lights and electronic ballasts used 75 percent less electricity than incandescent lights while providing equivalent illumination. Development of electronically adjustable speed drives for electric motors also occurred rapidly in the 1980s, yielding savings of 20 percent for this important component of commercial and industrial electricity consumption. And improvements in home refrigeration units provided great savings as well. While the average new refrigerator in 1971 used 1,726 kWh per year, a new one produced in 1980 used only 1,280 kWh. By 1992, a new refrigerator's average consumption had declined to about 690 kWh.[57] Surveying these and other commercially available energy-efficiency devices in late 1990, prominent energy analysts like Lovins concluded that electricity consumption could be reduced by 30–75 percent by the year 2010 without any changes in behavior or lifestyle.[58]

When employed by utility companies as part of regulator-mandated programs, these energy-efficiency technologies became known in the mid-1980s as part of an approach called "demand-side management" or "DSM." Demand-side management focused on customers' demand for useful energy services rather than on a utility's ability to supply power. Though DSM subsumed activities once called "conservation," many DSM advocates avoided that term because of the implication, left over from the days when a cardigan-clad President Carter admonished the American public to turn down their thermostats in the winter, that conservation meant deprivation. According to these advocates, effective DSM programs achieved energy savings in an economically efficient fashion. With a rebate from New York City's Consolidated Edison Company, for example, the American Express Company upgraded the lighting in its Manhattan office in 1992, resulting in better lighting and energy savings of about 4.5 million kWh annually, worth about $280,000.[59] Some people have argued that utilities have been in the DSM business for decades, initially because of their wish to *increase* customers' demand for electricity. But since the mid-1980s,

[57] Refrigerator usage data from Arthur Rosenfeld, Lawrence Berkeley Laboratory, letter to Hirsh. See also Mathew L. Wald, "Utilities Offer $30 Million for a Better Refrigerator," *New York Times,* July 8, 1992, p. 3.

[58] Arnold P. Fickett, Clark W. Gellings, and Amory Lovins, "Efficient Use of Electricity," *Scientific American* (September 1990): 66.

[59] The company benefited from the help of the Environmental Protection Agency's "Green Lights" program, which provides information about energy-efficient lighting systems and ways to finance lighting upgrades. See Environmental Protection Agency, *Green Lights Program Report* (Washington, D.C., 1992), p. 2, *Green Lights: The Second Year,* EPA 430-R-93-005 (Washington, D.C., 1993), p. 10, and *Green Lights: An Enlightened Approach to Energy Efficiency and Pollution Prevention,* EPA 430-K-93-001 (Washington, D.C., 1993), pp. 10–11.

the term has been used primarily to describe technologies and techniques that provide the same level (or a higher level) of electrical services while using less energy.

Though gaining momentum among conservation advocates and business people who saw economic benefits from reduced energy usage, the notion of demand-side management did not at first appeal to most electric utility managers. Some utilities, such as California's PG&E, were pushed into the conservation business by regulators. In the mid-to-late 1980s, the company's managers resisted energy-efficiency efforts because the utility owned excess generating capacity and because energy prices had declined substantially since earlier in the decade. Simply put, managers believed that conservation did not make as much economic sense for customers as when energy prices were high. Moreover, from a cultural point of view, some managers felt that their business was to sell electricity, not "unsell" it, as was required with DSM programs.[60]

Perhaps most important, utility managers resisted environmentalists' efforts to encourage DSM because traditional regulation allowed utilities to earn profits for shareholders only when they sold electricity. Though companies like PG&E had encouraged some conservation efforts since the energy crisis, thus deflating potential earnings, they did so because of the pressure to be good corporate citizens and because of prodding by their state's regulatory bodies.[61] But managers argued that by encouraging DSM efforts to the extent advocated by environmental activists, utilities would be damaging their companies' economic integrity and their ability to provide reliable service to customers in the future.

Utility executives' antipathy toward DSM efforts often revealed itself in contentious regulatory hearings, at which participants espousing environmental goals lashed out against companies' proposals for building new power plants. In some cases, regulators agreed with the intervenors, penalizing utilities that did not devote enough effort to DSM by reducing the allowed rate of return on their investments and by disallowing recovery of DSM expenses they incurred.[62]

[60] Richard F. Hirsh and Bettye H. Pruitt, "The Background, Origins, and Formative Phase of the Advanced Customer Technology Test (ACT^2) for Maximum Energy Efficiency" (report for Pacific Gas & Electric Co., San Francisco, 1993), p. 35.

[61] To be absolutely fair, one must recognize that some utilities objected less strenuously to conservation efforts because they reduced the companies' need to build new and expensive power plants. Still, utilities made money only when they sold power. When they had excess capacity, as many did in the mid-1980s, conservation efforts seemed to make little sense.

[62] Jonathan Raab, *Using Consensus Building to Improve Utility Regulation* (Washington, D.C., 1994), p. 100.

After enduring several of these fracases in the mid-1980s, a few system participants sought to develop "collaborative" efforts to achieve goals advanced by environmentalists and utilities.

In Massachusetts, the Conservation Law Foundation, a group that used litigation to achieve environmental objectives, and the New England Electric System worked together in 1988 to develop a comprehensive DSM program that also provided the company a financial incentive to participate. The onetime adversaries first persuaded regulatory bodies in three New England states to give up the traditional formula used to determine utility revenues, which was based on the number of kilowatt-hours sold. Instead, they agreed on a payment system that gave the company a portion of the economic savings realized by its customers who took advantage of aggressive DSM programs. The New England Electric System found the financial motivation sufficient to sustain its managers' interest. As the company's president acknowledged good-humoredly after entering into the agreement with the conservation group, "you have to give the rat some cheese."[63] For NEES, the cheese amounted to $15 million after spending $92 million on DSM programs in 1992.[64] In California, environmental litigators employed a similar approach. After embarrassing the state's utilities by demonstrating how their conservation efforts had lagged since 1985, the Natural Resources Defense Council worked with electric and gas utilities in a 1989 collaboration that allowed some utilities to retain 15 percent of the energy savings enjoyed by their customers, making energy conservation a lucrative business indeed.[65]

As an example of the eagerness some utilities demonstrated for DSM programs, consider Pacific Gas & Electric. Marking an about-face from the year before, when it claimed it had done all that was possible in the conservation realm, the firm announced in 1990 that it would displace 75 percent of about 2,500 MW of planned capacity additions for the decade with DSM measures. Besides pursuing aggressive educational efforts to make customers and building con-

[63]John W. Rowe, president of New England Electric System, "Making Conservation Pay: The NEES Experience," *Electricity Journal* 3 (December 1990): 19.

[64]John W. Rowe and Cheryl A. LaFleur, "Making Lemons into Lemonade," *Conservation Law Foundation Newsletter* (Fall 1992): 7. Since the $92 million investment was reimbursed to the company through customer rates, the company's return on investment was effectively infinite.

[65]The public utilities commission approved the plan in 1990. For the history of the collaborative process, see Jonathan Raab and Martin Schweitzer, "Public Involvement in Integrated Resource Planning: A Study of Demand-Side Management Collaboratives," Oak Ridge National Laboratory publication ORNL/CON-344, February 1992.

structors aware of new technologies, the utility offered generous rebates to offset the higher prices customers had to pay for more efficient appliances, lighting systems, motors, and the like. The company developed this plan, which boosted DSM spending from $84 million in 1989 to $150 million in 1991, as part of an overall environmental program that, it claimed, would benefit everyone.[66] As part of the bargain with regulators and public interest groups, PG&E would earn handsome returns even as it sold less electricity. Customers would see lower electricity bills due to their greater energy efficiency and because the utility would not need to raise rates as much as if it built costly new plants. Finally, society at large would benefit by experiencing reduced power-plant emissions in a state cursed with air pollution problems. The program also had more widespread consequences, the company claimed, since it would reduce the fossil-fuel combustion that contributes to the (still-debated) phenomenon of global warming. Pacific Gas & Electric thus embraced a policy that corporate executives had previously viewed as unacceptable. In response to the stress of the hitherto combative environmental movement, system participants had developed a collaborative strategy and agreed to seek regulatory change so that utilities could profit from DSM activities.[67]

The New England Electric System and PG&E were not alone in developing DSM programs. By the end of 1991, at least twenty-four utilities in ten states had participated in similar collaboratives, in which regulatory bodies changed rules so that utilities could profit from pursuing DSM programs.[68] In many states, regulators viewed what had occurred elsewhere favorably and started implementing similar rules for increasing DSM efforts without undergoing the collaborative process. It appears, then, that the goals of many environmental groups had been realized, with utility executives embracing DSM after they found they could make a good business out of it.

Conclusion: Radical Structural Change?

When comparing the attitudes and behaviors of power company managers in the early 1990s to those of two decades earlier, it seems that a large part of the utility system has undergone a radical change. This article has highlighted two features of that change. First, a system once characterized by centralized generation and managerial

[66] PG&E, "Annual Summary Report on Demand Side Management Programs" (March 1990 and March 1992).

[67] On collaboratives and PG&E, see Hirsh and Pruitt (n. 61 above), pp. 33–38.

[68] Raab and Schweitzer, p. 1.

control opened up to allow decentralized generating technologies, such as wind turbines, as well as a set of entrepreneurial human actors who had little interest in maintaining the previously monolithic utilities. Second, managers who formerly viewed the environmental movement as a destructive countercultural force had incorporated some of the movement's goals, such as new approaches to demand-side management, into their corporate strategies. It appeared, then, that the utility system had undergone dramatic changes in technology, source of human control, and even guiding strategies and attitudes.

These changes suggest that Hughes's conditions for radical change—"a confluence of contingency, catastrophe and conversion"—have been met in the American electric utility system. The market for electricity did receive a serious jolt in the early 1970s after the energy crisis caused prices to escalate and the growth rate in usage to decline precipitously. And while the California power industry has not had a highly publicized catastrophe, the possibility of such a disaster was in the news, particularly when it was discovered that the Diablo Canyon nuclear reactor straddled the San Andreas Fault and, in addition, had been built from reversed blueprints.[69] And, in a sense, there has been an apparent conversion in American attitudes toward energy and the environment, as demonstrated by some utility companies' acceptance of DSM measures.

Utility managers themselves harbored no doubt that their industry had undergone cataclysmic changes. For instance, in the watershed year of 1987, CEO Richard Clarke of the Pacific Gas & Electric Company complained bitterly that an incoherent legal environment forced his company to operate as a traditional, regulated utility in its core residential and commercial markets, while obliging it to compete for industrial customers against new power-producing entities benefiting from regulation that discriminated against conventional utilities. Meanwhile, Clarke's colleagues bemoaned the disintegration of their closed and fraternal management culture. Edwin Lupberger, chairman and president of Middle South Utilities, sighed that utility CEOs had lost the authority to "sit down in a room" and

[69] PG&E, "1981 Annual Report," p. 3; Bryce Nelson, "Nuclear Panel Suspends Diablo Plant License," *Los Angeles Times,* November 20, 1981; p. I1; "Diablo Canyon Loses Its License," *Newsweek,* November 30, 1981, p. 98; and "Utility Apologizes to NRC for Nuclear Plant Problems," *Los Angeles Times,* November 11, 1982, p. I3. For review of the problems encountered by PG&E with its Diablo Canyon plant, see Tom Redburn, "PG&E's Nuclear Hope Turns into a Disaster," *Los Angeles Times,* March 16, 1982, p. I1.

implement the policies they judged optimal for the electric system. "The times," Lupberger conceded, "have changed."

By 1990, utility managers diverged in their outlook. Some looked enthusiastically to a new era of financial opportunity in which lean, flexible companies would thrive. Others remained recalcitrant, like CEO and chairman Walter McCarthy of the Detroit Edison Company, who warned that the benefits of the changes to the ratepayer would be "trivial." Still others, like Middle South's Lupberger, pragmatically tried to put behind them the "political infighting, animosities and regulatory problems" of the 1970s and 1980s, in order better to decipher the new operating environment. Almost unanimously, however, the executives agreed with Chairman Richard Flynn of the New York Power Authority that "the old world is definitely gone."[70]

But let us look again at these events. Did they cause deep structural changes, or did utility managers simply make tactical responses to system stresses, with the hope that they could continue to operate normally, or as close to normally as possible? We have already noted how managers seek to retain control over as many variables in their systems as possible. In the "golden age" before the 1970s, utility managers clearly controlled nearly all variables in the electric system. Because of their long success in reducing the marginal cost of electricity, regulators endorsed almost all the plans utilities made for constructing power plants, transmission lines, and other facilities. Customers, satisfied by the cheap, dependable service they received (at least after service reached the hinterlands, due to the efforts of the REA and other factors), assumed that state commissions adequately represented their interests, and they eventually ceased to participate actively in the operation of the system. Outsiders rarely interfered, and managers enjoyed carte blanche to run their systems as they saw fit. The stresses of deregulation and environmentalism, however, challenged the managers, who in response sought ways to retain control.

It is worth recalling Hughes's distinction between conservative and radical inventions. The former, he suggests, preserve technological systems; they frequently originate with scientists or engineers with a stake in the existing system, such as those employed by indus-

[70]Lou Iwler, "Issues for 1985: Competition Heats Up as New Plant Comes on Line," *Electrical World* (January 1985), pp. 17–24; "1987 to Spotlight Competition and Deregulation," *Electrical World* (January 1987), pp. 15–20; "The Big Issue Is Deregulation," *Electrical World* (January 1988), pp. 11–20; "The Political Train Is on the Track," *Electrical World* (January 1990), pp. 11–15; H. A. Cavanaugh, "The Big Issues for 1993," *Electrical World* (January 1993), pp. 9–23.

trial research laboratories. A radical invention, on the other hand, often comes from independent inventors. The existing system cannot incorporate the new technology, which may form the cornerstone of a new system.[71]

The giant wind turbines envisioned by aerospace and electrical-equipment manufacturing firms working on federal contracts were a typically conservative invention. Though development of the machines took place outside of the utility industry, the products of the Department of Energy's giant turbine effort embodied the management values and economic assumptions of the established utility system—in this case, enormity of scale. For reasons described earlier, utility managers had met growth in consumers' electricity demand during the postwar period by installing increasingly large generating plants. Not only were large generating plants thought to capture economies of scale; they also came to symbolize the "macho" outlook of the system's human managers. Proponents of large wind turbines in the utilities, the federal government, and the equipment manufacturing firms seems to have shared the notion that electricity demand would continue unabated, and they conceptualized additions to generating capacity, including "alternative" technologies, in the largest feasible units. Additionally, the industry designed and produced the giant turbines so they could be plugged into the slots in the existing grid that would otherwise be filled by conventional coal or nuclear plants. By using the turbines in this way, the machines meshed well with the goals, values, and processes of the existing system. To return to Hughes's terminology, the giant turbines were conservative inventions because they would preserve and expand the existing system.[72]

Did the windfarms then represent a radical invention? They were a significant departure from past practice. Most notably, independent entrepreneurs rather than utility companies owned and controlled these generating facilities, made up of small-scale turbines. With the introduction of windfarms, utility managers lost significant control over the timing, quantity, and cost of producing electricity. And while modern wind turbines are not precisely low-tech, they are cer-

[71] Hughes, *American Genesis* (n. 11 above), pp. 57–59.

[72] This may in fact account for the failure of the giant turbines to be adopted. By the mid-1980s, increases in demand and the pace of construction were less predictable, contingent on local needs. Windfarms, which can rapidly be constructed in sizes finely matched to utility needs, suited the energy market. Giant turbines, by contrast, shared the unwieldiness and inflexibility of the giant fossil-fuel and nuclear plants with which they were intended to compete. Perhaps the giants were simply inappropriate for the new energy market.

tainly more so than the nuclear reactors that many managers still looked to as the future of electricity production.

But the answer must still be no. As an example of what truly radical change in the system might have looked like, consider Lovins's vision of "soft" technology described earlier.[73] California does not exhibit, either as cause or effect of the wind boom, the social change on the part of electricity users that Lovins propounded. The home wind movement typified by the household machine described in *Mother Earth News* stressed the lifestyle changes necessary for such fundamental social change to occur. Because of the small power output of a home wind turbine, it might, for example, be necessary to change the times and places of housework to take advantage of natural light and reduce reliance on electric lighting, or even to adjust cultural standards of cleanliness. By contrast, the California windfarms are transparent to electricity users: that is, they require no lifestyle changes at all on the part of consumers. One strongly doubts whether Altamont Pass would sport a single turbine if incorporation of wind power into the electric system required PG&E's customers to alter their daily routines.

With passage of the National Energy Policy Act of 1992, even the prevailing pattern of windfarm ownership stands in question. One important provision of that law revises the Public Utility Holding Company Act of 1935, allowing utilities to own unregulated power-generating subsidiaries. As a result, utilities can now own windfarms, and the time since passage has seen a flurry of interest in wind development by a number of utilities, chiefly in the upper midwestern states, whose wind resources are the best in the world.[74] We suspect that the next decade may bring the demise of venture capital in the wind industry, as utilities finance and operate their own windfarms.[75]

[73] Of course, Lovins represents only one possible version of radical change. Some Progressives of the early 20th century advocated a large-scale government-managed public power system that, while quite unlike Lovins's proposals, would have produced an electric system equally different from present reality.

[74] Brower et al. (n. 53 above), p. 12.

[75] In addition to the ownership of windfarm projects, renewable energy companies themselves reflect the growing interest of conventional sources of capital in this purportedly radical business. For example, the Kenetech Windpower Corporation, once America's most successful windfarm developer and wind-turbine manufacturer, listed as major investors the Allstate Insurance Company, the F. H. Prince Companies, and the Hillman Companies. Although PURPA prohibits utilities from owning wind farms, corporations providing debt or equity to Kenetech include subsidiaries of the Florida Power & Light Co., Niagara Mohawk Power, PacifiCorp, and the Iowa-Illinois Gas & Electric Co., in addition to Aetna, John Hancock Insurance, Chrysler Financial, the Security Pacific Bank, and Westinghouse. Taylor Moore, "Excellent Forecast for Wind," *EPRI Journal* 15 (June 1990): 23.

If that turns out to be the case, the existing electric utility system will have completely digested the threat to it presented by the most successful renewable energy resource technology. To be sure, certain benefits will accrue to American society. Wind turbines do not produce particulate pollution, and they reduce reliance on Middle Eastern oil and politically unpopular technologies such as nuclear power plants. (Wind-power developers still encounter some resistance, however, on the basis of visual pollution and danger poised to birds by the large turbine blades.)[76] Still, wind power has developed as a conservative invention, rather than the basis for radical change in the existing system.

Meanwhile, how do we interpret the apparently radical change in some managers' attitudes, as expressed by their adoption of DSM? Why are they so eager to cede some control to outsiders—especially outsiders who had, until recently, been their staunch opponents? The simplest, most straightforward reason is that utilities need to remain in the good graces of regulatory commissions. Trying to mitigate the price hikes of the 1980s, regulators looked to DSM as a way to reduce pressure to build expensive new power plants. Though resisting, utilities went along with regulators to obtain different benefits (such as a financially satisfying rate of return on other investments).

The case of Central Maine Power in the 1980s provides a dramatic example of this behavior. Seeking rate increases to pay for its share of the long-delayed Seabrook nuclear power plant in New Hampshire, and hoping for approval of plans to establish unregulated subsidiaries, the company lost credibility with Maine commissioners when a vice president admitted to perjury in a hearing. The event caused a corporate shakeout and, in 1984, John Rowe, a Wisconsin lawyer with no utility experience, took control of the firm. Seeking to placate regulators, he embraced DSM and cogeneration technologies—favored by the commissioners but dismissed by previous company executives. While Rowe appreciated how DSM could fit into

[76] Early champions of wind power presented it as an electric supply option with zero environmental cost. Neighbors of early turbines hotly disputed that assessment, pointing to noise and television interference. These problems have been reduced, but protests on the basis of the turbines' visual impact continue. More recently, influential environmentalists, such as the chief scientist of the National Audubon Society, have called for a moratorium on turbine construction because of the birds of prey killed by flying into the rotor blades. See Ellen Paris, "Palm Springs and the Wind People," *Forbes* 135 (June 3, 1985): 170–71; P. B. Bosley and K. W. Bosley, "Risks and Benefits of Wind Generated Electricity: Facts and Perceptions," *Energy Sources* 14 (January–March 1992): 1–9; and Teresa Tamkins, "Tilting at Wind Power," *Audubon* 95 (September–October 1993): 24, 26, 28.

his company's strategy for avoiding participation in new power-plant construction programs, he nevertheless expected to score valuable points by complying with the commissioners' desires. As a result, he rebuilt the trust and cooperation between the company and the regulatory body, which proved vital for the utility's financial turnaround.[77] While unique, this story illustrates the role DSM can play in reestablishing good will between antagonistic power executives and regulators.

Utility managers may have accepted DSM for other reasons too. Some critics of utility DSM programs charge that managers are abdicating control over some parts of their business in order to retain control over others. In particular, managers may be embracing conservation on the demand side of their business to limit competition with other power generators on the supply side. They accomplished this goal, the argument goes, by urging the states' regulatory commissions to make DSM part of their forward-looking energy resource plans. In California, for example, the Public Utilities Commission (with the approval of the California Energy Commission) approved plans for the state's utilities to meet a large fraction of the anticipated demand for power during the 1990s with DSM programs. While some of the programs would be pursued by independent energy service companies, most would be undertaken by the utilities themselves. Regardless of who performs the DSM work, however, the net effect is to reduce the need for supply-side resources. Hence, the state regulatory agencies, by approving DSM plans, have restricted the amount of business that independent power producers can gain. This outcome suits utility managers, as it allows them to retain technological and financial control over "their" system.[78]

Critics maintain that DSM inhibits independent power producers in another way. It may happen that, later in the 1990s, utilities will find that their projections for replacing generation supply with DSM have fallen short. At that time, the companies will argue that they can refurbish old, utility-owned plants and provide power at a lower cost than could competitors. As noted by Jan Smutney-Jones, executive director of the state's Independent Energy Producers Association, "there's a high degree of suspicion from the independent suppliers of electricity that DSM is actually being used as sort of a marketing tool, at least in the short run to suppress the [indepen-

[77]Julie Lanza, "John W. Rowe: Creative Energy," *Boston Business Journal* 12 (February 12, 1993), sec. 1, p. 19.

[78]Karen Morris, "California DSM Programs Anticompetitive," *DSM Quarterly* (Summer 1992): 6–9.

dent power] market. My suspicion is that [the utilities] are using DSM in the front part of the 90s to suppress the need for adding new independents, and then later in the decade will be in the mode of repowering their own plants." In sum, Smutney-Jones argues, "the utilities' strategy is to use the regulatory system to maintain market control."[79]

As historians, we find such an interpretation based on self-interest plausible, insofar as it outlines possible scenarios of the *effects* of DSM programs. If DSM makes new generating capacity unnecessary in the immediate future, independent producers, rather than utilities, will indeed be inhibited. Utility managers, then, will have found another way to retain control over the system. Nevertheless, we are cautious about imputing *conscious* political motives to utility managers who support DSM. Certainly, no executives have ever admitted to us that their companies pursue DSM merely to prevent independents from gaining market power. Nor are we sure what such an admission would mean.[80]

More important, historians must be careful not to view utilities as

[79] Ibid., p. 8. Utility managers are not the only actors who have the opportunity to retain control of the existing utility system in this DSM scenario. Some critics of utility DSM programs argue that environmental advocates and regulators also benefit. Environmentalists profit by having their goals endorsed by regulatory commissions, which gives them public credibility and wins them entrée into the boardrooms of utility companies. No longer dismissed as extremists, environmentalists now appear as insiders, part of the establishment, thus strengthening their positions in future policy debates. Meanwhile, regulators may see DSM as a way to retain their own power. As the industry becomes more competitive, the role of regulators as overseers of utility capital projects and ensurers of "fair" rates for customers diminishes. Because environmental activities such as conservation have widespread public support, DSM offers regulators a way to legitimize a new set of activities while they lose control over traditional utility actions. At the same time, some managers argue that customers will benefit through conservative change: not only is conventional utility management safer and more reliable than entrepreneurial activity, they suggest it is cheaper too, because utilities enjoy access to cheaper sources of long-term capital than do entrepreneurs. In short, adoption of radical-looking DSM programs offers a chance for many of today's participants in the utility community to maintain the system in a way that allows them to retain some elements of control. This self-interest interpretation is also known as a "rent-seeking" model of political economy. See James M. Buchanan, Robert D. Tollison, and Gordon Tullock, eds., *Toward a Theory of the Rent-Seeking Society* (College Station, Tex., 1980). For a critical view of the role of participants in the political economy of the utility industry, see Douglas A. Houston, *Demand-Side Management: Ratepayers Beware!* 2d ed. (Houston, 1993), pp. 36–41.

[80] The intentions of historical actors, or rather their first-person accounts of their intentions, offer problematic historical evidence at best. We doubt, in any case, that history should take as its primary goal the explication of individuals' intentions.

single historical actors with a single motive. At a finer grain of internal detail, utilities show a variety of tendencies, some of which conflict. For example, in 1984 Southern California Edison hired John Bryson, cofounder of the National Resources Defense Council, as executive vice president.[81] While Bryson may consider himself a business pragmatist, he also provides a strong voice for environmental values within the company, especially since becoming the company's chairman and CEO in 1991.[82] With regard to DSM, then, we doubt that utility managers speak with one voice, or that they act for a single reason, or even that they are aware of all the consequences of their actions.

Despite these reservations about the single-mindedness of historical actors, we argue that employment of wind turbines and DSM techniques constituted responses to conserve the utility system with as little change as possible. But beyond this conclusion, our study offers some useful methodological insights into the role played by technology (and its creators) for altering the momentum of systems. In particular, the examination of the threats posed by wind power and DSM to the utility system gives us a way to sharpen the methodological value of Hughes's notion of radical and conservative technologies. Hughes implies that negotiations concerning acceptance of a new technology hang on whether the invention is *essentially* radical or conservative. Yet, in the story of wind power and DSM, those attributes *resulted* from negotiations. Nothing inherent in the technologies determined that they would be used either to destroy the existing system nor enhance its stability.

In the terminology we have been using, the technologies could have been considered either conservative or radical. Some actors lobbied for utilities to adopt wind power, for example, as simply another supply option in the grid they owned and managed. Others pushed for privately owned windfarms, mostly outside the utilities' control. Still other actors envisioned wind power as a household or community-based energy option able to provide a radical alternative to the established electric system. Likewise, some advocates of energy efficiency expected that utilities would shy away from a business that appeared to have little to do with their core competency of generating and distributing an electrical commodity. But environmental activists, regulatory commissioners, and utility managers turned DSM

[81] Marc Reisner, "The Most Imaginative Power Company in the US," *Science Digest* (August 1986): 63–87.

[82] Peter Nulty, "Finding a Payoff in Environmentalism," *Fortune*, October 21, 1991, p. 79.

into a lucrative business that enhanced the political and institutional power utilities could wield. In other words, the technologies did not exert a one-way influence on their environment, as if a radical or conservative character was embedded in their design. Those qualities had to be determined as a result of negotiations between actors in the system.

This modification of Hughes's notion of conservative and radical inventions should prove useful to those who employ the systems approach. It will also benefit us as we continue to explore the possibility that the American utility system has been radically transformed. Though we have described here how power company managers have successfully digested the threats of wind power and DSM to the utility system, we realize that other events have occurred recently that continue to challenge the system's integrity. In particular, new small-scale technologies have been evolving that allow individual homes and businesses to produce power for themselves and sell excess capacity to other users, thus reducing the need for customers to be hooked up to the grid at all. At the same time, political pressure has intensified to eliminate the regulatory framework altogether and to allow greater competition among vendors of electricity—all enabled through the use of new technologies for generating power and for coordinating its movement through the grid or small, localized networks. Clearly, these events may constitute perils that eclipse the destabilizing effects of wind power and DSM. To determine whether the traditional stakeholders in the American electric power matrix have permanently succeeded in conserving their system, we will therefore need to focus on the ongoing negotiations among participants about the potentially radical nature of new technologies.[83]

[83] Richard Hirsh is currently completing a book manuscript that plans to offer a more definitive characterization of the contemporary electric utility system.

INDEX